HIGH-LET RADIATIONS
IN CLINICAL RADIOTHERAPY

HIGH-LET RADIATIONS IN CLINICAL RADIOTHERAPY

Proceedings of the 3rd Meeting on Fundamental and Practical Aspects of the Application of Fast Neutrons and other High-LET Particles in Clinical Radiotherapy, The Hague, Netherlands, 13–15 September 1978

SPONSORED BY THE RADIOBIOLOGICAL INSTITUTE OF THE ORGANIZATION FOR HEALTH RESEARCH TNO, RIJSWIJK, THE NETHERLANDS

Editors

G. W. BARENDSEN, *Rijswijk*

J. J. BROERSE, *Rijswijk*

K. BREUR, *Amsterdam*

PUBLISHED AS A SUPPLEMENT
TO THE EUROPEAN
JOURNAL OF CANCER

PERGAMON PRESS

OXFORD · NEW YORK · TORONTO · SYDNEY · PARIS · FRANKFURT

U.K.	Pergamon Press Ltd., Headington Hill Hall, Oxford OX3 0BW, England
U.S.A.	Pergamon Press Inc., Maxwell House, Fairview Park, Elmsford, New York 10523, U.S.A.
CANADA	Pergamon of Canada, Suite 104, 150 Consumers Road, Willowdale, Ontario M2J1P9, Canada
AUSTRALIA	Pergamon Press (Aust.) Pty. Ltd., P.O. Box 544, Potts Point, N.S.W. 2011, Australia
FRANCE	Pergamon Press SARL, 24 rue des Ecoles, 75240 Paris, Cedex 05, France
FEDERAL REPUBLIC OF GERMANY	Pergamon Press GmbH, 6242 Kronberg-Taunus, Pferdstrasse 1, Federal Republic of Germany

Copyright © 1979 Pergamon Press Ltd.

All Rights Reserved. No part of this publication may be reproduced, stored in a retrieval system or transmitted in any form or by any means: electronic, electrostatic, magnetic tape, mechanical, photocopying, recording or otherwise, without permission in writing from the publishers.

First edition 1979

British Library Cataloguing in Publication Data
Meeting on Fundamental and Practical Aspects of the Application of Fast Neutrons and Other High LET Particles in Clinical Radiotherapy, *3rd, The Hague, 1978*
High LET radiations in clinical radiotherapy.
1. Cancer – Radiotherapy – Congresses
2. Radiotherapy, High energy – Congresses
I. Title   II. Barendsen, G. W.   III. Broerse, J. J.
IV. Breur, K.
616.9'94'0642   RC271.R3   79-40323

ISBN 0-08-024383-5

Published as a supplement
to the *European Journal of Cancer*

Printed and bound in Great Britain by
William Clowes (Beccles) Limited, Beccles and London

Contents

Introduction to the Proceedings
G. W. Barendsen, J. J. Broerse and K. Breur — ix

Sessions I and II. REPORTS ON CLINICAL EXPERIENCE

Second preliminary report of the M. D. Anderson study of neutron therapy for locally advanced gynecological tumors — 3
L. J. Peters, D. H. Hussey, G. H. Fletcher and J. T. Wharton

Observations on the reactions of normal and malignant tissues to a standard dose of neutrons — 11
M. Catterall

Results of fast neutron radiotherapy at Amsterdam — 17
J. J. Battermann and K. Breur

Results of fast neutron beam radiotherapy pilot studies at the University of Washington — 23
T. Griffin, J. Blasko and G. Laramore

Results of clinical applications with fast neutrons in Edinburgh — 31
W. Duncan and S. J. Arnott

Fast neutron project at Fermilab — 37
G. A. Lawrence

Clinical observations of early and late normal tissue injury and tumor control in patients receiving fast neutron irradiation — 43
R. Ornitz, A. Heskovic, E. Bradley, J. A. Deye and C. C. Rogers

Results of clinical applications of fast neutrons at Hamburg–Eppendorf — 51
H. D. Franke

Results of clinical applications of negative pions at Los Alamos — 61
M. M. Kligerman, S. Wilson, J. Sala, C. von Essen, H. Tsujii, J. Dembo and M. Khan

Results of tumor treatments with alpha particles and heavy ions at Lawrence Berkeley Laboratory — 67
J. R. Castro, C. A. Tobias, J. M. Quivey, G. T. Y. Chen, J. T. Lyman, T. L. Phillips, E. L. Alpen and R. P. Singh

Results of clinical applications of fast neutrons in Japan — 75
H. Tsunemoto, Y. Umegaki, Y. Kutsutani, T. Arai, S. Morita, A. Kurisu, K. Kawashima and T. Maruyama

Five years of clinical experience with a combination of neutrons and photons — 79
H. J. Eichhorn, A. Lessel and K. Dallüge

Hyperbaric oxygen and hypoxic cell sensitizers in clinical radiotherapy: present state and prospects 83
S. Dische

Treatment at low dose rate, by low-LET radiation: present status and prospects 91
B. Pierquin, E. Calitchi and R. Owen

Combined chemo/radiotherapy of cancer: present state and prospects for use with high-LET radiotherapy 95
T. L. Phillips

Sessions III and IV. PHYSICAL ASPECTS AND RADIOBIOLOGY

Review of performance of high-LET radiation sources used in clinical applications 105
D. K. Bewley

Dose distributions of clinical fast neutron beams 109
B. J. Mijnheer and J. J. Broerse

Dosimetry intercomparisons and protocols for therapeutic applications of fast neutron beams 117
J. J. Broerse, B. J. Mijnheer, J. Eenmaa and P. Wootton

Characteristics of fast neutron sources 125
J. B. Smathers

CONTRIBUTIONS ON CHARACTERISTICS OF FAST NEUTRON SOURCES 129

Neutron dosimetry, radiation quality and biological dosimetry 147
J. Booz

CONTRIBUTIONS ON NEUTRON DOSIMETRY, RADIATION QUALITY AND BIOLOGICAL DOSIMETRY 151

Review of RBE data for cells in culture 171
E. J. Hall and A. Kellerer

RBE values of fast neutrons for responses of experimental tumours 175
G. W. Barendsen

Neutron RBE for normal tissues 181
S. B. Field and S. Hornsey

Fast neutron radiobiology 187
H. B. Kal and A. J. van der Kogel

CONTRIBUTIONS ON FAST NEUTRON RADIOBIOLOGY 193

Dosimetry and radiobiology of negative pions and heavy ions 209
M. R. Raju

CONTRIBUTIONS ON DOSIMETRY AND RADIOBIOLOGY OF NEGATIVE PIONS AND HEAVY IONS 213

The Committee on Radiation Oncology Studies plan for a program in particle therapy in the United States 237
J. R. Stewart and W. E. Powers

Contents

Sessions V and VI. EVALUATION OF PRESENT RESULTS AND FUTURE OF HIGH LET RADIOTHERAPY

Evaluation of clinical experience concerning evaluation of tumour response to high-LET radiation 243
J. Dutreix and M. Tubiana

Evaluation of normal tissue responses to high-LET radiations 251
K. E. Halnan

The application of RBE values to clinical trials of high-LET radiations 257
H. R. Withers and L. J. Peters

Doses and fractionation schemes to be employed in clinical trials of high-LET radiations 263
J. F. Fowler

Review of protocols for high-LET radiotherapy in the United States 267
S. Kramer

Clinical trials with fast neutrons in Europe 273
K. Breur and J. J. Battermann

Summary of discussion on multi-centre collaboration in clinical trials with high-LET radiations 277
G. W. Barendsen

Concluding remarks on the future of high-LET radiotherapy 281
G. W. Barendsen, K. Breur and J. J. Broerse

List of participants of the 3rd meeting on Fundamental and Practical Aspects of the Application of Fast Neutrons and other High-LET Particles in Clinical Radiotherapy 283

Introduction to the Proceedings

G. W. BARENDSEN,* J. J. BROERSE* AND K. BREUR† (*Organizers and Editors*)

*Radiobiological Institute TNO, Rijswijk, and †Antoni van Leeuwenhoek Hospital, Amsterdam.

INTEREST in the clinical applications of ionizing radiations that cause energy deposition in tissue through particles of high linear energy transfer has continued to increase during the past years. As a consequence of experimental evidence and increasing clinical experience, it became evident that high-LET radiations might provide a significant advantage for the treatment of some types of cancer, but not for all types and all sites. These different observations can be correlated partly to the fundamental radiobiological properties of these radiations, but a significant role can also be ascribed to dosimetric problems and technical disadvantages of the neutron therapy equipment employed. A comprehensive analysis of all problems associated with the clinical application of high-LET radiations was therefore considered to be important for the selection of future trials.

The programme of this meeting was developed on the basis of our judgement that clinical experience and its evaluation should be given priority. Thus reviews of results obtained in all centres which had treated a significant number of patients were presented on the first day of the meeting. In addition a few papers were presented on clinical results of treatments based on other fundamental radiobiological factors, namely the elimination of the influence of hypoxic cells through hyperbaric oxygen or hypoxic cell sensitizers, exploitation of the influence of repair of sublethal damage through low-dose-rate irradiation and the utilization of interaction of chemotherapeutic agents and radiation. Subsequent discussions during the following 2 days showed that these presentations of clinical data on the first day had made a significant impact. These papers on clinical experience constitute the first part of these proceedings.

During the second day attention was directed at technical and physical aspects of high-LET radiations and at new radiobiological data. The number of contributions submitted was so large that they have been reviewed and discussed in special reports, to point out agreements and significant differences among the results obtained. These reviews and short contributions constitute the second part of the proceedings.

During the third day future developments of high-LET radiotherapy have been discussed with special emphasis on international co-operation and on protocols for randomized clinical trials. These papers and discussions are presented in the last part of these proceedings.

The meeting was attended by 210 participants from fifteen countries and was financially supported by the Organization for Health Research TNO and The Ministry of Education of the Netherlands. Thanks are due to the many individuals who helped in the planning and organization of this meeting, in particular Miss M. C. von Stein and Mrs. H. Zandstra-Overvoorde. We hope that the publication of these proceedings will serve as a useful source of information and will contribute to international co-operation in this developing area of clinical radiotherapy.

References

Proceedings of the first meeting on Fundamental and Practical Aspects of the Application of Fast Neutrons in Clinical Radiotherapy, published in *Europ. J. Cancer* **7**, 97–267 (1971).

Proceedings of the second meeting on Fundamental and Practical Aspects of the Application of Fast Neutrons in Clinical Radiotherapy, published in *Europ. J. Cancer* **10**, 199–398 (1974).

SESSIONS I and II

Reports on Clinical Experience

Second preliminary report of the M. D. Anderson study of neutron therapy for locally advanced gynecological tumors

L. J. PETERS, D. H. HUSSEY, G. H. FLETCHER AND J. T. WHARTON*

*Department of Radiotherapy, *Department of Gynecology,
The University of Texas System Cancer Center,
M. D. Anderson Hospital and Tumor Institute,
Houston, Texas 77030, U.S.A.*

Abstract—*Between April 1973 and April 1978, a total of 530 patients with locally advanced tumors of all types were treated at the MDAH-TAMVEC facility. This report is based on ninety-six patients with advanced gynecological tumors treated in whole or part with fast neutrons. Of these, thirty-seven patients were treated with neutrons only or a neutron boost following photon irradiation. In the remaining fifty-nine patients a mixed beam technique was used, consisting of twice-weekly neutron doses of 65 rad nγ and three-times-weekly photon doses of 200 rad. The best therapeutic ratio was observed in this group, although the patients treated with neutrons only had, on average, more advanced disease. The local tumor-control rate achieved with mixed beam therapy was superior to that in a comparable group of forty patients treated during the same period with photons alone, but no improvement in survival has so far resulted, due to an excess of deaths from metastatic disease and other causes in the mixed beam group. The preliminary results for a subset of patients treated in a prospective randomized clinical trial comparing mixed beam and photon treatments show a similar trend.*

Introduction

In October 1972, M. D. Anderson Hospital (MDAH) initiated a pilot study of fast neutron radiotherapy using the cyclotron located at Texas A & M University. Initially, the clinical material was limited primarily to patients with head and neck or breast tumors because the 16-MeV$_{d \to Be}$ neutron beam then employed had poor skin-sparing and depth-dose properties. When 50 MeV$_{d \to Be}$ neutrons became available in April 1973, the study was extended to include patients with gynecological tumors.

Three neutron therapy treatment schedules have been employed for gynecological tumors: (1) 50 MeV$_{d \to Be}$ *neutrons only*, (2) 50 MeV$_{d \to Be}$ neutrons following initial treatment with 25 MV photons (*neutron boost*), and (3) combined MeV$_{d \to Be}$ neutrons and 25 MV photons in which patients were treated twice weekly with neutrons and three times weekly with photons (*mixed beam*). The cyclotron was not available for clinical use between July and December 1975 and between July and December 1976. During these periods, patients who would have been candidates for the neutron therapy pilot study were registered in a "control group" and treated conventionally with 25 MV photons. Since January 1977, all patients with locally advanced squamous carcinomas of the cervix have been entered into a prospective randomized trial comparing *mixed-beam* irradiation with 25-MV *photon* therapy.

The objective of this paper is to compare the results of the neutron therapy pilot study for gynecological cancer with those of the "control group" treated with photons. The preliminary results of the randomized trial for squamous carcinoma of the cervix are also discussed.

Energy and dosage conventions

The neutron treatments were given with the Texas A & M variable energy cyclotron using a neutron beam produced by 50 MeV deuterons incident on a thick beryllium target (50 MeV$_{d \to Be}$). The mean neutron energy of this beam is approximately 22 MeV. The photon treatments were delivered with 25 MV photons using an Allis Chalmers betatron or a Sagittaire linear accelerator.

The neutron beam doses have been determined by measurements with a tissue-equivalent ionization chamber immersed in tissue-equivalent liquid. The physical doses are expressed in rads including both the neutron and gamma components ($rad_{n\gamma}$). The *gamma dose* for the 50-MeV$_{d \to Be}$ beam is approximately 10 per cent of the total physical dose at 10 cm depth.

Equivalent doses (rad_{eq}) have been calculated by multiplying the physical dose delivered with neutrons by an RBE of 3.1 and adding this to the dose delivered with photons. The RBE of 3.1 was determined clinically by comparing the late effects of neutrons twice weekly with the late effects of photon irradiation delivered in fractions of 200 rad five-times weekly (Hussey, 1975). The complication rate seen with 2080 $rad_{n\gamma}$/13 fractions/6½ weeks with 50 MeV$_{d \to Be}$ neutrons is similar to that seen with a dose of 6500 rad/32 fractions/6½ weeks with photons. Radiobiology studies evaluating oromucosal necrosis in Rhesus monkeys and fibrosis in pigs' skin have shown that the dose response curve for late effects with neutrons is independent of the weekly fractionation schedule (Withers, 1977).

those of a photon beam generated by a 4-MV linear accelerator (Fig. 1). The depth of maximum build-up (D_{max}) is 0.8 cm, and a 10 × 10-cm beam is attenuated to 50 per cent of D_{max} at a depth of 13.8 cm. The dosimetric properties of 25 MV photons are far superior, with D_{max} occurring at 4.5 cm and the depth of 50 per cent attenuation at 24.3 cm.

The neutron treatments were delivered with the patient in a standing position using a *fixed horizontal beam*. This complicates field shaping and results in an increased patient diameter. Furthermore, the small intestine shifts into the pelvis when patients are standing, leading to an increased risk of bowel complications. A compression device has been designed to minimize these disadvantages by reducing the patient diameter and displacing the intestines out of the pelvis.

Materials and Methods

The analysis includes 136 patients with locally advanced gynecological cancers treated between April 1973 and April 1978. The data were

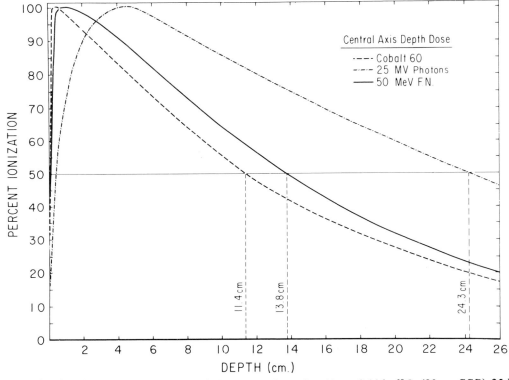

Fig. 1. *Surface build-up and central axis depth dose curves for a 10 × 10-cm field for ^{60}Co (80-cm SSD), 25 MV photons (100-cm SSD), and 50 MeV$_{d \to Be}$ neutrons (140-cm SSD). The depth of 50 per cent attenuation are shown for comparison. Courtesy: Caderao, Hussey, Sampiere, Johnson and Wharton (1976).*

Dosimetric properties and technical considerations

The 50-MeV$_{d \to Be}$ neutron beam has skin-sparing and depth-dose properties similar to

analyzed in August 1978. The clinical material is listed by site and clinical stage in Table 1.

The aim has been to deliver 5000 rad_{eq} to the whole pelvis in 5 weeks and re-evaluate the patient for completion of treatment using intracavitary radium or an external beam boost. In general, patients showing a good regression of parametrial disease completed treatment with intracavitary radium, whereas those with persistent disease at the pelvic wall completed treatment with an external beam boost. The treatment methods and tumor doses are listed in Table 2.

when the dosage schedules and treatment techniques were not well established. In general, the patients treated with *neutrons only* presented with more advanced disease than those in the other treatment categories (Table 1). Fifteen patients (88 per cent) had stages IIIB or IV carcinomas of the cervix or massive tumors of other sites. Approximately half presented with abnormal intravenous pyelograms and half of those

Table 1. Summary profile of the total patient population (April 1973–April 1978)

	Neutrons only	Neutron boost	Mixed beam	Photons
No. treated	17	20	59	40
Mean age ±1 S.D.	57±13 yr	53±16 yr	55±12 yr	50±12 yr
Site and stage[a]				
IIB CA Cx	0	2 (10.0%)	4 (6.8%)	4 (10.0%)
IIIA CA Cx	2 (11.8%)	6 (30.0%)	24 (40.7%)	17 (42.5%)
IIIB CA Cx	7 (41.2%)	8 (40.0%)	23 (39.0%)	15 (37.5%)
IV CA Cx	2 (11.8%)	1 (5.0%)	3 (5.1%)	3 (7.5%)
Other sites[b]	6 (35.3%)	3 (15.0%)	5 (8.5%)	1 (2.5%)
Abnormal IVP	9/17 (52.9%)	2/20 (10.0%)	11/59 (18.6%)	14/40 (35.0%)
Regional metastasis[c]	7/13 (53.8%)	6/14 (42.9%)	19/52 (36.5%)	10/36 (27.8%)

[a] MDAH staging system: stage IIB—lateral parametrial involvement, or massive involvement of the corpus (barrel-shaped); stage IIIA—involvement of one pelvic wall or lower third of the vagina; stage IIIB—involvement of both pelvic walls, or one pelvic wall and the lower third of the vagina; stage IV—biopsy-proven rectal or bladder involvement (IVA), or distant metastasis (IVB).
[b] Other sites: 9 endometrium, 4 vagina, 2 urethra.
[c] Regional node metastasis on lymphangiogram or at lymphadenectomy.

Table 2. Summary of treatment received[a]
(Total population)

	Neutrons only	Neutron boost	Mixed beam	Photons
External beam only				
No. patients	17	20	28	15
Mean equivalent dose[b] ±1 S.D. (range)	6050± 550 (rad_{eq}) (4090–6480)	6660± 360 (rad_{eq}) (5880–7010)	6530± 290 (rad_{eq}) (5990–7000)	6550± 670 (rad_{eq}) (4600–7490)
External beam + radium				
No. of patients	0	0	31	25
Mean external dose[b] ±1 S.D. (range)	—	—	5100± 520 (rad_{eq}) (4000–6040)	5050± 710 (rad_{eq}) (4000–6020)
Mean radium dose ±1 S.D. (range)	—	—	4220±1470 mg hr (720–6720)	4700±1460 mg hr (1000–6840)

[a] Abbreviations: S.D. = standard deviation.
[b] Total "equivalent" dose based on RBE=3.1 for photon fraction size of 200 rad.

Treatment categories

Neutrons only. Seventeen patients were treated only with neutrons using fractions of 160 $rad_{n\gamma}$ delivered twice weekly. These patients were treated in the initial phase of the program

who had *lymphangiograms* or selective lymphadenectomies had regional nodal metastasis. None of the patients treated with *neutrons only* showed sufficient regression of parametrial disease to justify completion of treatment with intracavitary radium.

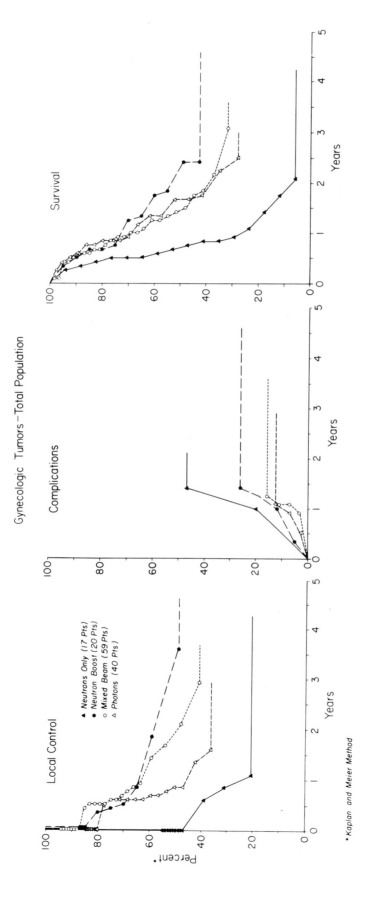

Fig. 2. Actuarial local control, complications, and survival curves (Kaplan and Meier method (1958)). The interval from start of radiation therapy to the date of analysis from 45 to 62 months (mean=54 months) for the neutrons-only group, from 29 to 57 months (mean = 46 months) for the neutron-boost group, from 4 to 46 months (mean = 25 months) for the mixed-beam group, and from 4 to 52 months (mean = 22 months) for the photon group.

Neutron boost. Twenty patients were treated with a *neutron boost* following initial treatment with 25-MV photons. The majority of these patients presented with slightly less advanced disease than those in the other treatment categories but were referred for completion of treatment with neutrons because of persistent or progressive disease. These patients received 4000 to 5000 rad in 4 to 5 weeks with photons followed by a *neutron boost* to the residual disease. The *neutron-boost* dose ranged from 640 $rad_{n\gamma}$/4 fractions/2 weeks to 960 $rad_{n\gamma}$/6 fractions/3 weeks (2000 rad_{eq}/2 weeks to 3000 rad_{eq}/3 weeks).

Mixed beam. The *mixed-beam* schedule was adopted in September 1974 because twice weekly fractionation with *neutrons only* had resulted in a relatively high local failure rate and a significant number of complications. The aims were: (1) to improve the dose distribution by delivering part of the treatment with 25 MV photons and (2) to improve the therapeutic ratio by utilizing five-times weekly fractionation. *Mixed-beam* treatments were given twice weekly with neutrons and three-times weekly with photons. The neutron fraction size was 65 $rad_{n\gamma}$ (200 rad_{eq}).

to allow the use of intracavitary radium (Table 2).

Photons. Forty patients were treated conventionally with 25 MV photons. Twenty of these patients presented at times when the cyclotron was not available for clinical use and twenty were allocated to photon therapy in the randomized trial. The stage distribution of the *photon* group is similar to the stage distribution of the *mixed-beam* group. Sixty per cent of the patients treated with photons completed treatment with intracavitary radium (Table 2).

Results

Actuarial local control, complication, and survival curves (Fig. 2) have been utilized to analyze the results because the patients treated with the *mixed beam* or *photons only* have not been followed as long as those treated with *neutrons only* or a *neutron boost*.

The results with *neutrons only* are inferior to those achieved with other treatment schedules. The local control and survival rates with the *neutron-boost* group are slightly better than those in the other treatment categories although the complication rate for the *neutron-boost* group is high.

The local control rate for patients treated with

Table 3. Current status of the total patient population[a]
(April 1973–April 1978)
Unlimited follow-up—analysis August 1978

	Neutrons only	Neutron boost	Mixed beam	Photons
Dead	16	11	26	18
of local failure (LF)	12 (2 c̄ DM)	6 (2 c̄ DM)	16 (1 c̄ DM)	15
of distant metastasis (DM) (s̄ LF or complications)	2	2	3	2
of intercurrent disease	1	1	4 (1 c̄ DM)	1 (1 c̄ comp)
of complications (comp)	1	2	3	0
Alive	1	9	33	22
NED	0	5	24	15
with local failure (LF)	0	2	7 (1 c̄ comp)	6
with distant metastasis (DM)	0	0	2	0
with complications (comp)	1	2	0	1

[a] Abbreviations: c̄ = with; s̄ = without, NED = no evidence of disease or complication; LF = local failure; DM = distant metastasis; comp = complications.

Fifty-nine patients were treated with *mixed-beam* irradiation – thirty-four in the *pilot study* and twenty five in the randomized trial. The stage distribution of the *mixed beam* group is similar to that of the *photon* group, but less advanced than the stage distribution of the *neutrons only* group. Approximately half the patients treated with the *mixed beam* showed sufficient tumor regression

mixed beam irradiation is better than the local control rate for patients treated with *photons*. The *survival rates* are no different for the two groups because a greater number of *mixed-beam* patients died of distant metastasis, intercurrent disease, or complications (Table 3). Two of the *mixed-beam* patients died of pulmonary emboli following radium applications.

The results are analyzed by site and stage in Table 4. For moderately advanced disease (stages IIB and IIIA carcinomas of the cervix) the local control rate with *mixed-beam* irradiation (61 per cent) is similar to that achieved with *photons* (57 per cent). For massive tumors (stages IIIB and IV carcinomas of the cervix and massive tumors of other sites), the local control rate with *mixed-beam* irradiation (61 per cent) is superior to that achieved with *photons* (37 per cent). The survival rates for the *mixed beam* and photon groups are the same when analyzed by site and stage.

squamous carcinomas of the cervix were randomized to receive treatment with mixed beam irradiation or photons. Twenty-five were treated with the *mixed beam* and twenty with *photons*. The *stage distribution* was the same for both groups.

The preliminary results are listed in Table 5. With a limited duration of follow-up, the local control rate is superior with *mixed-beam* irradiation as compared to *photon* irradiation, but the survival rates are no different. Seventy-two per cent of patients treated with the *mixed beam* have had local control of their disease as

Table 4. Preliminary results for the total population by site and clinical stage
(April 1973–April 1978)
Unlimited follow-up–analysis August 1978

	Neutrons only	Neutron boost	Mixed beam	Photons
Local control	5/17 (29.4%)	11/20 (55.0%)	36/59 (61.0%)	19/40 (47.5%)
IIB CA Cx	—	1/2	4/4	2/4
IIIA CA Cx	0/2	3/6	13/24	10/17
IIIB CA Cx	1/7	5/8	14/23	6/15
IV CA Cx	1/2	1/1	2/3	1/3
Other sites	3/6	1/3	3/5	0/1
Survival	1/17 (5.9%)	9/20 (45.0%)	33/59 (55.9%)	22/40 (55.0%)
IIB CA Cx	—	1/2	4/4	3/4
IIIA CA Cx	0/2	2/6	13/24	9/17
IIIB CA Cx	0/7	3/8	14/23	9/15
IV CA Cx	0/2	1/1	1/3	1/3
Other sites	1/6	2/3	1/5	0/1
Complications	2/17[a] (11.8%)	4/20[b] (20.0%)	4/59[c] (6.8%)	3/40[d] (7.5%)

[a] 1 sigmoiditis, 1 small bowel obstruction.
[b] 1 proctitis, 2 small bowel obstructions, 1 small necrosis.
[c] 1 proctitis, 2 small bowel obstructions, 1 small bowel necrosis.
[d] 1 vault necrosis, 1 sigmoiditis, 1 small obstruction.

Complications. Thirteen patients (9.6 per cent) developed *major complications* (Table 4). The incidence of small bowel complications was greater with neutron therapy than with photons, possibly because more small bowel is irradiated when patients are treated in a standing position with a fixed horizontal beam. The complication rate was also influenced by recent surgery, since 15 per cent (5/33) of patients who were treated initially with selective lymphadenectomies developed major complications as compared to only 8 per cent (8/103) of patients who had not had recent *pelvic surgery*.

Preliminary results of the randomized trial

Between January 1977 and April 1978 forty-five patients with stages IIB, III and IVA

compared to 55 per cent in the *photon* group. Eighty per cent of the patients in each group were alive at the time of analysis. One patient in the *mixed-beam* group died of a pulmonary embolus at the time of radium insertion. Actuarial local control, complication, and survival curves are shown in Fig. 3.

Discussion

The crude results for patients treated with *neutrons only* are poorer than those for the other treatment categories. However, the patients in the *neutrons-only* group presented on average with more advanced disease and were treated during the early phase of the program when the dosage schedules were not well established and before the compression device had been developed. The highest actuarial local control rate was achieved with the *neutron-boost* technique (Fig. 2). However, patients in this group presented with less advanced disease than those in the other treatment categories. This group also had the highest incidence of complications (Table 4), probably because all received high doses of external-beam irradiation, using opposed antero-posterior portals.

because of an excess of deaths from distant metastases, complications, or intercurrent disease in the *mixed-beam* group (Table 3). The same trend is apparent in the randomized trial, even though no complications have occurred to date in the *mixed-beam* group (Fig. 3). These results point out the inadequacy of survival as an indication of local treatment effectiveness in patients with far advanced cancer, but in view of

Table 5. Preliminary results of the randomized trial in squamous carcinoma of the cervix

(January 1977–April 1978)
Analysis August 1978

Clinical Stage[a]	Local Control		Survival		Complications	
	Mixed beam	Photons	Mixed beam	Photons	Mixed beam	Photons
IIB	3/3	1/2	3/3	2/2	0/3	0/2
IIIA	5/10	5/8	8/10	7/8	0/10	0/8
IIIB	8/10	5/8	8/10	7/8	0/10	1/8
IVA	2/2	0/2	1/2	0/2	0/2	0/2
Total	18/25 (72.0%)	11/20 (55.0%)	20/25 (80.0%)	16/20 (80.0%)	0/25 (0%)	1/20 (5.0%)

[a] MDAH staging system.

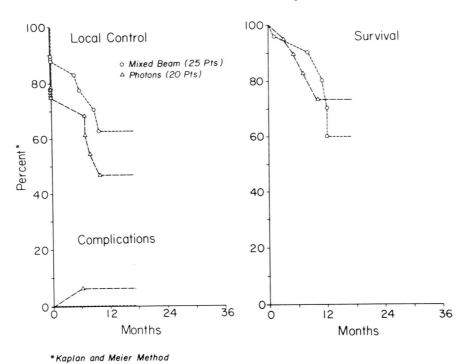

Fig. 3. Actuarial local control, complication, and survival curves (Kaplan and Meier method (1958)) for patients treated in the randomized trial.

The best therapeutic ratio overall was achieved with the *mixed-beam* technique. In relation to an evenly matched control group treated with photons alone, the *mixed-beam* group had a higher local control rate with a similar incidence of complications. In spite of the improved local control rate, no improvement in survival was noted with *mixed-beam* therapy

the subjective element in assessment of local tumor control, we believe that the absolute end point of survival should be included in all data reporting. It may well be that until effective therapy for metastatic disease is developed, significantly improved survival as a result of improved local disease control will not be realized.

In interpreting the results of this study, one

must keep in mind the dosimetric and technical disadvantages of neutron therapy equipment. Although the TAMVEC beam is one of the most energetic in clinical use, the dose distribution for neutron therapy is distinctly inferior to that which can be achieved with conventional 25-MV X-ray techniques, and the added constraint of a fixed horizontal beam represents a considerable disadvantage for pelvic irradiation since the intestines fall into the pelvis when the patient is standing.

It is clear from the results of the various pilot studies that prospective randomized clinical trials are required to determine the true value of fast neutron therapy. These should be carried out with hospital-based high-energy neutron therapy machines that are as versatile as modern-day photon therapy machines, so that the potential radiobiological advantages of neutron therapy are not obscured by the technical disadvantages of the equipment. Finally, in view of the relatively modest gains achieved in most of the trials reported, attention should be directed towards identifying those tumors most likely to respond selectively to neutron therapy (Withers, this meeting).

Acknowledgment—This investigation was supported in part by Grant CA 12542, awarded by the National Cancer Institute, DHEW.

References

CADERAO, J. B., D. H. HUSSEY, G. H. FLETCHER, V. A. SAMPIERE, D. E. JOHNSON and J. T. WHARTON (1976). Fast neutron radiotherapy for locally advanced pelvic cancer. *Cancer* 37, 2620-2629.

HUSSEY, D. H., G. H. FLETCHER and J. B. CADERAO (1975). A preliminary report of the MDAH–TAMVEC neutron therapy pilot study. In O. F. Nygaard, H. I. Adler, W. K. Sinclair (Eds), *Proceedings of the Fifth International Congress of Radiation Research,* Radiation Research, Academic Press, New York, pp. 1106–1117.

KAPLAN, E. L. and P. MEIER (1958). Nonparametric estimations from incomplete observation. *Amer. Statist. Assoc. J.* 53, 457–480.

WITHERS, H. R., B. L. FLOW, J. I. HUCHTON, D. H. HUSSEY, K. A. MASON, G. L. RAULSTON and J. B. SMATHERS (1977) Effect of dose fractionation on early and late skin responses to γ-rays and neutrons. *Int. J. Radiat. Oncol. Biol. Phys.* 3, 277–233.

Observations on the reactions of normal and malignant tissues to a standard dose of neutrons

MARY CATTERALL

Fast Neutron Clinic, Medical Research Council Cyclotron,
Hammersmith Hospital, London W12 0HS

Abstract—*A standard dose of neutrons (1560 rad given in 12 treatments over 26 days) has been given to nearly all of the 800 patients treated on the M.R.C. Cyclotron at Hammersmith Hospital. The effects of this dose on the skin, bone, cartilage, larynx and intestine are described in this paper. The results of the treatment of tumours of the head and neck and breast are also given and complications are discussed. Attention is drawn to the inadequacies of all neutron sources at present used for treating patients and to the possible dangers of attributing treatment failures to neutrons themselves, without recognizing the importance of poor penetration, wide penumbra and other physical characteristics. Better neutron machines at least equal to Co-60 gamma-sources are required before proper comparisons of neutron and photon treatments can be made in most sites and most tumours.*

Introduction

The clinical investigation of cancer treatments with fast neutrons from the Medical Research Council cyclotron at Hammersmith Hospital was designed to assess the effects of neutrons on patients with malignant disease as precisely as possible. The neutrons used were of 7.5 MeV energy and had a poor penetration into the tissues and a wide penumbra, being in fact equivalent to 250-kV X-rays and markedly inferior to megavoltage photons. This neutron energy resulted in uneven dose distribution throughout all the tumours, except those which were superficial and small. With the establishment of a satisfactory dose régime (1560 rad in 12 treatments over 26 days), this became the standard and was applied without alteration in 90% of the cases. But because of the low neutron energy, doses of only 900 rad were received by parts of some large tumours while the wide penumbra resulted in high spots of 2000 rad in some normal tissues. Neutrons were not combined with photons in any treatment. The effects were observed on a wide variety of tumours and all irradiated normal tissues in 800 patients during the years 1970–1978. One hundred and sixty of the patients have been observed for 2 to 8 years.

Effects on normal tissues

The effects on all normal tissues and the tolerance doses of specific structures such as the spinal cord, the eye, the stomach and the oropharynx are described and discussed fully elsewhere (Catterall and Bewley, 1979). In this paper some observations are given on the following.

The skin

This is a very complex organ and its structure and physiology differ widely in different parts of the body. Additional important factors influencing radiation effects are infection, trauma, age and nutrition. But the two most important factors in our experience are, firstly, the size of the area treated and, secondly, the site.

The effect of size is illustrated by two patients who both had tumours of the oropharynx and both were treated with parallel opposed fields. However, in one patient each field measured 14 × 8 cm and produced a confluent skin reaction over the front of the neck, measuring in all 420 cm². This resulted in marked fibrosis of the skin, subcutaneous tissue and the sternomastoid muscles. These were, however, painless and symptomless. The patient also had dysphagia for a time after treatment and this was treated by gastrostomy to give the throat a complete rest. Normal swallowing returned after 13 months and the gastrostomy was closed. In the other case each field measured 8 × 7 cm and each was surrounded by normal skin and blood vessels. This patient therefore had two separate areas each of 56 cm². Very little telangiectasis appeared and, on palpation, the skin felt soft and normal. Both were followed for more than 3 years after treatment.

Bone

In contrast to fat, the kerma for neutrons is low in bone and we have seen no necrosis of normal bone in any site following our standard dose.

Cartilage

Despite its relative hypoxia, we have seen no necrosis of normal cartilage nor have there been

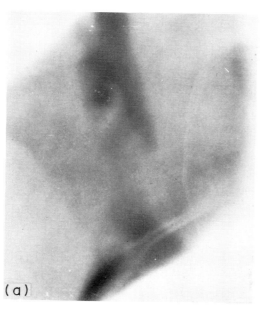

Fig. 1a. Neutrogram T_3N_3 carcinoma of larynx at first treatment.

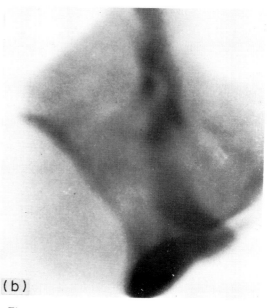

Fig. 1b. Neutrogram of tumor shown in Fig. 1a at end of treatment.

In areas where there is ample fat, for example the buttocks, the submandibular region and the breasts, firm fibrosis resulted, but this was quite painless, unless it became infected or unless trauma precipitated necrosis.

The main reason for the production of fibrosis is that in soft tissue hydrogen is responsible for the major part of energy transfer. Fat contains a high proportion of hydrogen and therefore absorbs a relatively large quantity of energy from fast neutrons (Bewley, 1970). Bewley points out that because of the high kerma for fat, the dose may be increased by 10–15%. With photons, on the other hand, fat receives a slightly lower dose than other soft tissues. Despite this difficulty with neutrons, we observed that our standard dose can be comfortably tolerated in all sites if the area does not exceed 100 cm². For larger areas, the factors already mentioned must be taken into consideration.

signs or symptoms of necrosis after treatment of advanced carcinoma of the larynx in which the cartilage was involved with tumour.

The larynx

Although the larynx is composed of mucosa, cartilage and skin and has a good blood supply, its anatomy is so complex that it will not tolerate 1560 cGy (rad) in 12 treatments over 26 days without producing severe symptoms. This conclusion is, of course, made on observations of the treatment of very advanced tumours.

Some tumours regressed quickly during treatment and in Figs. 1a and 1b the neutrograms taken at the first and last treatment show how a laryngeal tumour and the enlarged nodes regressed during treatment leaving significant air gaps. These caused inaccuracies in the prescribed dose and caused more radiation to be given to the mucosa and adjacent structures than had been intended. These errors can be avoided by repeating neutrograms and simulator check films at every fourth treatment and measuring the additional space which appears if the tumour is regressed quickly. Additional pieces of bolus can then be made to compensate for the gap, and fixed on to the original bolus in the appropriate places, thus restoring the dose distribution to that originally planned.

tumours with, therefore, smaller volumes irradiated and less damage done by the tumours. It is not yet known whether this dose will be adequate to sterilize the tumours.

The intestine

The patients treated so far had already had extensive and usually repeated surgery and, or, diathermy; infection and sepsis were also present in some. The dose distribution was very

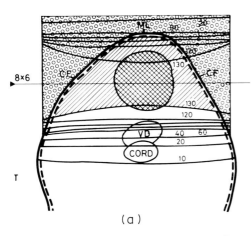

Fig. 2a. *Original plan for carcinoma of larynx, showing tumor within 130% isodose.*

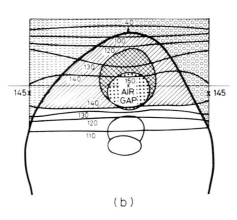

Fig. 2b. *Showing how the appearance of an air gap increased the dose to 150%.*

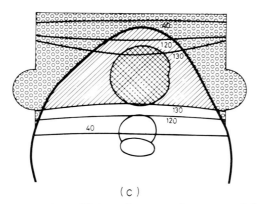

Fig. 2c. *The addition of accurately measured bolus restored the tumour dose to 130%.*

Figs. 2a and 2b show the original plan and how the dose distribution was changed by the air gap. Figure 2c shows how the compensators restored the isodoses to those intended at the start of treatment.

On the evidence of twelve patients followed for 7 to 36 months, it does appear that 1560 rad can be tolerated if the overall time is prolonged to 33 days and the fractions increased to 15 and by compensating for air gaps produced by rapidly regressing tumours. However, it must also be noted that these twelve patients had smaller

unsatisfactory and the wide penumbra caused the irradiation of an unacceptably large amount of normal bowel and vasculature. It is therefore impossible, at this stage, to give a tolerance dose. We believe that more penetrating, more sharply defined beams and earlier cases are required for this information.

Effects on tumours

On the evidence available the standard dose of 1560 cGy (rad) given in 12 treatments over 26 days is required to cause clinically complete regression with little chance of recurrence in adenocarcinomas, squamous cell carcinomas, soft tissue sarcomas and salivary gland tumours. Gliomas Grades III and IV completely regressed in most cases with the standard dose.

Table 1 summarizes the results of the controlled clinical trial of advanced tumours of the head and neck (Catterall, Sutherland and Bewley, 1975; Catterall, Bewley and Sutherland, 1977).

characteristics of the beam with its poor penetration and wide penumbra and its fixity in the horizontal position and, secondly, the very advanced stage of most of the tumours which in many cases had themselves caused irreparable damage to the normal tissues or required very large areas to be irradiated, thus making healing practically impossible.

These factors, in greater or less degree, apply also to all centres where neutrons are being investigated clinically. So marked are the

Table 1. Advanced tumours of the head and neck prospectively randomised controlled clinical trial. All sites

	Total	Complete regression	Recurrent	Persisting control	Significance	Total complications	Complications orolaryngo-hypopharynx
Neutron	71	55	1	54	p 0.001	12	8
Photon	63	27	15	12		2	1

Eight of the twelve complications were in the larynx or oropharynx and these appear to have been reduced by the measure described.

In Table 2 are the results of all tumours treated in the paranasal sinuses, oral cavity and salivary glands, not only those in the controlled clinical trial. These were all extremely advanced and would have had a zero to 30% chance of cure by photon radiation, according to the results of treatment reported in the literature of comparable tumours.

differences between the available neutron beams and the deeply penetrating, sharply defined isocentric beams of megavoltage machines that they risk masking the real effects of neutrons on tumours and normal structures. Despite this, many so-called "controlled" clinical trials are being undertaken to compare the treatment of tumours where the inequality of neutron and photon treatment plans is unacceptably great. One such site is the lower end of the oesophagus. At first sight, because neutrons of 7.5 MeV

Table 2. Noteworthy sites summary of all cases (head and neck) 1970–1978

Site	Total treated	Complete regression	Recurrence	Persisting control	Complication
Nasopharynx and antra	27	23	1 (4%)	22 (81%)	6 (22%)
Oral cavity	39	29	1 (3%)	28 (72%)	1 (3%)
Salivary glands	34	30	1 (3%)	29 (85%)	7 (4 had complete recovery) (9%)

In Table 3 are shown the results of a group of patients with advanced but mainly localized tumours of the breast. The numbers are small but complete regression of all tumours was achieved without complications and without recurrences for periods ranging from 8 months to 8 years.

energy are less attenuated by lung tissue and corrections can be made for the amount of lung traversed, it seems possible to deliver a considerably higher dose to the tumour than to the surrounding lung. At the central level a treatment plan looks satisfactory. However, with

Table 3. Advanced breast cancer

Stage	No. of patients	Complete regression	Duration (years)	Recurrence	Complications
$T_{4C}N_0M_0$	5	5	8, 6, 4, 3, (8 months)	0	2 Traumatic ulcers. Both healed
$T_{4C}N_{1b}M_0$	4	4	5, 3, 2, 1½	0	0
$T_{4C}N_{1b}M_1$	2	2	3 1	0	0
Total	11	11		0	0 in long term

Complications

Before ascribing complications exclusively to neutrons, two essential factors have to be considered. These are, firstly, the physical

a field length of 14 cm, the advantage given by the presence of lung tissue is lost because of the presence of the heart and diaphragm. When activation dose meters in a naso gastric tube are placed throughout the tumour volume, the dose at the ends of the treated area is found to differ by a factor of 2 (Catterall and Bewley 1979). Thus with the central part receiving 1560 rad, the top end received 1780 rad and the lower end 900 rad. It is clearly inappropriate to compare such unevenness of dose with a nearly uniform one which could be achieved with megavoltage photons.

All neutron beams, even of very high energy, will deposit more energy than from megavoltage photons in superficial structures and therefore, so long as very large fields of radiation are used, complications are likely to be more frequent. Premature clinical trials risk giving neutrons a further unmeritedly adverse report, such as the one received in 1945 and from which it took 20 years to reopen the clinical investigation.

Furthermore, in critically observing the effects of neutrons, the complications produced must be compared with those produced by surgery in the treatment of much smaller tumours. For example, fibrosis of the neutron-treated breast should be compared with a mastectomy, fibrosis at the site of a sarcoma in a limb, with amputation, reduced mouth opening with severing of the 7th nerve, and loss of an eye with exenteration of the orbital cavity and maxillectomy. It also has to be remembered that in the treatment of the tongue and the floor of the mouth, neutrons result in virtually no complications compared with surgical excision of the tongue, the floor of the mouth and the mandible.

The contrast in energy absorption between neutrons and photons in fat and bone is sometimes forgotten. But it does exist and in the writer's opinion makes the use of a single RBE for the interchange of neutron photon doses inappropriate. Nor does the term "rad equivalent dose", which appears in some protocols, appear to be justified.

Conclusion

The consensus of many experienced clinicians, based on the evidence from prospectively controlled trials and retrospective comparisons, is that fast neutrons, in the treatment régime used at Hammersmith Hospital, promise a significant advance in the treatment of some tumours. It must be pointed out, however, that extreme care has been taken at all stages in localization, planning, treatment and follow-up of the patients. Although much of this care was taken to try to deal with the physical inadequacies of the neutron beam, effective treatment with neutrons and the reduction of complications will always demand meticulous attention to detail at all stages.

There is an urgent need to continue and to expand the investigation of fast neutrons in the treatment of solid tumours. For such clinical trials to be undertaken, however, machines must be made available which produce neutrons with clinically acceptable physical characteristics. These include a reliable beam which is as penetrating and sharply defined as that of ^{60}Co with an equally good output and an isocentric head. Such machines should be sited within selected hospitals.

References

BEWLEY, D. K. (1970) *Fast Neutron Beams for Therapy in Current Topics in Radiation Research*, North Holland Publishing Co., p. 249.

CATTERALL, M. I. SUTHERLAND, and D. K. BEWLEY (1975) First results of a randomised clinical trial of fast neutrons compared with X or gamma rays. In *Treatment of Advanced Tumours of the Head and Neck. British Medical Journal* 2, 653.

CATTERALL, M., D. K. BEWLEY and I. SUTHERLAND (1977) Second report of a randomised clinical trial of fast neutrons compared with X or gamma rays. In *Treatment of Advanced Tumours of the Head and Neck. Brit. Med. J.*, 1, 1642.

CATTERALL, M. and D. K. BEWLEY (1979) *Fast Neutrons in the Treatment of Cancer*, Academic Press.

Results of fast neutron radiotherapy at Amsterdam

J. J. BATTERMANN AND K. BREUR

Department of Radiotherapy, Antoni van Leeuwenhoek Hospital, Amsterdam, The Netherlands

Abstract—*An analysis of the results of fast neutron radiotherapy at Amsterdam is given. Although the local tumour control rate is high for head and neck tumours (about 80%) and for bladder and rectal tumours (72%), the complication rate was higher than after photon treatment (in the head and neck area 6% complications, in the pelvic area 20%). The results for brain tumours are disappointing.*

An RBE value of 15 MeV neutrons relative to ^{60}Co gamma-rays is estimated at 2.8 for damage to subcutaneous tissue and at 3.5 for damage to intestinal mucosa, at doses of 80–90 rad ($n + \gamma$) per fraction.

Introduction

Clinical interest in the use of fast neutron therapy is due to the fact that, even with modern high-voltage machines, the local cure rate for some tumours remains poor. As hypoxia could be one of the main reasons for radioresistance to X-rays, the lower OER for neutrons would be a great advantage. In addition a diminished influence of repair of sub-lethal damage for neutrons in comparison with X-rays might provide an advantage for the treatment of some tumours (Barendsen, 1966). A possible therapeutic gain due to these properties has been investigated first for treatment of advanced, localized tumours. The encouraging results, obtained in a controlled clinical trial on head and neck tumours by Catterall, Sutherland and Bewley (1975, 1977) were a stimulus to other centres to start clinical studies, also on other tumour sites (Parker et al., 1977).

In The Netherlands a 14-MeV d-T machine was installed in 1975 in the Antoni van Leeuwenhoek Hospital (Netherlands Cancer Institute) and is used routinely since April 1976.

The effects of fast neutrons on tumours and normal tissues, observed in more than 200 patients that have been treated till June 1978, will be discussed.

Material and methods

Neutron beam

In our institute we use a 14-MeV fast neutron beam from a sealed d-T tube, developed by the Philips Research Laboratories, Eindhoven. The output of the apparatus is about 10^{12} n/sec, giving a dose rate of 6-8 rad/min in air at 80 cm SSD. More detailed physical information has been given elsewhere (Broerse and co-workers, 1977).

Clinical material

Only patients for which conventional treatment methods could not offer a reasonable chance of cure were selected for our pilot studies. Most patients had very advanced tumours, making them unsuitable for radical surgery and often they were in a poor general condition or of very advanced age.

More than 200 patients have been treated since the installation of the machine. In May 1978 a controlled clinical trial was started for head and neck, rectal and bladder tumours.

In Table 1 a summary is given of the tumour sites treated.

Treatment methods

The patients were treated five times per week with fast neutrons only. Occasionally a neutron boost was given in addition to a photon dose of 3000–4000 rad, but these results will not be discussed here.

Both in the head and neck area and in the

pelvis the dose given was 1720* rad neutrons (midline) in 20 fractions over 26 to 28 days. Most patients were treated with two opposing fields, applying both fields daily. During the study wedge filters of 45° became available, to be used in appropriate situations. On the basis of clinical observations during the study, the total dose and treatment technique in the pelvic area had to be modified. The dose was subsequently changed to either 1600* rad/4 weeks or to 1750* rad/5 weeks (midline). By means of a six-fields technique a more homogenous dose distribution in the target area and a reduction of the dose in the vulnerable normal tissues could be obtained. This was facilitated by the use of a computer program, developed by Drs. v. d. Laarse.

Table 1. Tumour sites treated with fast neutrons

Head and neck		53
Parotid glands		11
Pelvis	bladder	29
	rectum	34
	female genital	7
Brain		17
Soft tissue		15
Lungs	Pancoast	3
	studies on metastases	27
Miscellaneous		19
Total		215

Results

The data were analysed in July 1978. The interval from the end of treatment till the date of analysis ranged from 3 to 30 months. The number of patients, alive and without evidence of tumour in the treated area at the time of analysis, was scored as NED.

Table 2. Number of patients treated in the head and neck area

Total number of patients	53
Treatment stopped, due to poor condition	5
Combined treatment with photons	2
Deceased within one month after treatment	3
Evaluable number of patients	43

For a few tumour sites sufficient experience has been obtained to allow an analysis of tumour regression rate, normal tissue reactions and complication rate.

Head and neck tumours

In the head and neck area fifty-three patients have been irradiated with fast neutrons. Table 2 shows the numbers of patients suitable for evaluation.

Tables 3 and 4 present the results obtained up till now. Although all patients had advanced tumours (mostly T3, T4), often with regional nodes, thirty-four out of forty-three patients (79%) had complete clinical tumour regression. The observation period, however, is rather short in most cases.

Necropsy could confirm the clinical observations in eight cases (three with and five without local recurrences). Most recurrences were seen in patients with very advanced tumours of the tongue (T4, N3). As could be expected in view of the advanced stage of the primaries and the age of the patients a number of them died of distant metastases or of intercurrent diseases.

Most patients developed a moderate degree of edema of the larynx as reaction on neutron irradiation; in one patient a tracheostomy was needed because of severe edema. Subcutaneous

Table 3. Results of fast neutron treatment of advanced head and neck cancer

Site	Number	Local NED[a]	Persistent tumour[a]	Local recurrence[a]	Deceased[a]
Nasopharynx	1	1(24)	—	—	—
Oropharynx	9	7(3,4,5,7,12,14[b],24)	2(2,3)	—	7(2,3,3,4,5,12,14)
Tongue	15	9(½,1,4,4,6,6,9,14,16)	2(3[b],3)	4(1,8[b],8,10)	10(½,1,1,3,4,6,6,10,14,16)
Hypopharynx	5	5(½,5,6,8,10[b])	—	—	3(½,8,10)
Larynx	8	7(?,7,8,8,8[b],14[b],18)	—	1(20[b])	7(?,7,8,8,8,18,20)
Neck nodes	2	2(4,6)	—	—	—
Miscellaneous	3	3(7,13,24)	—	—	1(7)
Total	43	34(79%)	4(9%)	5(12%)	28(65%)

[a] In brackets the follow-up time is given in months after the end of treatment.
[b] Confirmed at necropsy.

*Just recently we had to conclude that our given dose was about 8% higher than the value derived initially. The given dosages are corrected values.

fibrosis was seen in practically all patients; the degree of fibrosis was correlated with the volume treated. To decrease this fibrosis we now reduce the dose with 5% when the field sizes are more than 80 cm^2.

Tumours in the pelvis

Patients were treated for inoperable bladder carcinoma (T4), rectum carcinoma (both primaries and recurrences) and gynaecological tumours.

Table 4. Cause of death in head and neck patients

Site	Number	Deceased			
		without evidence of local recurrence metastases/	intercurrent	with evidence of local tumour	of severe complications
Nasopharynx	0/1	–	–	–	–
Oropharynx	7/9	2	3	2	–
Tongue	10/15	1	6	3	–
Hypopharynx	3/5	2	1	–	–
Larynx	7/8	2	2	1	2
Neck nodes	0/2	–	–	–	–
Miscellaneous	1/3	–	–	–	1
Total	28/43	7	12	6	3

Three patients died with complications; in one patient a very severe fibrosis developed after 1730 rad/20 fractions/4 weeks with a field size of 160 cm^2. In the other patient the total dose was raised to 1900 rad as there was hardly any mucosal reaction after 1720 rad. This patient died of a bronchus carcinoma but also had a necrosis of the larynx. The third patient developed a moderate fibrosis and edema of the larynx and died of pneumonia.

The local results in tumours of the *parotid glands* are promising (Table 5). A complete tumour regression was seen in ten of eleven patients with inoperable primaries or local recurrences.

Table 5. Results of fast neutron treatment for parotid gland tumours

Total treated	11
Alive, NED:	4 (2,7,13,20 months)
Alive, local cured, metastases	1 (24 months)
Deceased	6 (<1,4^a,5,7,15,17^b months)
complications	2
metastases	1
intercurrent	3

^{a} Skin necrosis,
^{b} Myelopathy.

Two patients died of complications. Both, however, had a recurrence after a high dose supervoltage X-rays to the same area. At necropsy no tumour was found.

Table 6. Number of patients treated in the pelvic area

Total number of patients:	70
Treatment stopped, due to poor condition	5
Combined treatment with photons	6
Deceased within 1 month after end of treatment	2
Evaluable number of patients	57

Table 6 gives the total number of patients treated and Tables 7 and 8 show the local results.

As can be seen from Table 8 the complication rate in this group is high. Almost 15% (8/53) died of peritonitis, due to an excessive dose in the intestine. Also some of the patients that are alive have a severe irritation of the bowel, making a colostomy necessary in four of them.

The early reactions after irradiation of the pelvis did not suggest such a high late complication rate. The initial complaints usually were less than after a dose of 5000 rad photons in 4 weeks. However, after 4 to 10 months, many patients developed severe bowel complications. In eight cases even a necrosis of the intestine was found in the maximum dose area. We were thus forced to reduce the maximum dose in the treated area. This is done by using six fields, to improve the dose distribution, and by a reduction of the total dose to 1620 rad. Three patients were without complications for 6–9 months after the end of treatment, using this new régime, but one died of distant metastases.

All patients with severe complications died of a peritonitis due to intestinal necrosis. At necropsy no tumour was found in the treated area. In total necropsy was done in twelve patients, only in one patient tumour was found in the treated area.

Brain tumours

Seventeen patients with advanced tumours of the brain (astrocytoma grade III, glioblastoma multiforme) were treated with a combination of photons (3060 rad/17×/23 days) on the whole brain and a boost of fast neutrons (1120 rad/16×/22 days) on the target area. This

Table 7. Results of fast neutron treatment of advanced tumours in the pelvis

Site	Number	Local NED[a]	Persistent tumour[a]	Local recurrence[a]	Deceased[a]
Bladder	23	17 (2,3[b],4,4,4,5, 5,6[b],7,8[b],8,9,9, 14,14,28)	5(1,3,4,5,6)	1 (4)	16 (1,2,3,3,3,4,4,4,5, 5,6,6,7,8,9,14)
Rectum	27	19 (0,3[b],3,4[b],4,4,5, 6[b],6[b],6,8,9,10[b], 13,15[b],15,17,20[b],20)	6(1,3,3,4,6,6)	2(3[b],8)	16 (0,3,3,3,4,4,4,5,6, 6,6,6,9,10,15,20)
Female genital	7	5 (4,5,7,8,9[b])	2(2½,4)	–	5 (2½,5,7,8,9)
Total	57	41 (72%)	13 (23%)	3(5%)	37(65%)

[a] In brackets the follow-up time is given in months after the end of treatment.
[b] Confirmed at necropsy.

Table 8. Cause of death of patients with pelvic tumours

Site	Number	Deceased			
		without evidence of local recurrence	metastases/intercurrent	with evidence of local tumour	of severe complications
Bladder	15/23	3	5	2 + 3	2
Rectum	16/27	4	1	2 + 4	5
Gynaecological	5/7	2	1	1	1
Total	36/57	9	7	4 + 8	8

scheme was chosen because of the poor results described by Catterall (1975) and by Parker (1976) when only neutrons are used. However, as Table 9 shows, the results in our series are not better than the results in other series. The historical control results obtained with an almost equivalent scheme of photon irradiation (4000 rad whole skull, boost of 2000 rad, total in 7 weeks) are the same. The mean survival time in this group was 8 months, in the neutron-treated group 6 months.

Table 9. Results of fast neutron treatment for advanced brain tumours

Total treated	17
Alive, NED	5 (4,6,7,10,10 months)
Alive, recurrence	2 (11,16 months)
Deceased	10 (1,1,2,5,6,7[b],8[a],8[b],8,12 months)

[a] No recurrence at necropsy, focal necrosis.
[b] Recurrence at necropsy.

Sarcomas

Of fifteen patients treated for very advanced soft tissue sarcomas only six showed a good tumour regression. Perhaps this is due to the fact that many of these patients died within a few months of distant metastases, already present at the moment of first treatment. A better selection and a longer follow-up are needed to obtain more detailed results.

Estimations of RBE values

About 10 years ago patients were treated for T4 bladder carcinoma with ^{60}Co gamma-rays, using two opposing fields. In 4 weeks a total dose of 5200 rad to 6200 rad (midline) was administered to the pelvis. The higher the dose used, the better the local control rate. As the depth dose distribution of ^{60}Co gamma-rays is approximately comparable with the 14-MeV neutron beam, the maximum doses were used to compare skin and subcutaneous tissue damage, caused by gamma-rays and fast neutrons, respectively, using a scoring system as given in Table 10. Although physical measurements

Table 10. Scoring system used for skin and intestinal damage as late effect of neutron and gamma radiation

Skin	Intestine
0–2: mild reaction	0–2: mild reaction
3: fibrosis	3: persistent diarrhoea
4: severe fibrosis, shrinking	4: colostomy
5: necrosis	5: necrosis

show some build-up in the 14-MeV neutron beam the clinical importance of this skin sparing is minimal in fast neutron therapy.

Figure 1 shows the skin and intestinal complications seen with different maximum doses of fast neutrons and ^{60}Co gamma-rays. Both in ^{60}Co gamma irradiations and in fast neutron irradiations the dose maximum is 10% to 25% higher than the dose in the target area, depending on field sizes and patient diameter. Apart from the neutron dose we have to take into account a gamma component of the neutron beam of 200–300 rad depending on field size and depth.

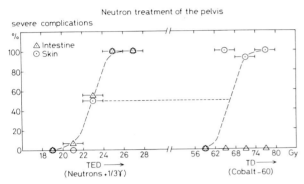

Fig. 1. *Relations of fractions severe complications and total doses in skin and intestine of patients treated with 15 MeV neutrons and ^{60}Co gamma radiation, respectively. At the 50% level a RBE of 2.8 is obtained for skin.*

Although we did not have the same scoring possibilities in the ^{60}Co γ-ray group, as most patients died within $\frac{1}{2}$ to 1 year after treatment and were not scored in those days, we could derive from the records that there was a moderate to severe fibrosis in all patients without signs of skin necrosis. None of the patients had severe intestinal complications, but some of them suffered from a persistent diarrhoea.

Using the severe complication rate (score 4) for skin and intestine (score 3) we can estimate RBE values of, respectively, 2.8 and 3.5 relative to ^{60}Co gamma-rays. Complete tumour regression for more than 4 months with ^{60}Co irradiation was 33% (6200 rad/4 weeks). For fast neutrons the local tumour control rate was about 80% (1720 rad/4 weeks). Insufficient data are available yet to determine RBE values for bladder tumour responses. The presented data suggest a RBE of about 4 or possibly even higher.

By using six fields we obtain a better dose distribution and can avoid maximum dose spots of more than 1800 rad. With these doses no severe complications from skin or intestine have been seen up till now.

Similar local results were achieved in *rectal cancer*. From our centre Tierie (1978) described sixty-six patients who received 6500 rad photons from an 8-MeV linear accelerator in 6 weeks. It was shown that no patient was locally cured. In the fast neutron group most patients remained without local recurrences up till now, but the complication rate was much higher, as compared with the X-ray treatments.

Conclusions

The first results show high tumour control rates in most sites of advanced cancer. Our estimations for subcutaneous and intestinal damage show a RBE relative to ^{60}Co gamma-rays of 2.8 and 3.5. The local tumour control rate, however, seems to be much higher with fast neutrons than can be achieved with photons. The need for controlled clinical trials in head and neck, bladder and rectal cancer is obvious. In May 1978 we started such trials. Particularly favourable effects were seen in inoperable salivary gland tumours. On the contrary there was no apparent advantage for malignant gliomas. Although the d-T machine in our institute has proved to be very reliable, the output is too low to treat a sufficient number of patients daily. International co-operation, e.g. in the EORTC Fast Particle Working Group, is therefore necessary to obtain within a short time conclusions of a more final nature.

References

BARENDSEN, G. W. (1966) Possibilities for the application of neutrons in radiotherapy: Recovery and oxygen enhancement of radiation induced damage in relation to linear energy transfer. *Europ. J. Cancer* **2**, 333–345.

BROERSE, J. J., D. GREENE, R. C. LAWSON, and B. J. MIJNHEER (1977) Operational characteristics of two types of sealed-tube fast neutron radiotherapy installations. *Int. J. Radiat. Oncol. Biol. Phys.* **2**, (suppl. 2), 361–365.

CATTERALL, M. (1975) Personal communication.

CATTERALL, M., I. SUTHERLAND and D. K. BEWLEY (1975) The first results of a randomized clinical trial of fast neutrons compared with X- or gamma-rays in treatment of advanced tumours of the head and neck. *Brit. Med. J.* **2**, 653–656.

CATTERALL, M., I. SUTHERLAND and D. K. BEWLEY (1977) Second report on results of a randomized clinical trial of fast neutrons compared with X- or gamma-rays in treatment of advanced tumours of the head and neck. *Brit. Med. J.* **1**, 1642.

PARKER, R. G., H. C. BERRY, A. J. GERDES, M. D. SORONEN and C. M. SHAW (1976) Fast neutron beam radiotherapy of glioblastoma multiforme. *Amer. J. Roentgenol.* **127**, 331–335.

PARKER, R. G., H. C. BERRY, J. B. CADERAO, A. J. GERDES, D. H. HUSSEY, R. ORNITZ and C. C. ROGERS (1977) Preliminary clinical results from fast neutron teletherapy studies. *Cancer*, **40**, 1434–1438.

TIERIE, A. H. (1978) Radiotherapy in marginal resectable and non-resectable rectum cancer. *Radiologica Clin.* **47,** 222–227.

Results of fast neutron beam radiotherapy pilot studies at the University of Washington*

T. GRIFFIN, J. BLASKO AND G. LARAMORE

Division of Radiation Oncology, University of Washington Hospital, Seattle, Washington 98105, U.S.A.

Abstract—*Fast neutron beam clinical trials were started at the University of Washington in 1973. Since that time, thirty-seven patients have been treated for Grades III and IV astrocytomas; thirty-six of these patients have died. The mean survival was 10.8 months for patients with Grade III lesions and 7.5 months for patients with Grade IV lesions. Fifteen of these patients had autopsies and identifiable cancer was seen in only one instance. One hundred and thirteen patients were treated for metastatic cervical adenopathy from squamous cell cancers of the head and neck. With a mean follow-up time of 14.5 months, 54% of the total group remained locally free of disease. Thirty-seven per cent of patients treated with neutrons only and 61% of patients treated with mixed beam irradiation maintained a complete remission at the sites of their adenopathy ($p<.025$). Fewer complications were seen in the mixed-beam irradiated group. One hundred and twenty-six patients were treated for primary squamous cell carcinomas of the head and neck region. With a mean follow-up time of 1.5 years, the local control rate for the entire group was 51%. Twenty-eight per cent of those treated with neutrons alone and 63% of those treated with mixed beam were locally controlled. Mixed-beam therapy appears to be superior to neutrons only, both in terms of complication rates and local tumor control in these tumor systems.*

Introduction

A National Cancer Institute-supported study of fast neutron beam radiation therapy of human malignancies was begun at the University of Washington in 1971. Approximately 2 years later, after a medical treatment beam was developed and biological characterization of that beam including RBEs for several tissues and OERs at several sites in an absorber were determined, clinical trials were started. Between 10 September 1973 and 13 May 1977, over 200 patients were treated in pilot studies, most with advanced cancers of the head and neck region including Grades III and IV astrocytomas. The purpose of this paper is to report on the results of these studies in terms of local tumor control, normal tissue effects and patient survival.

Neutron beam characteristics

The neutron beam used in these studies was generated using the University of Washington cyclotron to accelerate deuterons to 21 MeV which then impacted on an intermediate thickness, water-cooled, beryllium target (Wootton, 1975). At the nominal treatment distance of 150 cm from the collimator apparatus, the average treatment dose rate was 40 rad/min. Estimated energy distributions for the output neutron beam indicate a broad peak centered at approximately 8 MeV with a range of the 50% level between 3–14 MeV. Isodose measurements in a TE phantom show the 50% level at a depth of approximately 9 cm at the center of the beam. Neutron-nuclei interactions produce photons and because the neutron flux and energy distribution varies with depth, so does the photon contaminant in the beam. For the field sizes used in these studies, we estimate the photon dose fraction to be < 10% at a depth of 10 cm^2. The phantom measurements include this photon flux as well and it is implicitly included when we use

---

* Supported in part by National Cancer Institute Grant CA 12441.

the term "neutron rad" ($rad_{n\gamma}$) (Weaver et al., 1979).

Dose-fractionation patterns

Prior to the institution of clinical trials, the relative biologic effectiveness (RBE) of neutrons relative to photons was estimated using animal models (Wootton, 1975; Geraci et al., 1974; Geraci et al., 1975; Geraci et al., 1975; Nelson et al., 1975). Single dose experiments on various tissues with doses in the range of 25 to several hundred rad resulted in RBE values ranging from 1.0 to 3.0. Studies with increased fractionation for early skin damage and late foot deformity in mice revealed RBE values in the range of 2.5–3.0; however, these dose-fractionation schemes did not accurately correspond to the more highly fractionated schemes used on patients. For general purposes of comparison, we assumed an RBE = 3 relative to ^{60}Co for our neutron beam. It may be that the actual value for some tissues is higher than this.

In an attempt to adapt standard photon treatment policies at the University of Washington to neutron therapy, total photon equivalent doses of 900 rad/week were given in these clinical trials. Photon equivalent doses were calculated by multiplying the neutron rad dose by an RBE of 3, and then adding this to the photon dose.

Four patterns of treatment were investigated during the period of study: (1) 150 $rad_{n\gamma}$ two times per week for approximately 6–7 weeks; (2) 100 $rad_{n\gamma}$ three times per week for approximately 6–7 weeks; (3) 75 $rad_{n\gamma}$ four times per week for approximately 6–7 weeks, and (4) 60 $rad_{n\gamma}$ on Monday and Friday plus 180 rad ^{60}Co γ-rays on Tuesday, Wednesday and Thursday (mixed beam) for 6–8 weeks. In addition, a few patients were treated with a neutron boost after various doses of photon irradiation. Early in the study, patients were randomly assigned to either the 2- or 3-neutron-fractions-per-week treatment pattern. Based on Dr. Rasey's studies in our laboratory suggesting an increased therapeutic ratio for two neutron fractions plus three fractions of conventional photons per week (Rasey et al., 1977), the mixed beam treatment option was added. A 4-neutron-fraction-per-week treatment pattern was later adopted by all neutron programs in the United States as the standard neutron-only fractionation scheme. Total photon equivalent doses for patients who completed their planned course of treatment ranged from 5100 to 7020 rad (Table 1).

Material, methods and results

The analysis of data generated by this clinical trial will be presented in four categories: (1) Grades III and IV astrocytomas; (2) metastatic cervical adenopathy from squamous cell carcinomas of the head and neck region; (3) primary squamous cell carcinomas of the head and neck region; (4) carcinomas of the major salivary glands. With few exceptions, patients entered on these studies were judged to have less than a 10% chance of survival with conventional cancer therapy.

Grades III and IV astrocytomas

Thirty-seven patients with biopsy-proven glioblastoma multiforme were accepted for fast neutron radiation therapy. Tissue specimens were reviewed by neuropathologists and were graded III or IV according to the classification of Kernohan et al. (1952). Fifteen patients had Grade III astrocytomas and twenty-two had Grade IV lesions. The mean age of those patients with Grade III lesions was 47.3 years while the mean age of those with Grade IV lesions was 49.9 years. The location of the tumors is shown in Table 2. Several patients had tumor present in

Table 2. Tumor location for Grades III and IV astrocytomas

Location	Grade III	Grade IV
Frontal lobe	8	4
Parietal lobe	3	11
Temporal lobe	5	8
Occipital lobe	1	z3

more than one lobe and these multiplicities of tumor location are reflected in the figures shown in the table. In no case was there evidence of tumor involvement below the tentorium. Regarding their initial surgery, nine patients had only a biopsy or very limited resection of their tumor, eleven patients had a moderate tumor resection and sixteen patients had an extensive resection of their tumor. All patients referred to us during the period of study were accepted for

Table 1. Fractionation schemes and photon equivalent dose ranges used in the fast neutron beam radiotherapy pilot studies

Fractions[a]/week	Total photon equivalent dose[b]
2 neutron	5400–6600
3 neutron	5100–6600
4 neutron	5400–7020
2 neutron and 3 photon	5760–7020

[a] All fractionation schemes give a total photon equivalent dose of 900 rad/week
[b] Photon equivalent dose = neutron rad dose × RBE of 3 + photon dose.

treatment regardless of their condition in order to avoid biasing the results. Pretreatment evaluation consisted of history and physical examination with assessment of neurological function according to the classification system of Order et al. (1968) (Table 3). At the initiation of treatment, thirteen patients were functional Class I, fifteen patients were functional Class II, six patients were functional Class III and three patients were functional Class IV.

Table 3. Functional classification of patients after Order et al. (1968)

Class	Definition
I	Intellectually and physically able to work; neurological findings minor or not present.
II	Intellectually intact and physically able to be home although some nursing care may be required; neurological findings present but not a major factor.
III	Major neurological findings requiring hospitalization and medical care and supervision.
IV	Requires hospitalization and is in serious physical and neurological state.

Initially, patients were treated with neutrons only, either 2 or 3 fractions per week. Twenty-one completed their planned course of treatment to their entire brain to doses of 1550–1850 $rad_{n\gamma}$. Five patients were unable to complete their planned course of therapy because of progressive neurological deterioration. They received doses ranging from 250–1250 $rad_{n\gamma}$. Ten patients were treated with mixed-beam irradiation to their whole brain with or without mixed-beam boosts to the tumor-bearing volume. They received whole-brain photon equivalent doses of approximately 5000 rad with boost photon equivalent doses of approximately 1500 rad. One patient recieved 5000 rad ^{60}Co whole-brain irradiation and a 450 rad_n boost. He was treated on a randomized prospective protocol sponsored by the Radiation Therapy Oncology Group, and although he is currently alive and clinically free of disease, he will be excluded from further consideration as he was not part of our original pilot study.

All thirty-six patients treated in our pilot studies are now dead. Survival times for patients with Grade III and Grade IV lesions are outlined in Table 4 where they are compared with survival times for a control group of patients who were previously treated with conventional photon irradiation in our institution. The mean survival after completing treatment was 10.8 months for patients with Grade III lesions and 7.5 months for patients with Grade IV lesions. Particularly for Grade III lesions, the average survival was appreciably less than that of patients treated with photons alone. The neutron-treated patients also had no noticeable improvement in quality of survival when compared with the results of conventional photon irradiation. Figure 1 graphically compares the survival of patients treated with neutrons only, mixed beam and photon controls. There was no significant difference in survival between the group treated with neutrons only or mixed beam.

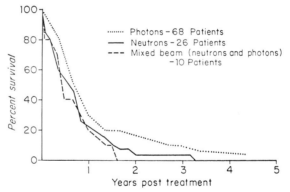

Fig. 1. Survival of patients with Grades III and IV astrocytomas after photon, neutron or mixed-beam irradiation

Autopsy data is available on fifteen of the thirty-six patients, and in only one instance could

Table 4. Comparative survival of patients with astrocytomas Grades III and IV

	Grade	No. patients	Survival (months) Average	Range
Photon	III	15	26.0	2.5–55
	IV	53	9.9	1 –38
	Combined	68	13.3	1 –55
Neutron	III	11	12.6	0.5–39
	IV	15	7.0	0.5–18
	Combined	26	9.4	0.5–39
Mixed beam (neutron/ photon)	III	4	6.0	0.5–16
	IV	6	8.7	1 –19
	Combined	10	7.6	0.5–19

gross tumor progression be documented (this patient also had liver metastases). The results for the fourteen remaining patients showed that for both groups of patients—neutrons alone and mixed-beam therapy—the bulk of the tumor had been replaced by a localized mass of coagulation necrosis. In general, there were some abnormal cells intermixed with the regions of coagulation but there was no real evidence of residual or recurrent tumor. The neuropathologists currently feel these abnormal cells are reactive astrocytes. A radiation-induced diffuse gliosis and white matter demyelination were found in regions far from the tumor volume and are thought to be related to the ultimate cause of death in these patients.

Metastatic cervical adenopathy from squamous cell carcinomas of the head and neck region

One hundred and thirteen patients with metastatic cervical adenopathy from squamous cell carcinomas of the head and neck region were treated with fast neutron beam irradiation between September 1973 and March 1977. Many patients presented with multiple neck nodes on both sides of the neck. They were classified according to the maximum dimension of the largest node mass. Twenty-eight patients presented with adenopathy measuring 3 cm or less in maximum dimension; sixty patients presented with single or multiple nodes measuring from 3–6 cm in maximum dimension, and twenty-five patients presented with adenopathy measuring greater than 6 cm (Table 5). The largest palpable mass measured 12 cm in maximum dimension. Patients were treated with the fractionation patterns and total doses previously outlined in Table 1. Thirty-three were treated with neutrons only and 80 were treated with mixed beam irradiation. All patients were evaluated at least once weekly during treatment and periodically thereafter. Node sizes were recorded and response rates graded at each examination.

Table 5. Size of neck nodes in maximum dimension

	Size in cm		
	3	3–6	6
Number of patients	28	60	25

With a maximum follow-up time of 49.9 months (mean 14.5 months), 54% of the total group remain disease free at the sites of their cervical adenopathy. Thirty-seven per cent of the patients treated with neutrons only and 61% of those treated with mixed beam irradiation maintained a complete remission. Mean follow-up times for the two groups are comparable.

Of the twenty-eight patients with adenopathy measuring 3 cm or less in greatest dimension, four were treated with neutrons only and have been followed for up to 22.4 months (mean 14.1 months). Twenty-four were treated with mixed beams with a mean follow-up time of 14.4 months (maximum 22.8 months). Three of the four (75%) neutron-only patients remained clinically disease free at the sites of their adenopathy either until the present time or until the time of their death due to uncontrolled primary disease. Eighty-eight per cent (21/24) of the patients treated with mixed beams maintained a complete remission. Many of these patients had multiple and/or bilateral adenopathy (Table 6).

Table 6. Per cent of patients remaining free of disease at the sites of their adenopathy. Four neutron-only patients were in the ≤3-cm group, twenty-one in the 3–6-cm group and 8 in the >6-cm group, twenty-four patients treated with mixed beam were in the <3-cm group, thirty-nine in the 3–6-cm group and seventeen in the 6-cm group

	Size of adenopathy		
	≤3 cm	3–6 cm	>6 cm
Neutrons only	75%	38%	13%
Mixed beam	88%	59%	29%

Of the sixty patients who had single or multiple nodes measuring from 3–6 cm in maximum dimension, twenty-one were treated with neutrons only and thirty-nine were treated with mixed beams. The maximum and mean follow-up times for the neutron-only and mixed-beam groups were 49.9 months maximum and 14.8 months mean, and 43.8 months maximum and 14.8 months mean, respectively. Thirty-eight per cent of the neutron-only group and 59% of the mixed beam group have maintained a complete remission of their adenopathy.

Of the twenty-five patients who had adenopathy measuring greater than 6 cm, eight were treated with neutrons only and seventeen were treated with mixed beams. With a maximum follow-up time of 30.7 months (mean 12.7 months for the neutron-only group and 14.3 months for the mixed-beam group), 13% of the neutron-only treated patients and 29% of the mixed-beam treated patients maintained a complete remission of their adenopathy.

If the possibility of further seeding of cervical lymph nodes is eliminated by excluding patients whose primary tumors were uncontrolled at the time of tumor regrowth in the neck, the overall control rate for the remaining group of patients was 72%. One hundred per cent of the nodes 3 cm or less were controlled, 82% of nodes measuring from 3–6 cm were controlled and 42% of the nodes greater than 6 cm were controlled. Of patients treated with mixed beams, 96% of nodes measuring from 3–6 cm were controlled and 50% of nodes measuring greater than 6 cm were controlled. Table 7 compares these results with those reported from the M. D. Anderson Hospital after conventional photon irradiation (Schneider et al., 1975).

Table 7. Per cent of patients with primary tumor control remaining free of disease at the sites of their adenopathy. The neutron-treated group includes both neutrons only and mixed beam; twenty-four were in the ≤3-cm group, thirty-six were in the 3–6-cm group and fifteen were in the >6-cm group. Twenty-one patients treated with mixed beam were in the ≤3-cm group, twenty-four were in the 3–6-cm group and ten were in the >6-cm group. The group of M. D. Anderson photon patients reported after exclusions retrospectively eliminated patients thought to have either inadequate doses or inadequate radiation portals.

	Size of adenopathy		
	≤3 cm	3–6 cm	>6 cm
All M.D.A. photon patients[a]	81%	66%	—
M.D.A. photon patients after exclusions[a]	91.5%	78.5%	—
All neutron-treated patients[b]	100%	82%	40%
Mixed-beam-treated patients[b]	100%	96%	50%

[a] Excluding patients with fixed nodes.
[b] Including patients with fixed nodes.

No difference in tumor response was apparent between any of the neutron-only fractionation schemes.

Acute treatment reactions in the skin and mucous membranes after treatment with neutrons alone were brisk, but only slightly more severe than those seen with conventional photon irradiation of large tissue volumes. Mixed-beam irradiation produced less severe reactions than the neutron-only schemes.

Late treatment complications were seen in soft tissues, bone, cartilage and spinal cord. Subcutaneous fibrosis was more common after irradiation with neutrons only than it was after treatment with mixed beams. Three patients developed cartilage necrosis, two developed mandibular necrosis and two developed signs and symptoms of cervical spinal-cord damage after treatment with neutrons only. One patient developed mandibular necrosis after irradiation with mixed beam: however, this complication was associated with recurrent tumor invading bone (Table 8).

Table 8. Severe late complications of fast neutron beam therapy seen in patients treated for metastatic cervical adenopathy

	Neutrons only	Mixed beam
Mandibular necrosis	2	1
Radiation myelopathy	2	0
Cartilage necrosis	3	0

Primary squamous cell carcinomas of the head and neck region

One hundred and twenty-six patients presenting with advanced primary squamous cell carcinomas of the head and neck were treated with fast neutrons during the period of study. Thirty-five presented with carcinomas of the oral cavity, nine presented with carcinoma of the nasopharynx, forty-nine had carcinomas of the oropharynx, seventeen had carcinomas of the hypopharynx, thirteen presented with carcinomas of the larynx and three had carcinoma of the maxillary sinus. Nine patients who presented with metastatic cervical adenopathy with an unknown primary site presumably in the head and neck region are excluded from this analysis.

Forty patients were treated with neutrons only, either 2, 3 or 4 fractions per week, and eighty-four patients were treated with mixed-beam irradiation as outlined in Table 1. The mean follow-up time for the entire group at the time of analysis was 1.5 years. The local control rate for the overall group was 51%. The local control rate for patients treated with neutron only was 28% and for patients treated with mixed beam irradiation it was 63% (Table 9).

Table 9. Local control rate after fast neutron beam treatment of primary squamous cell carcinomas of the head and neck region

	Number	Local control*
All patients	126	64 (51%)
Neutrons only	40	11 (28%)
Mixed beam	84	53 (63%)

The stress imposed by surgery on neutron-irradiated tissues can serve as an unique test of normal tissue tolerances to fast neutron beam therapy, and twenty-five of these patients underwent a major surgical procedure for persistent or recurrent cancer. The overall major complication rate in this group of patients (fistula formation, necrosis of skin flap, infection, wound dehiscence, skin necrosis and non-union of mandible) was 52%. The major complication rate for patients who had planned preoperative

Table 10. Complication rate for major
surgery after fast neutron beam treatment

Type of combined treatment	Complication rate
Planned surgery after irradiation	4/5
Salvage surgery after irradiation	9/20
Total	13/25 (52%)

irradiation to photon equivalent doses of 4500–5000 rad was 4/5. The major complication rate for salvage surgery after full-course neutron irradiation was 9/20 (Table 10). The complication rate was much better after mixed-beam than after neutron irradiation only—11% vs. 73% (Table 11). The time interval between irradiation and surgery seemed to have no influence on the complication rate.

Table 11. Complication rate after various
types of preoperative fast neutron beam
irradiation

Type of irradiation	Complication rate
Neutrons only	11/15 (73%)
Mixed beam	1/9 (11%)
Neutron boost	1/1

Major salivary gland tumors

Eleven patients were treated with neutrons for tumors of the major salivary glands between August 1974 and April 1976. Five patients were treated with neutrons alone and six patients were treated with mixed-beam irradiation. Patients were usually treated with various combinations of wedged pairs. Eight presented with tumors of the parotid gland and three presented with tumors of the submaxillary gland. Five patients had adenocystic carcinoma; two had mucoepidermoid tumors; one had a malignant mixed tumor; one had an adenocarcinoma; one had squamous cell carcinoma and one had an undifferentiated tumor. All tumors were high grade except one massive low-grade mucoepidermoid carcinoma.

Table 12. Local control rates after
irradiation of malignant salivary gland tumor

	Local control rate		
Tumor size	< 3 cm	3–6 cm	> 6 cm
Neutron*	3/3 (100%)	4/4 (100%)	0/4 (0%)
Photon	10/10 (100%)	2/6 (33%)	0/3 (0%)

*Includes neutrons alone and mixed beam.

Patients were grouped according to the maximum dimension of tumor mass. The results of treatment were then compared to the results obtained in a control population with similarly sized tumor masses previously treated with conventional photon irradiation (Table 12). The three neutron patients with disease less than 3 cm are alive with no evidence of disease, with a maximum follow-up time of 32 months (mean 27 months). All ten patients in this category treated with photons alone achieved local control. Three of these ten are dead—one of unrelated disease, one of progressive tumor after a marginal recurrence and one of pulmonary metastases. The remaining seven are alive with no evidence of disease, with maximum follow-up of 120 months (mean 36 months). All four neutron-treated patients in the 3–6 cm category achieved local control with maximum follow-up of 18 months (mean 12 months). One failed in the unirradiated neck and died with distant metastases (12 months) with no evidence of disease at the primary site. One is alive with distant metastases (7 months). The other two are alive with no evidence of disease (12 months, 18 months). Two of six patients treated with photons only achieved local control (maximum follow-up 48 months, mean 26 months). Seven patients were referred to us with disease greater than 6 cm in maximum dimension and all patients in this category (four neutron, 3 photon) failed to achieve local control. Most of these patients had massive disease and never achieved a complete response.

The number of patients in this series is too small to evaluate differences between treatment with neutrons only and mixed-beam irradition. There have been no long-term complications noted in this group of patients to date.

Conclusions

1. Although local tumor control was achieved in 14/15 (94%) of the autopsied patients with Grades III and IV astrocytomas treated in this pilot study, all neutron-irradiated patients died after variable courses of progressive neurological deterioration. The cause of death was probably related to radiation-induced diffuse white-matter deterioration. Hopefully, current trials which limit the whole-brain neutron dose, such as the RTOG protocol 76–11, will avoid this problem and still allow for good tumor control.

2, Metastatic cervical adenopathy serves as a near ideal *in vivo* test site for evaluating fast neutron beam teletherapy because the tumor volume is easily measured, the results of treatment can be determined accurately, and there is a reasonable data base describing the results of conventional treatment with which to compare any new findings. In this tumor system, local tumor control with mixed-beam irradiation was superior to that achieved with neutrons only ($p < .025$). Also, fewer treatment complications were seen with the mixed-beam modality suggesting an increased therapeutic ratio over neutrons alone. This conclusion is supported by data published by Dr. Rasey. Finally, although follow-up times were not strictly comparable, comparisons with published results of conventional photon therapy suggest an advantage for neutron treatment, especially in the category of disease measuring 3–6 cm, and especially for the mixed beam treatment program.

3. Results of treatment of primary tumors in the head and neck region including salivary gland tumors support the conclusions made in the metastatic cervical adenopathy test system.

4. Major surgery performed in a previously neutron-only irradiated field is hazardous, as is demonstrated by our major complication rate of 73%. The major complication rate of 11% after mixed beam treatment is no greater than that seen after comparable conventional photon therapy.

References

GERACI, J., K. JACKSON, G. CHRISTENSEN, R. PARKER, M. FOX, and P. THROWER, 1974) The relative biological effectiveness of cyclotron fast neutrons for early and late damage to the small intestine of the mouse. *Europ. J. Cancer* **10,** 99–102.

GERACI, J., K. JACKSON, G. CHRISTENSEN, P. THROWER, and M. FOX, (1975a) Cyclotron fast neutron RBE for various normal tissue. *Radiology* **115,** 459–463.

GERACI, J., K. JACKSON, P. THROWER, and M. FOX, (1975b) An estimate of the patient risk in cyclotron neutron radiotherapy using mouse testes as a biological test system. *Health Physics* **29,** 729–737.

KERNOHAN, J. and G. SAYRE, (1952) Tumors of the central nervous system. In: *Atlas of Tumor Pathology*, Sec 10, Fasc. 35 and 37. Armed Forces Institute of Pathology, Washington D.C., pp. 17–42.

NELSON, J., R. CARPENTER, and R. PARKER, (1975) Response of mouse skin and the C3HBA mammary carcinoma of the C3H mouse to X-rays and cyclotron neutrons: effect of mixed neutron-photon fractionation schemes. *Europ. J. Cancer* **11,** 891–901.

ORDER, S., S. HELLMAN, C. VON ESSEN, and M. KLIGERMAN, (1968) Improvement in quality of survival following whole brain irradiation for brain metastases. *Radiology* **91,** 149–153.

RASEY, J., R. CARPENTER, N. NELSON, and R. PARKER, (1977) Cure of EMT–6 tumors by X-rays or neutrons: effect of mixed fractionation schemes. *Radiology* **123,** 207–212.

SCHNEIDER, J., G. FLETCHER, and H. BARKLEY, (1975) Control by irradiation alone of non-fixed clinically positive lymph nodes from squamous cell carcinoma of the oral cavity, oropharynx, supraglottic larynx and hypopharynx. *Amer. J. Roentgenol* **123,** 42–48.

WEAVER, K., H. BICHSEL, J. EENMAA, and P. WOOTTON, (to be published, 1979) Measurement of photon dose fraction in a neutron radiotherapy beam. *Medical Physics*.

WOOTTON, P., K. ALVAR, H. BICHSEL, J. EENMAA, J. S. R. NELSON, R. PARKER, K. WEAVER, D. WILLIAMS, and W. WYCKOFF, (1975) Fast neutron beam radiotherapy at the University of Washington. *J. de l'Ass. Can. Rad.* **26,** 44–53.

Results of clinical applications with fast neutrons in Edinburgh

W. DUNCAN and S. J. ARNOTT

*Department of Radiotherapy, University of Edinburgh,
Western General Hospital, Edinburgh, Scotland*

Abstract—*The fast neutron therapy facilities based on a CS30 Cyclotron are described. One hundred and twenty-four patients with various forms of advanced cancer were treated in the first year of clinical studies. Randomly controlled trials have begun in "head-and-neck" cancer. The immediate local tumour control rate (75%) and morbidity (24%) are similar for both neutron- and photon-treated patients. Twenty-three patients have been included in the trial of brain tumours, but assessment of relative survival cannot be made. Patients with advanced cancer of the gastrointestinal tract do not appear to show any increase in the rate of tumour regression.*

The clinical facility in Edinburgh is based on The Cyclotron Corporation's CS30 cyclotron from which is obtained a deuteron beam with an energy just in excess of 15 MeV. This deuteron beam is directed on to a thick beryllium target producing neutrons with a similar energy to that of the Hammersmith Hospital, London (Williams *et. al.,* 1977; Field and Morris, 1977). The unit is provided with two treatment rooms, one accommodating a fixed horizontal beam (Fig. 1) and the other an iso-centric neutron therapy machine (Fig. 2). The cyclotron has been used for clinical studies since March 1977 in association with the fixed horizontal beam. The deuteron current has been kept at about 70% of maximum output giving a dose rate of about 30 rad per minute for a 10 × 10-cm field at an FSD of 125 cm. The cyclotron has operated splendidly and only 5 half days have been lost due to minor problems in the 15 months since clinical work began. We would pay tribute to Mr. D. D. Vonberg and Mr. T. E. Saxton of the MRC Unit for such a wonderful record of performance. Since April 1978 we have been able to use the iso-centric machine for static field therapy. Although we are not entirely satisfied by the design of its primary shield, we have been monitoring its operation at 50% of the maximum beam current under conditions of continuous clinical practice.

The clinical experience reported is of the first year during which patients were treated only on the fixed horizontal beam. A set of wooden applicators provides a range of forty-two field sizes from 6 × 6 cm to 21 × 21 cm. The facility is provided with an optical range finder, optical front and back pointers and illuminated field definition. In regular use is a set of three electronically interlocked wedge filters of 25°, 35° and 45°.

Most patients are treated lying in the supine position but some patients with "head-and-neck" tumours have been treated sitting. In both cases great care is taken in the accuracy of beam-direction techniques and in maintenance of the patient's position. Field selection is always performed by a senior specialist and dose distribution optimized by a computerized treatment planning system.

Initially a group of patients with various advanced cancers was irradiated as part of dose-ranging studies and also to obtain experience of the beam-direction techniques to be applied on this new facility. In our clinical work we normally specify dose in terms of what we call the "Total Effective Dose". This is the measured neutron dose together with one-third of the gamma radiation. The factor of 3 we refer to as a "notional scaling factor", a term preferable to RBE in circumstances where we do not actually know the relative effectiveness of the neutron beam which varies with fraction size and with different normal tissues and tumours being considered.

All our patients have been treated daily and a radical course of treatment is given in 20 frac-

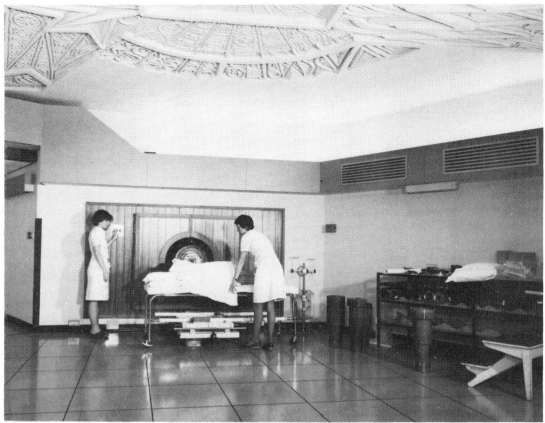

Fig. 1. *The Edinburgh fixed horizontal neutron beam facility.*

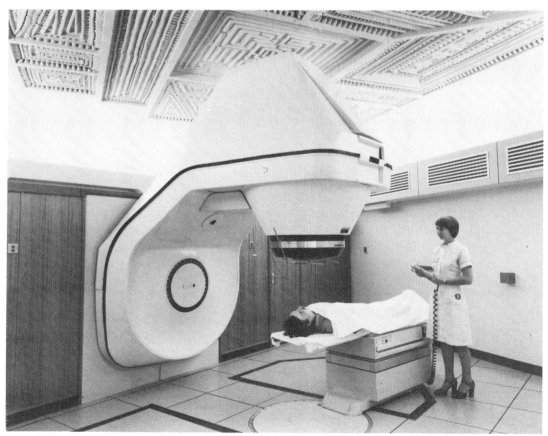

Fig. 2. *The iso-centric neutron therapy machine.*

tions over 1 month (26 to 28 days). Based on our initial experience of skin and mucosal reactions we concluded that a total dose of 1600 rad in 20 fractions (1560 rad neutrons) gave immediate reactions similar to our radical doses of megavoltage X-rays (5600 rad). In the case of central nervous system irradiation by neutrons we concluded from animal work and the clinical studies already reported that our maximum total effective dose with neutron therapy should be 1300 rad in 20 fractions (1260 rad neutron dose). Quite quickly after beginning clinical studies we felt able to proceed with randomly controlled trials in the "head and neck" and in the brain, using of course the fixed horizontal-beam facility.

Table 1. First year's clinical experience

Head and neck cancer		30
Oral and oropharynx	23	
Hypopharynx	4	
Brain tumours		13
Gastro-intestinal cancers		40
Stomach	15	
Rectum	15	
Metastatic lymph nodes		12
Metastatic bone lesions		9
Miscellaneous		20
Melanoma	5	
Sarcoma	5	
Salivary tumours	3	
Total number of patients		124

We have in the first year completed treatment and assessment of 124 patients who may be categorized in six main groups (Table 1). Forty patients with "head-and-neck" cancers have been treated, most of whom have been included in a randomized controlled trial. Patients with brain tumours, principally astrocytomas Grades III and IV, have received neutron therapy, the majority again within the protocol of a randomly controlled trial. Experience has also been gained in irradiating with fast neutrons a number of patients with advanced gastro-intestinal cancers, inoperable metastatic lymph nodes, metastatic bone lesions and a group of miscellaneous tumours including malignant melanoma, soft tissue sarcomas and salivary gland cancers.

The trial of "head-and-neck" cancers includes males and females under the age of 79 years with histologically confirmed squamous cell cancer of oral cavity and larynx and pharynx, but excluding the nasopharynx. There must be no evidence of metastatic disease beyond the cervical lymph nodes. Some less advanced cancers are excluded as shown in Table 2, for which we

Table 2. Head-and-neck trial, Site and staging of tumours

Site	Stages (all Mo)
Oral cavity and oropharynx	Excluding only T_1 No and in the case of tongue T_2 No when suitable for implant
Larynx	Excluding only T_1 No and in the case of glottis T_2 No
Hypopharynx	No exclusions

consider the results of megavoltage X-ray therapy techniques to be excellent. Twenty-nine patients (Table 3) have been included in the trial and are able to be assessed at least 2 months after completion of therapy. It should be remembered in considering this preliminary evaluation that the maximum length of follow-up is only 1 year. The site and stage distribution is satisfactory in the two treatment groups. The assessment of local control is given in Table 4 and it is evident that at present there is no difference between the two treatment groups. The overall local control rate is 75.8%. The majority of these patients had advanced disease and so it is not surprising that the radiation morbidity is high—an overall rate of 24.1% (Table 5). It will be seen that there is no real difference in the incidence of high dose effects between the neutron-irradiated or photon-treated groups. However, in the small group of oral and oropharangeal cases the time to complete healing of the mucosal reactions was 3.5 weeks in the neutron treated group compared to 6.5 weeks in the photon group. We therefore feel that there is a need to make some adjustment to the doses we are using, as the photon dose may be just a little high compared to the neutron dose. No distinction can be made between the two groups by measuring the height of normal tissue reactions as in all cases confluent fibrinous reactions were observed.

Table 3. Head-and-neck trial, Site distribution

Site	Neutrons	Photons
Oral and oropharynx	8	8
Larynx and hypopharynx	6	7
Total	14	15

At present twenty-three patients have been entered into the brain trial and again it is too early to make any real assessment of the results. The maximum length of follow-up is again only 1 year, but only three out of the ten patients treated by neutrons are alive while eleven out of thirteen patients treated by photons are surviving. In this case, although stratification is involved before random allocation of patients into the two

Table 4. Head-and-neck trial, Local tumour control

Site	Neutrons		Photons	
	Patients	Number	Patients	Number
Oral and oropharynx	8	5	8	5
Larynx and hypopharynx	6	6	7	6
Total	14	11	15	11

Overall control rate = 75.8%.

Table 5. Head-and-neck trial, High dose effects

Site	Neutrons		Photons	
	Patients	Number	Patients	Number
Oral and oropharynx	8	3	8	2
Larynx and hypopharynx	6	0	7	2
Total	14	3	15	4

Overall morbidity rate = 24.1%.

treatment options, it may be seen that a disproportionate number of patients aged over forty years and who carry a less favourable prognosis are in the neutron group, and relatively more of this group are more disabled (Karnofsky grades IV and V) than the photon group of patients. (Table 6) (Karnofsky, 1948). This disparity will, of course, be removed when greater numbers of patients have been recruited. We have not been aware of any serious morbidity and we have not seen the syndrome of progressive dementia and rapid death described by other investigators as occurring 2 or 3 months after completion of neutron therapy. It has been stated that nine of these patients have died following treatment, and it has been possible to obtain autopsies on six. The whole brains are carefully fixed and later sliced before microscopic sections are chosen. In two of these brains, both treated by neutrons, we have seen evidence of focal demyelinization. In all six cases there was gross residual tumour.

We were particularly interested to examine the response of tumours arising in the gastro-intestinal tract because of the opinions expressed that neutrons may also be more effective than X-rays in this group of cancers. All these patients had very advanced disease and neutron therapy was given electively. Normally the dose of 1600 rad in 20 fractions was given to these tumours and it will be seen (Table 7) that in only one case was complete tumour regression seen. There was no appreciable morbidity associated with this treatment but we have to say that we would expect in our experience a higher tumour control rate following X-ray therapy. The results of treating patients with inoperable cancer of the stomach are very similar (Table 8). Of fifteen patients treated by fast neutrons, in two there was considered to be complete tumour regression. In this group one patient suffered osteo-necrosis of the lower rib margin following gastrectomy. However this patient is otherwise well for no tumour was seen in the operative specimen—there was a tiny area of ulceration near the fundus which histologically was involved with non-specific inflammatory changes.

Table 6. Randomly controlled trial—Cerebral astrocytoma, Distribution by prognostic criteria

Criteria	Neutrons	Photons
Age		
Under 40 years	1	6
Over 40 years	9	7
Physical status		
Grade II	2	6
Grade III	3	5
Grade IV	5	2
Tumour circulation		
Vascular	4	5
Avascular	6	8
Total	10	13
Surviving patients	3	11

Table 7. Management of recurrent rectal carcinoma

Total patients	15
Complete tumour regression	1
Partial tumour resolution	12
No improvement	2
Morbidity	0

Table 8. Management of inoperable gastric carcinoma

Total patients	15
Complete tumour regression	2
Partial tumour resolution	7
No improvement	6
Morbidity	1

Now that we have the iso-centric neutron therapy machine we have begun randomly controlled trials in patients with rectal and gastric cancer in the hope of assessing the relative effectiveness of neutrons and photons in treating adenocarcinoma of the gastro-intestinal tract.

It will be realized that our clinical experience of fast neutron therapy is only of 1 year's duration. The results presented here must be regarded as provisional and any definitive evaluation of our data would be inappropriate and injudicious. The quantitative and qualitative differences between neutrons and photons are unlikely to be substantial in our opinion from our initial clinical impressions. But differences there are, and their evaluation will hopefully proceed in a well-integrated programme of international research so that if neutrons have a role in cancer management it will be accurately defined within the context of the exciting parallel developments that are now being explored to enhance the effectiveness of photon therapy.

Acknowledgements

We would acknowledge the collaboration of our clinical colleagues who have referred patients for these studies, and particularly Dr. J. McLelland and Dr. J. A. Orr who have participated in the management and assessment of patients in the randomly controlled trials.

References

FIELD, S. B. and CAROLINE C. MORRIS, (1977) Comparison of the RBE of fast neutrons at Edinburgh and Hammersmith. Brit. J. Radiol. **50,** 923.

KARNOFSKY, D. A., W. H. ABELMANN, L. F. CRAVER and J. H. BURCHENAL, (1948) The use of the nitrogen mustards in the palliative treatment of carcinoma. *Cancer (Philad.)* **1,** 634.

WILLIAMS, J. R., J. LAW, D. E. BONNETT and C. J. PARNELL, (1977) Initial experiences with the fast neutron facility in Edinburgh. *Proc. of Third Symp. on Neutron Dosimetry in Biology and Medicine,* Munchen.

Fast neutron project at Fermilab

GILBERT A. LAWRENCE

Cancer Therapy facility,* Fermilab,† Batavia, Ill., U.S.A.

Abstract—*The Cancer Therapy Facility (CTF) at Fermilab has been using fast neutrons in cancer treatment since 1976. The neutrons produced by a proton beryllium interaction [p(66 MeV)Be (2.21 cm)] have a mean energy of 25 MeV. The depth dose distribution resembles that of a 4-MeV photon beam. The patients mainly had advanced and recurrent cancer of the head and neck or glioblastoma multiforme. The therapy was by neutrons alone or neutron boost. The normal tissue reaction was similar to that expected from a radical course of photons. The tumour response is analysed by site. The past experiences and ongoing studies are discussed.*

Introduction

The Cancer Therapy Facility (CTF) at the Fermi National Accelerator Laboratory, Batavia, Illinois, started treating patients with fast neutrons in September, 1976. The laboratory is primarily a high-energy physics research center. The proton accelerator is composed of four accelerators each injecting into the following one to accelerate protons to a final energy of 400 GeV. The first three known as the injector (Cockcroft-Walton, Linac, Booster) are required for about 1 second out of every 10 to 16 seconds for high-energy physics. Nine to 15 seconds of each work cycle are available for medical research. This unique arrangment allows harmonious coexistence of the high-energy physics and medical research programs. For neutron production protons are extracted from the middle of the linac. The extracted proton beam has an energy of 66 MeV. The protons strike a beryllium target 22 mm thick to produce neutrons. The neutron energy spectrum extends to 64 MeV, with an estimated mean of 25 MeV. The characteristics of the beam for a 10 × 10-cm² field at a target axis distance of 153.2 cm are: D_{max} at 1·5 cm, 50% isodose at 14·9 cm and the dose rate at D_{max} typically 45 rad per minute. The physical and radiobiological characteristics of the beam have been published in previous reports (Amols *et al.*, 1977; Cohen and Awschalom, 1976). The dose distribution for various field sizes resemble that of a 4-MeV X-ray beam with similar skin sparing (Fig. 1).

Materials

This center is the first high-LET particle facility for cancer therapy in the mid-west. It is situated 60 km west of Chicago, from where the majority of patients are referred. Chicago, the second largest city in the United States, has a large number of cancer-treatment centers with distinguished surgical, medical, and radiation oncologists. Many of the early patients referred to the CTF were for recurrences following radical treatment with surgery, radiotherapy, or chemotherapy. Those who had no previous treatment were generally patients with extensive disease and poor nutritional status.

The initial period was spent treating pilot cases with advanced diseases in an attempt to develop treatment techniques, observe tumor and normal tissue response, and to establish an RBE and an iso-effect time dose fractionation relationship for the beam. After careful observation of skin and mucosal reaction and tumor response the data in Table 1 was determined. An RBE of 3 is used as a basis of adapting neutron treatments to conventionally adopted photon doses for tumor control and normal tissue tolerance, except for the central nervous system when for reasons of safety an RBE of 4 is assumed.

The following is a report on the first 105 patients treated at the CTF up to the end of 1977. Our main efforts to evaluate the effects of fast neutrons were directed at head and neck tumors. All but one of the patients had massive T_3T_4,

---

* Funded by NCI Grant No. CA18081-04.
† Operated by Universities Research Association, Inc., under contract with the U. S. Department of Energy.

N_2N_3 lesions. Thirteen patients did not complete their planned treatment. Only sixteen patients were eligible to enter into the RTOG (Radiation Therapy Oncology Group) protocols (four head and neck, twelve brain).

treatment is usually carried out with the tumor at the axis of rotation of the chair. The source axis distance (SAD) is 153.2 cm. However, for larger field sizes or field sizes which do not match the collimators, the SAD may be varied or the

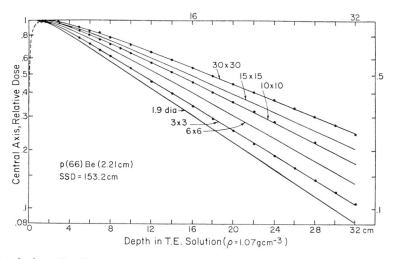

Fig. 1. Depth dose distribution for Fermilab's fast neutron beam in tissue equivalent (T.E.) solution. Field size in centimeters.

Neutron therapy techniques

The Fermilab neutron beam is fixed horizontally with a wide range of fixed size collimators made of polythelene concrete. The collimators permit the use of wedges, shields and boluses. The collimator angle can be adjusted coaxially with the central axis of the beam.

treatment done at a similar source skin distance (SSD). The back and the headrest of the chair are made of aluminium and lucite. Various back and side supports are available to minimize the interactions of the beam with these fixtures. Lesions in the upper torso are treated with the patient sitting, those in the lower half are treated with the patient standing.

Table 1. Fermilab neutron dosage options (6 weeks' treatment)

Fractions per week	4	3	2	1
Nominal weeks treatment	6	6	6	6
Number of fractions—min.	25	20	13	7
max.	27	21	14	8
Actual treatment days—min.	43	44	43	43
max.	47	48	49	50
Dose per fraction	80	100	150	250
Total dose (rad)—min.	2000	2000	1950	1750
max.	2160	2100	2100	2000
Photon equiv. (1900 ret)	6500	6100	5600	4900
Virtual RBE	3.0	2.9	2.7	2.5

The patient sits in a chair or stands on a pedestal which can move in three dimensions and can rotate about a vertical axis through 360 degrees. The point of intersecion of the vertical axis of rotation and the central axis of the beam provides a treatment isocenter analogous to that of a conventional rotation therapy machine. The

The patient planning, simulation and setup are done in the same room as patient therapy, above the treatment level (Fig. 2). The patient is immobilized using the conventional "lite cast" method for head and neck and nylon straps for the rest of the body. Laser beams and X-ray confirmation are used in treatment planning and setup. The axis of rotation of the chair, the X-ray and the laser beams meet at the planning isocenter. The elevator is then lowered until the patient comes in front of the neutron beam. Laser beams at the treatment level meet at the treatment isocenter and help in making any final adjustments. Neutrograms are used to confirm beam placement and direction in the treatment position.

1. Fast neutron therapy alone (18/70).
2. Fast neutron boost to the region of gross tumor following photon irradiation to the area of clinical and subclinical disease (32/70).
3. Fast neutron therapy for recurrence following a radical course of photon irradiation (20/70).

Tumor doses, when neutrons were used alone,

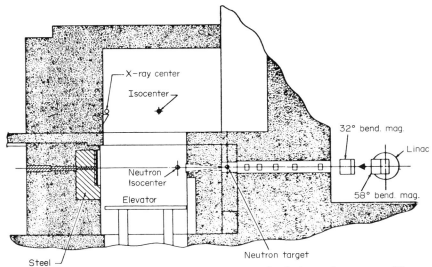

Fig. 2. Design of the planning and treatment room with relation to the target and linac.

The treatment planning is done on a PDP-10 computer, programmed with capabilities to calculate and plot iso-dose distribution for fast neutrons and photon beams. The latter is especially important when patients are being treated with a mixed beam or neutron boost. Electron beam planning capability is still under development. The photon radiation is delivered by the radiotherapist at the referring institution. There is excellent and close cooperation and coordination between the radiotherapists at the referring institutions and those at the CTF. The success of the CTF is largely due to the active participation of various radiotherapists in this project. The patients are subsequently followed up by all clinicians participating in the treatment.

Dose

Total doses herein reported as "neutron rad" are in fact neutron + gamma doses, following the current American practice. No more than 5% of the doses described are attributable to the gamma component.

Head and neck tumors

For head and neck tumors (seventy patients) fast neutron therapy was used as follows:

ranged from 2000 to 2400 neutron rad over a period of 6 weeks. The fractionation varied from two to four fractions a week. Clinically uninvolved nodes in the neck, supraclavicular region, and superior mediastinum received 1650 neutron rad in 5 weeks. Every effort was made to keep the spinal cord dose below 1250 neutron rad.

When neutrons are used as a boost the radiation is restricted to the gross tumor. The dose given was 700–800 neutron rad over a period of 2 weeks following 4000–5000 photon rad. For treatment involving a mixture of photons and neutrons, the maximum permissible dose to the spinal cord and brain stem is 5000 photon rad equivalent. The formula for determining safe tolerance limits to spinal cord is $D_\gamma + 4D_n \leq 5000$ rad.

When neutrons are used for recurrence following a radical course of photon irradiation, the dose varied from 700 to 2100 neutron rad. This being related to the previous dose of photon radiation, time gap, field size and the presence of neutron-sensitive structures like the spinal cord and base of brain in the treatment field.

Glioblastoma multiforme

Sixteen patients with Grade III and IV astrocytomas have been treated by a neutron boost (400–600 neutron rad) following a course of 5000–6000 photon rad to the whole brain (1000 rad /5 fr/wk). The time interval between the two courses of radiation has varied from 3 to 15 days.

Results

Normal tissue tolerance

Fifteen patients lived more than 10 months to permit the study of late effects after neutron radiation to the head and neck. The skin and subcutaneous tissue showed some degree of induration and fibrosis. These were no more severe than that following a radical course of photon irradiation. However, these reactions were moderately severe when the patient was treated by fast neutrons for recurrence following photon irradiation. It must be noted that the neutron beam has a skin sparing action (Fig. 1) similar to a 4-MeV photon beam.

All patients who had dryness of mouth and alteration in taste during treatment had these symptoms return to normal or markedly improved. No dental problems were encountered. No cases of soft tissue or osseous necrosis were seen. One patient had transient (1 month) symptoms of cord damage. He was treated for repeated recurrences following multiple surgeries by 700 neutron rad followed 2 weeks later with 3000 photon rad. The cervical cord received a dose of 700 neutron rad and 2000 photon rad. Two patients who received 1300 and 1500 neutron rad, respectively, have shown no evidence of cord damage at 1 year after exposure.

Tumor response in head and neck region

In fifty-five cases, the tumors were squamous cell carcinoma of the upper aero-digestive tract. Fourteen patients had carcinoma of the major or minor salivary glands. Local control is defined as absence of clinical disease in the treated area at the time of last evaluation or at the time of death. Post-neutron induration and tumor fibrosis sometimes makes evaluation of tumor control debatable. Table 2 is an analysis of the response to therapy.

Table 2. Results of fast neutron therapy for advanced carcinoma of head and neck

Type of Radiation	Site of tumor	No. of patients	Local N.E.D.	Persistent tumor	Local recurrence	Edge recurrence
Neutrons alone	Oral cavity	4	0	2	2	0
	Oropharynx	1	1	0	0	0
	Nasopharynx	2	0	1	0	1
	Hypopharynx	3	2	0	0	1
	Larynx	0	0	0	0	0
	Antrum	1	1	0	0	0
	Orbit	2	2	0	0	0
	Salivary gland	5	3	1	1	0
	Total	18	9	4	3	2
Photons followed by neutron boost	Oral cavity	14	4	7	2	1
	Oropharynx	9	3	4	2	0
	Nasopharynx	2	1	0	0	1
	Hypopharynx	1	0	0	0	1
	Larynx	1	1	0	0	0
	Antrum	1	1	0	0	0
	Orbit	0	0	0	0	0
	Salivary gland	4	2	2	0	0
	Total	32	12	13	4	3
Neutrons for photon recurrences	Oral cavity	6	3	3	1	0
	Oropharynx	2	0	1	1	0
	Nasopharynx	2	0	1	1	0
	Hypopharynx	3	1	1	1	0
	Larynx	1	1	0	0	0
	Antrum	2	0	2	0	0
	Orbit	1	0	1	0	0
	Salivary gland	2	2	0	0	0
	Total	20	7	9	4	0

The median follow-up time is 9 months. A number of patients with persistent tumor died because of distant metastasis with the treated lesion regessing or stable. Others are long-term survivors with the tumor being stable after an initial regression. In these patients, evaluation is difficult. In this series any doubt in local tumor control is labeled as persistent disease. Two such patients came to radical neck dissection. Histopathological examination of the node mass in both cases revealed no viable tumor present.

Glioblastoma multiforme

The evaluation of brain tumor response to therapy has been difficult. Even with the help of a CAT scan, persistent tumor, necrosis, edema and haemorrhage are not always distinguishable. The mean survival time is 8 months. Many of these patients were part of the RTOG protocol study, where a neutron boost was compared with a photon booster (Table 3). The short survival time would tend to suggest recurrent tumor rather than radiation induced necrosis.

Table 3. Results of neutron boost for glioblastoma multiforme.

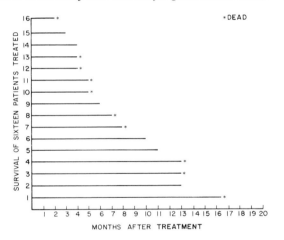

Miscellaneous tumors

As pilot studies, a large number of miscellaneous advanced lesions have been treated. These include carcinoma of the bladder, cervix, pancreas, colon and rectum. While the response has been satisfactory, the group is too inhomogeneous to draw definite conclusions.

Discussion

It is our belief that for neutrons, the alteration of fractionation (two to four fractions a week) does not require alteration of the overall dose when given over the same period of time. Fast neutrons being a localized form of therapy, local tumor response and control must be the criteria of effectiveness. However, given the size of the tumor in the present series and the presence of sensitive structures like the spinal cord, eye and base of brain, treatment planning becomes a test of skill on the part of the radiotherapist and the physicists. The challenge has sometimes been too great as measured by the presence of edge recurrence when the bulky disease in the center of the field has been well controlled.

Lack of significant local complications strongly suggests that one may cautiously increase the tumor dose. This increase will hopefully decrease the incidence of persistent disease after a radical course of fast neutron therapy. It should be noted, however, that the present time interval is too short to observe all late effects. While longer follow-up periods will show more recurrence and complications, it will also help reclassify some patients from the persistent tumor group to the locally controlled group. It must be noted that the majority of these patients were inoperable or had recurrences following surgery. The only other alternative form of treatment was chemotherapy. The 24% (7/29) neutron salvage for recurrences following radical photon treatment is encouraging.

Neutron boost, in our opinion, may well be as effective as giving the whole course of radiation with neutrons. Theoretically, if hypoxic cells are the reason for tumor resistance, it appears that neutrons are neither needed for microscopic disease nor for oxygenated cells. Hence, the neutron boost to the area of gross disease after 5000 photon rad to the clinical and subclinically involved region would presumably have a significant effect only on the residual core of hypoxic cells. Our results are not a randomized study and may well be a biased group of patients who have been referred for neutron boost because of an unsatisfactory response to 5000 photon rad. We have hence embarked on a randomized study comparing a photon boost to a neutron boost to the area of gross tumor.

The use of neutrons in glioblastoma multiforme needs re-evaluation. Our limited experience combining photons with neutrons show no improvement in patient survival. A possible combination of hypoxic cell sensitizers like Misonidazole with a sub-radical dose of neutron radiation would seem to be an appropriate next phase in evaluating the role of neutron therapy. While this paper analyzes the clinical experience of the first 16 months of the Fermilab CTF, the most important information was the development of treatment techniques and establishment of treatment policies. It may be useful, however, to add our impressions of 1978. As part of the RTOG protocol for head and neck tumors, we have treated fourteen patients with a mixed beam using two fractions of neutrons and three fractions of photons per week. The total tumor dose is 7000 rad equivalent over 7 weeks in the region of gross tumor and 5000 rad equivalent over 5 weeks to the area of subclinical disease. The patients tolerate this régime very well and initial tumor regression and control has been satisfactory.

Fourteen cases of pancreatic carninoma have been treated with neutrons alone. All these patients have tolerated a radical course of neutrons (2100 neutron rad) without interruption and with minimal nausea and loss of weight. The survival and tumor response are too early to evaluate. In future cases, however, we plan to decrease the dose by 7% (1950 neutron rads) and add chemotherapy pre-or post-treatment.

Eight cases of carcinoma of the esophagus have been treated to 2000 neutron rad. There has been no evidence of acute radiation pneumonitis or any other untoward side effects.

We are now in the process of conducting a feasibility study on using fast neutrons with chemotherapy in non-oat cell carcinoma of the lung. The addition of chemotherapy, however, requires a cautious approach. Severe acute and late pulmonary reactions may outweigh the benefit of treatment.

It is important that clinical oncologists working on new modalities of treatment establish a TNM classification for advanced tumors. The present staging system is good to distinguish between operable and inoperable cases. For example, in head and neck cancer the N_2 lesions are contralateral and N_3 are fixed neck nodes irrespective of size. To state success or failure without reference to tumor volume is inadequate and confusing. This will be an important factor in the comparison of results at different centers. Another important facet is the need to develop a system to evaluate post-operative high-LET radiation.

With an increasing number of high-LET particle therapy units now coming into use, it is important that we develop an accepted terminology. Confusion exists on the concept of RBE, the exact definition of dose, tumor control and its relation to tumor volume. It is important when comparing results across national borders, to appreciate the subtleties of treatment philosophy and accepted beliefs and practices in cancer therapy. The American philosophy of using radiotherapy for sub-clinical disease and the British belief that operable neck nodes are best controlled by surgery, affects type of case accession. While the exact role of fast neutrons is not yet established, it is our opinion that the present results are encouraging enough to continue enthusiastically with randomized clinical trial for various types of tumors.

References

AMOLS, H. I., J. F. DICELLO, M. AWSCHALOM, and colleagues (1977) Physical characterization of neutron beams produced by protons and deuterons of various energies bombarding beryllium and lithium targets of several thicknesses. *Medical Physics,* **4,** no. 6, 486, 493.

COHEN, L. and M. AWSCHALOM, (1976) The cancer therapy facility at the Fermi National Accelerator Laboratory, Batavia, Illinois: a preliminary report. *Applied Radiology*, November–December 1976, pp. 51–60.

Clinical observations of early and late normal tissue injury and tumor control in patients receiving fast neutron irradiation*

R. ORNITZ, A. HERSKOVIC, E. BRADLEY, J. A. DEYE AND
C. C. ROGERS

Division of Radiology, Oncology and Biophysics, George Washington University Medical Center, Washington, D.C., U.S.A.

Abstract—*The clinical experience of the first 211 patients treated at MANTA from October 1973 to May 1978 is described. Acute cutaneous, mucosal, gastrointestinal reactions and late effects including myelitis, damage to brain, bowel, soft tissue and mandibular necrosis are described. A review of tumor response data is also submitted.*

Introduction

Ever since Stone condemned the excessive late normal tissue injury observed in seventeen patients exposed to 8 MeV neutrons, interest has been focused on the early and late normal tissue responses in patients undergoing fast neutron irradiation (Stone, 1948). From October 1973 to May 1978, 211 evaluable patients have been irradiated at MANTA utilizing a 15-MeV cyclotron-produced neutron beam. One hundred and seventy-nine patients were treated with 320 (neutron + gamma) rad, four times per week for 7 weeks or less, for squamous cell carcinoma of the upper aerodigestive tract, glioblastoma multiforme, salivary gland tumors, adenocarcinoma of the pancreas and stomach and soft tissue sarcomas. It is the purpose of this report to analyze the early and late normal tissue reactions and local tumor control in these patients and to suggest possible directions for future investigation.

Materials and methods

Energy and dosage nomenclature

The Naval Research Laboratory cyclotron is a 76-inch sector focusing isochronous accelerator which produces a mean neutron energy of 15 MeV by bombardment of a thick beryllium target with 35 MeV deuterons. The NRL beam characteristics are nearly identical to those of Co-60 at 80 cm source-skin-distance. MANTA utilizes a fixed horizontal beam which is shaped by the use of Benelex (pressed wood) collimator inserts and tungsten blocks suspended from the collimator.

From October 1973 to May 1978, 211 patients were irradiated in clinical pilot studies to assess normal tissue tolerance and local tumor control in a variety of anatomical locations. Two different fractionation patterns were utilized during this phase and are summarized in Table 1. All doses quoted in this paper will refer to the total neutron plus gamma dose (rad $n + \gamma$), the latter measured to be 2% at D_{max} in tissue equivalent fluid (Smathers *et al.*, 1975). The photon equivalent dose (rad eq) is derived by multiplying the neutron physical dose by a minimum RBE value of 3.0 for the 15-MeV neutron beam at MANTA. This concept was developed as a convenience in order to discuss dose prescription with referring physicians not familiar with fast neutron irradiation. Typical neutron dose prescriptions and their photon equivalents are also presented in Table 1.

*This investigation was supported by Grant No. 1P01–CA–17465–03, awarded by the National Cancer Institute, DHEW.

Table 1. Manta neutron dosage schedules

Site	Fraction size[a]	FX. no.	Elapsed time (days)	Total dose(rad $n\gamma$)	Photon equivalents (rad eq)[b]
October 1973 to December 1974 (N=49)				2 Fractions/week	
Upper aerodigestive	140	14	49	1960	7000/7 weeks
Broncheogenic	140	12	42	1680	6000/6 weeks
Glioblastoma multiforme	140	14	49	1960	7000/7 weeks
Spinal cord tolerance	140	10	35	1400	5000/2 weeks
Renal tolerance	140	4	14	560	2000/2 weeks
January 1975 to May 1978 (N=162)				4 Fractions/week	
Upper aerodigestive	80	28	49	2240	7000/7 weeks
Broncheogenic/ esophagus	80	24	42	1920	6000/6 weeks
Glioblastoma multiforme	80	28	49	2240	7000/7 weeks
Soft tissue sarcoma	80	28	49	2240	7000/7 weeks
Pre-op. bladder	80	20	35	1600	5000/5 weeks
Definitive bladder	80	28	49	2240	7000/7 weeks
Adenoidcystic carcinoma	80	26–28	45–49	2080–2240	6500–7000/ $6\frac{1}{2}$–7 weeks
Spinal tolerance	80	20	35	1600	5000/5 weeks
Renal tolerance	80	8	14	640	2000/2 weeks
Hepatic tolerance	80	12	21	960	3000/3 weeks
Small bowel tolerance	80	22	42	1760	5500/$5\frac{1}{2}$ weeks

[a] Daily fraction size reduced from 80 rad $n\gamma$ to 78 rad $n\gamma$ after January 1977.
[b] Equivalency based on megavoltage X-ray or cobalt-60 fractionated five times weekly utilizing a daily dose of 180–200 rad/fraction.

Response criteria

Cutaneous and mucosal reactions were scored by both a radiation oncologist and radiation biologist according to a modified grading system developed by Fowler as depicted in Table 2 (Fowler, 1975). Acute gastrointestinal tolerance was necessarily scored by both subjective and objective criteria which are presented in Table 3.

Late normal tissue injury was assessed according to the nature of the anatomic and physiological lesion. The clinical diagnosis of transverse myelitis was established after exhaustive neurological evaluation including myelography, CT scanning, tomography, and spinal fluid analysis failed to document a tumor-related anatomical lesion in the cervical spinal cord or the presence of cerebral metastases. Bowel toxicity and brain necrosis were established by histopathologic review of resected specimens. The assignment of excessive subcutaneous fibrosis was made by the clinical evaluation of a minimum of two radiation oncologists.

Results

Acute reactions

Mucosal reactions. Mucosal reactions were scored in twenty-two patients receiving irradiation to the head and neck. Peak erythema was observed at 15.7 days and subsided by 44.9 days. Maximum erythema was quite mild as evidenced by a maximal score of 1.16. The development of mucositis was observed to be maximal at 20.6 days and waned by 40.5 days. The oral reactions were unusually mild peaking with a score of 1.56. Not a single case of confluent mucositis with pseudomembrane formation has been observed. A few patients have developed a focal punctate mucositis which was not time-dose limiting. A concurrent candida infestation could not be documented in these patients. Peak mucosal reactions appeared to occur 1 week earlier than those reported at the University of Washington where 8 MeV neutrons were fractionated either twice or three

Table 2. Scoring system for oral cavity and cutaneous reactions

	Developing mucositis	Fading mucositis
	DM 0—No reaction	FM 0—No reaction
	DM 1—Minimal mucositis	FM 1—Scant mucositis
	DM 2—Patchy mucositis	FM 2—Moderate mucositis
	DM 3—Confluent mucositis	FM 3—Severe mucositis
Site	*Developing erythema*	*Fading erythema*
S1 Soft palate	DE 0—No reaction	FE 0—No reaction
S2 Hard palate	DE 1—Slight erythema	FE 1—Slight erythema
S3 Oropharynx	DE 2—Moderate erythema	FE 2—Moderate erythema
S4 Tongue	DE 3—Severe erythema	FE 3—Severe erythema
S5 Buccal mucosa		
S6 Gingiva		
S7 Floor of mouth		
	Developing desquamation	*Fading desquamation*
	DD 0—No desquamation	FD 0—No desquamation
	DD 1—Slight desquamation	FD 1—Slight desquamation
	DD 2—Moderate desquamation	FD 2—Moderate desquamation
	DD 3—Severe desquamation	FD 3—Severe desquamation

times weekly with 300 rad $n + \gamma$ for 6 weeks (Parker, 1976).

Cutaneous reactions. Cutaneous reactions were scored in sixty-one patients at various anatomical sites (scalp, head and neck, abdomen, and pelvis). Peak erythema for scalp reactions occurs at 28 days in contrast to 35 days at the University of Washington. In general, peak cutaneous erythema scores are higher than mucosal erythema scores (1.90 vs. 1.16) and several cases of partial field moist radiodermatitis have been observed, some severe enough to require patient treatment interruption. However, all reactions have completely healed without complications. There appears to be both an accelerated and augmented cutaneous response when either actinomycin or adriamysic have *previously* been administered.

Gastrointestinal response. A total of nineteen patients have received fast neutron irradiation for adenocarcinoma of the pancreas (11) and stomach (8). Ten of the eleven patients (10/11) with pancreatic lesions also received concurrent 5-fluorouracil (375-350 mg/M^2 IV) during the first three and last three treatment fractions while six of eight patients (6/8) with gastric adenocarcinoma received the identical regimen. Maintenance chemotherapy consisted of a cyclical 8-week program of 5-fluorouracil, adriamycin and mitomycin C.

While nearly all patients receiving fast neutron irradiation to the upper abdomen were receiving chemotherapy with known gastrointestinal toxicity, acute reactions were decidedly mild although persistent leukopenia and thrombocytopenia were observed following treatment,

Table 3. Gastrointestinal toxicity scoring system

Score	Definition
0	No nausea, vomiting, diarrhea, or abdominal cramping. Diet unrestricted. Weight stable or increased during treatment
1.0	Mild nausea ± vomiting. Stool frequency increased without diarrhea Antiemetics not required. Weight loss 2% or less of pre-treatment weight
2.0	Moderate nausea and vomiting controlled with antiemetics Moderate diarrhea controlled with opiates No evidence of dehydration. Weight loss 5% or less of pre-treatment weight
3.0	Severe nausea and vomiting not controlled with antiemetics and opiates No evidence of dehydration. Weight loss 5% of pre-treatment weight Treatment interruption required
4.0	Hospitalization required for intravenous feeding and supportive care

often delaying the initiation of maintenance chemotherapy. The cumulative gastrointestinal toxicity score for the combined pancreatic-gastric group was 1.61. No significant difference in tolerance could be observed between the pancreatic patients or gastric patients as evidenced by respective scores of 1.45 and 1.63. Not a single case of radiation nephritis or radiation hepatitis has been observed when limiting renal dose to 624 rad $n\gamma$ and hepatic dose of 936 rad $n+\gamma$. The average weight loss for the entire group of eighteen patients during treatment was 5.43 lb without a significant difference demonstrated between the pancreatic and gastric patients (5.73 vs. 5.13 respectively).

Late effects

Late normal tissue injury responses have been documented in the large and small intestine, bone, soft tissues, cervical spinal cord and brain. A total of twenty-six late normal tissue injuries have been observed in 211 patients irradiated between October 1973 and May 1978. In eight cases, the injury was judged either to be directly or indirectly fatal. However, in two of the eight fatalities, a combined modality approach proved to be intolerable whereas in the remaining six cases, no other treatment other than fast neutron irradiation was administered. Thus, an over-all fatal complication rate of 2.8% can be attributed to fast neutron irradiation alone in a patient population with extremely advanced disease. The following anatomical or physiological lesions have been documented: transverse myelitis of the cervical spinal cord, mid-brain and frontal lobe necrosis, radiation enteritis and colitis, soft tissue necrosis, osteoradionecrosis and severe subcutaneous fibrosis. A summary of all late complications observed as of May 1978 are summarized in Table 4.

Table 4. Complications December 1974 to May 1978

Total patients evaluated	211
Total late complications	26
Brain necrosis	3
Transverse myelitis	4
Bowel necrosis[a]	3
Hemorrhagic gastritis	3
Soft-tissue necrosis[b]	3
Mandibular necrosis	1
Excessive fibrosis	4
Cataract formation	1
Laryngeal edema/fibrosis	4

[a] Includes two patients undergoing pre-operative irradiation for carcinoma of the bladder.
[b] Two of three patients had post-operative irradiation.

CNS:

Spinal cord. A total of four cases of transverse cervical myelitis have been observed. Two cases of transverse myelitis have been documented at the identical dose of 1543 rad $n+\gamma/28$ fractions/49 days including scatter. One patient was treated for a T3N3 B M0 squamous cell carcinoma of the pharyngeal tongue while the other patient was irradiated for an extra-osseous well-differentiated chondrosarcoma overlying the mandible. The latent period for the development of neurological deterioration was 8 and 4 months respectively.

Five months post-completion, the third patient developed Lhermitte's sign which progressed by 9 months to involve the right lateral spinothalamic tract sparing the posterior columns. By 12 months, the patient developed a dense right hemiplegia which has stabilized by 22 months.

The fourth case occurred at a dose of 1600 rad $n+\gamma/20$ fractions/34 days following irradiation for a recurrent squamous cell carcinoma of the larynx. Twenty-three months post-irradiation, the patient presented with progressive left hemiplegia. The four cases of transverse myelitis are summarized in Table 5.

Brain. Three cases of pathologically confirmed brain necrosis have been observed and in two, only limited cerebral portals were utilized. Two of the three cases occurred below the 2000 rad $n+\gamma$ level, utilizing limited fields, while the third case received 1655 rad $n+\gamma$ to the whole brain and a limited field boost totalling 2155 rad $n+\gamma/28$ fractions/52 days.

Gastrointestinal. As contrasted to the excellent acute gastrointestinal tolerance to fast neutron irradiation, several late bowel injuries have been documented. In two of the six cases, a combination of pre-operative pelvic irradiation and total cystectomy was not tolerated whereas in the third case, definitive pelvic irradiation was utilized. Pre-operative irradiation consisted of delivering 1600 rad $n+\gamma/20$ fractions/35 days to the whole pelvis utilizing a four-field "Box" technique with all fields treated daily. Both patients expired secondary to bowel necrosis post-operatively (1, 4 months). An additional three cases of hemorrhagic gastritis have occurred when concurrent 5-fluorouracil was administered in the pancreatic and gastric pilot study, but not when chemotherapy was withheld. The three cases occurred between 3 months and 8 months post-completion. Calculated tumor dosage ranged from between 1722–1760 rad $n+\gamma/22$ fractions/38 days. Two of the three cases healed with conservative management while the third required partial gastrectomy.

Table 5. Radiation myelitis

Group parameters:	Patients treated from December 1974 to August 1977 who survived a minimum of 3 months post-completion of neutron irradiation alone (excludes mixed beam and boost)
Fractionation:	80 rads $n+\gamma$ four times weekly for 6 to 7 weeks
	Spinal cord reduction at 1600 rad $n+\gamma$ 20 fractions/35 days
Diagnosis:	All patients have undergone a minimum of one myelogram, CT scan, cervical spine series, and CSF cytology
Patients at risk:	40
Myelopathy:	4
Incidence:	10%
Cervical cord dosage:	Case 1 .... 1543 rad $n+\gamma$/28 fractions/49 days Case 2 .... 1543 rad $n+\gamma$/28 fractions/49 days Case 3 .... 1575 rad $n+\gamma$/30 fractions/50 days Case 4 .... 1600 rad $n+\gamma$/28 fractions/49 days

Analysis: May 1978

Soft tissue and skeletal effects. Three cases of soft tissue necrosis and one case of complicated mandibular osteonecrosis have been documented. In two of the three soft tissue injuries, previous surgery had been performed whereas in the third case the necrosis was spontaneous 30 months after completion of irradiation.

A total of four cases of unacceptable subcutaneous fibrosis have been documented, three of which occurred in the head neck region following 2240 rad $n+\gamma$/28 fractions/49 days. Additional sequelae have included progressive trismus, reduced cervical range of motion and severe epiesophageal stenosis.

Clinical results

Head and neck

In general, neutron radiotherapy has been reserved for extremely advanced disease. The first forty-nine cases were retrospectively considered to have been under-dosed during the dose-searching pilot phase from October 1973 to December 1974. For oral cavity primary sites, all but one patient had greater than T_2 or N_1 disease or had surgical recurrence. Most of these lesions were either in the anterior tongue or floor of the mouth. Nevertheless, of those patients receiving radical neutron irradiation, ten of twenty-two patients (45%) had a complete response. Some patients are still alive without evidence of disease. The numbers of mixed-beam patients or neutron-boost patients are much smaller, but four of these nine (44%) also received a complete response. Only three of the fourteen patients (21%) with complete responses recurred locally. In the oropharynx, 91% patients were at least T_3 of whom 12/24 (50%) had N_3 neck nodes. The majority were from the pharyngeal tongue or the tonsil. Again, the complete response rate is encouraging with eleven of twenty-three (48%) complete responses. Although three patients recurred locally (27%) in the larynx, of sixteen patients, only one patient had less than T_3 or N_3 disease. Most of the lesions were divided primarily between larynx and supraglottis although there were four epiglottic primary lesions. The neutron only complete response rate was ten out of twelve (83%) and mixed beam complete response rate was two out of three (66%) although there were five local recurrences in these patients (41%).

In the hypopharynx, the predominant site of origin was the pyriform sinus. Only one patient had less than T_3 or N_3 disease with two surgical recurrences. There was a 50% (5/10) complete response rate and five of the mixed-beam or boost patients had a complete response. Only one local recurrence could be documented (20%).

Our salivary gland data consisted of eight patients fairly evenly divided between parotid gland, submandibular gland and minor salivary

Table 6.

	Pt. no.	Response			Survival by response (median survival in months and range)			Recurrences		
		Complete	Partial	None	Complete	Partial	None	Local	Regional	Distant
ORAL CAVITY:										
Neutrons alone	22	10	12	0	9(3–29+)	5(2–24)	—	2	2	2
Neutron boost	4	1	2	1	6	8(8)	?	1	0	0
Mixed beam	5	3	1	1	4(4–6)	3(3)	2	0	0	0
OROPHARYNX:										
Neutrons alone	23	11	12	—	7(5–42)	4(3–9)	—	3	2	2
Neutron boost	3	1	2	—	11	5(2–14)	—	—	—	1
Mixed bead	2	—	2	—	—	4(4)	—	—	—	—
LARYNX:										
Neutrons alone	12	10	2	—	9(3–36)	3(3–5)	—	5	2	1
Neutron boost	1	—	—	1	—	—	(6)	—	—	—
Mixed beam	3	2	1	—	7.5(7–8)	2	—	—	—	1
HYPOPHARYNX:										
Neutrons alone	10	5	5	—	6(5–8)	4(3–9)	—	—	—	2
Neutron boost	2	2	—	—	7(6–7)	—	—	—	—	—
Mixed beam	3	3	—	—	9(8–11)	—	—	1	—	—

gland. Six of the eight cases had adenoid cystic carcinoma. There were three complete reponses out of five patients with adenoid cystic disease who had complete radiotherapy with one local recurrence due to lack of adequate bolus. Both the mixed malignant and clear cell carcinoma patients had complete responses, although the former locally recurred.

In summary, of sixty-seven patients receiving neutrons alone for squamous cell carcinoma of the upper aerodigestive tract, 36/37 (53.7%) have achieved a complete response although 10/36 (27.7%) have had in-field recurrence (Table 6). Of sixty-two metastatic cervical nodes irradiated with neutrons alone, 31/62 (50%) have been locally controlled with only a 3% in-field recurrence rate. Forty-five per cent of the nodes (28/62) were judged to be fixed and 11/28 (39%) of these nodes were locally controlled.

Glioblastoma multiforme

Twenty-six patients with glioblastoma multiforme were treated with a fairly even distribution between those patients graded as 3, 4 and not specified. Comparatively more patients with lower grades received neutrons alone than photons plus neutron boost. Nevertheless, the mean survival was identical for grade 4 lesions whether neutrons alone or photons plus neutron boost was administered. There did not appear to be a significant increase in survival between those patients receiving neutrons alone above 1800 neutron rad. The best mean and median survival for neutrons alone was at the 1800 rad level which was approximately $2\frac{1}{2}$ months longer than photons plus neutron boost, (Table 7).

Gastrointestinal

Three areas were irradiated as part of pilot projects: five esophageal patients, eight gastric patients and seventeen pancreatic patients.

Only three out of five esophageal patients completed a planned course of radiotherapy. Two out of the three who were treated with combinations of photons and neutrons had a complete response and two are alive without disease at approximately 6 months. Eight patients with locally advanced gastric carcinoma were treated, six of which received combination chemotherapy. Of these, two had a complete response and one patient is still alive apparently free of disease. Seventeen patients with locally advanced pancreatic carcinoma were treated, twelve with combination chemotherapy. The distinction between partial and complete response is quite subjective, allowing assessment largely in terms of crude survival rate. Additional controlled clinical trials will be required to assess a potential gain factor for pancreatic carcinoma.

Sarcomas

Twenty-one patients were treated with either bony or soft tissue sarcoma, the majority of which were either of low or unspecified grade. Two of these patients had stage 4 disease. Thirteen of the patients had a subtotal resection whereas only one patient had a total resection. Response did not seem to vary with grade or resection except for the chondrosarcomas, the majority of which had a complete response and had a subtotal resection. There were seven patients with osseous sarcomas receiving

Table 7. Glioblastoma multiforme (N=26)

	Survival						
	Neutrons alone		Boost				
			Not planned			Planned	
Median (range)	15 (1–23)		4 (4–14)			8.5 (4–12)	
Grade	N	Mean[a] Median[a]	N	Mean[a]	Median[a]	N Mean[a]	Median[a]
III	6	11.7 10.5(6–23)	1	4	4(4)	—	—
IV	3	8.3 11 (3–11)	1	4	4(4)	4 8.3	8.5(4–12)
Glioblastoma	6	6.0 6 (1–11)	2	10	10(6–14)	—	—

Dose (rads N+γ)		Mean[a]	Median[a]
1639–1680/12/42	(5)	5.0	3(1–10)
1830–1885/13/56, 44	(2)	11.0	11(11)
2000–2200/26/49	(3)	10.0	4(3–23)
2240/28/49	(6)	9.5	10(6–11)

[a] In months.

neutrons alone. Six had a complete response (86%). Two of the three patients receiving photons plus neutron boost had a complete response. In the soft-tissue sarcomas, four out of seven patients with neutron boost had a complete response. Control and survival data for the sarcomas are reviewed in Table 8.

reported was commensurate with excessive dose delivery and should not contraindicate a properly conducted controlled clinical trial (Sheline et al., 1971). Only one such study has been performed to date by Catterall et al. which has demonstrated clearly superior local control with neutron irradiation for squamous cell carcinoma

Table 8. Sarcoma

Histology	Neutron alone			Neutron boost		
	Progression	Partial response	Complete response	Progression	Partial response	Complete response
Osteosarcoma	—	—	1(18[a])	—	—	—
Chondrosarcoma	—	1(6[a])	5(32+[a])	—	1(17[a]) (no change)	2(28[a]) (1 patient recurred local and distal)
Rhabdomyosarcoma	—	1	—	—	—	—
Leiomyosarcoma	—	1(7[a]) (disseminated disease)	1(2+[a])	—	—	—
Fibrous sarcoma (one patient inadequate)	—	—	1(5+[a])	—	1	—
Angiosarcoma	—	—	1(21+[a])	—	—	—
Myxoliposarcoma	1	—	—	—	—	—
Synovial sarcoma	—	—	1(20+[a])	—	—	—
Lymphangiosarcoma	—	—	—	—	—	1(18) (recurrence)
Embryonal Cell Carcinoma	—	—	—	—	1(1[a])	—
Totals						
Boney sarcomas	0	1	6	0	1	2
Soft-tissue sarcomas	1	2	4	0	2	1

[a] In months.

Discussion

The acute and late normal tissue responses to fast neutron irradiation have been of interest to both biologists and clinicians since Stone's initial report on what was felt to be *excessive* late damage (Stone, 1948). In 1971 Sheline et al. reviewed the original Berkeley data and concluded the excessive late injury potential of the upper aerodigestive tract (Catterall et al., 1975, 1977). However, in the most recent British report, there appears to be a somewhat higher complication rate in the neutron group.

The experience at MANTA since October 1973 raises many fundamental questions. Inasmuch as there is a paucity of information regarding *measured* RBE data for normal human tissues and organ systems at clinically used fractionated doses, the dose prescriptions and concept of rad equivalency may be misleading for many normal tissues with the exception of skin. The spectrum of late tissue injury ranges from post-mitotic neural tissue to rapidly proliferating intestinal parenchyma. While neurotoxicity has been observed both by the British and American groups and is undoubtedly radiation-related, many of the late injuries observed at MANTA have occurred in disturbed normal tissue due to previous surgery (soft-tissue necrosis) or have occurred following preoperative irradiation. In other cases, patients appear to have received supra-tolerance normal tissue doses based on acute RBE data derived from both *in vitro* cell studies and human skin measurements.

The clinical experience to date emphasizes a definite uncertainty factor with regard to the relationship between acute and late normal tissue response to fast neutron irradiation. The late effects reported were not preceded by exaggerated acute toxicity and tended to occur somewhat earlier than would be expected with conventional irradiation. These observations strongly suggest the urgent need for continued large animal research investigating the late RBE function in a wide range of normal tissues at clinically used fractionated doses as well as in compressed fractionation schedules.

With regard to local tumor control, the results for upper aerodigestive lesions have not been as good as reported by the British group and in-field recurrences have been more frequent. At MANTA we have cleared 53.7% of advanced squamous lesions with a 27.7% recurrence rate. Control of metastatic cervical adenopathy has been encouraging with the data very close to that reported at the University of Washington, although mixed-beam irradiation found superior to neutrons alone at Washington was not investigated in the pilot study.

In summary, the ultimate demonstration of the potential efficacy of fast neutron irradiation must be measured by local control without complications. Inasmuch as acute normal tissue reactions have not been dose-limiting, this end-point of effectiveness must be judged after the late effect RBE for normal tissue has been established and dose prescriptions based on a tolerable level of injury.

Survival data for glioblastoma multiforme remains poor which appears to be the universal experience to date despite tumor eradication at post-mortem examination. Salivary gland and sarcoma control rates are encouraging at MANTA and warrant additional trials.

The severe limitations afforded by fixed horizontal beam treatment make comparison with modern megavoltage accelerator treatment difficult in that limitation of dose to normal surrounding tissue is often impossible with the available collimators, attendant wide penumbra and patient positions. In that the final evaluation of the role of fast neutron irradiation must be measured by uncomplicated local control in prospective clinical trials, there is urgent need for continued neutron clinical investigation utilizing the best available neutron delivery systems.

References

ATTIX, F. H., L. S. AUGUST, P. SHAPIRO and R. B. THEUS, (1975) The Physics and dosimetry of fast neutrons for radiotherapy. NRL Memorandum, Report 3123, September.

CATTERALL, M., I. SUTHERLAND and D. K. BEWLEY, (1975) First results of a randomized clinical trial of fast neutrons compared with X or gamma rays in treatment of advanced tumors of the head and neck. *Brit. Med. J.* **2**, 653–656.

CATTERALL, M., D. K. BEWLEY and I. SUTHERLAND, (1977) Second report on results of a randomized clinical trial of fast neutrons compared with X or gamma rays in advanced tumors of the head and neck. *Brit. Med. J.* **1**, 1642

FOWLER, J. F., K. KROGT, R. E. ELLIS, P. J. LINDOP and R. J. BERRY, (1965) The effect of divided doses of 15 MeV electrons on the skin response of mice. *Int. J. Radiat. Biol.* **9**, 241–252.

PARKER, R. G., H. C. BERRY and A. J. GERDES, (1976) Early assessment of normal tissue tolerance of fast neutron beam radiation therapy. *Cancer* **38** (3), 1118–1123.

SHELINE, G. E., T. L. PHILLIPS, S. B. FIELD, J. T. BRENAN and A. RAVENTOS, (1971) Effects of fast neutrons on human skin. *Amer. J. Roentgenol.* **III**, 31–41.

SMATHERS, J. B., F. H. ATTIX, P. WOOTEN, D. K. BEWLEY, *et al.* (1975) Dosimetry intercomparisons between fast neutron facilities. *Med. Phys.* **2**, 195–200, July/August.

STONE, R. S. (1948) Neutron therapy and specific ionization. *Amer. J. Roentgenol.* **59**, 771–785.

Results of clinical applications of fast neutrons at Hamburg–Eppendorf

H. D. FRANKE

Tumor Center Hamburg, Radiotherapy Department of the Radiological Clinic (Director: Professor H. D. Franke), University Hospital Hamburg–Eppendorf, D-2000 Hamburg 20, Federal Republic of Germany

Abstract—*Results are presented of fast neutron treatments given to 232 patients between February 1976 and September 1978. The 14-MeV (DT) neutron therapy facility features an isocentric treatment system, a small contribution from low-energy neutrons and good dose distribution characteristics. Good results were obtained for soft-tissue sarcomas and differentiated thyroid carcinomas. Because of gut sensitivity, carcinoma of the bladder, prostate and rectum in advanced stages are treated with mixed photon–neutron schedules. In general differentiated tumours react to increasing portions of fast neutrons with better remission and the reduction of recidives; exceptions are glioblastomas owing to the high sensitivity of the normal brain structures to fast neutrons. This is in accordance to own histological findings in the brain of guinea pigs. The mixed photon–neutron schedule of glioblastoma seems to have no better results than the therapy with photons only. Disappointing was the effect on very big, not homogeneously irradiated and very differentiated tumours.*

Introduction

The technical and physical characteristics of the AEG/RDI neutron therapy facility in Hamburg have been described earlier (Franke *et al.*, 1974, 1978).

Specific features are:

1. The neutron therapy unit is installed within the radiotherapy department (Franke, 1974, 1976).

2. The isocentric treatment system allows precise reproduction of positioning, selection of beam direction and optimal dose distributions using arc and multiple port therapy.

3. The set-up is equivalent to telecobalt systems, thirty-five collimators facilitate adaption of the field size to the tumour size of each patient.

4. With a new target the neutron dose rate $(D_{n+\gamma})$ is about 20 rad/min at SSD = 80 cm (measured at the maximum of the build-up dose at 2.5 mm depth in a TE phantom) (Franke *et al.*, 1974, 1978; Franke, in press).

5. Depth dose characteristics are slightly inferior to that of Co-60 gamma rays, but better than that of cyclotron neutrons generated by up to 21-MeV deuterons (Franke *et al.*, 1978).

6. Neutron beam isodose profiles have been measured and are used for computerized treatment planning as exemplified in Fig. 1.

7. Energy distributions for neutrons and gamma-rays have been measured showing only a small proportion of low-energy neutrons (Schmidt *et al.*, 1977 a, b). This might contribute to the observation that skin reactions are only slightly more intense than with telecobalt therapy.

Clinical Applications

Measurements of physical parameters were started in 1974 and patient treatments commenced in Feburary 1976 (Franke *et al.*, 1978; Franke in press). Radiobiological studies are in progress, directed at obtaining insight in special clinical problems (Franke *et al.*, in press; Lierse and Franke, in press; Zywietz *et al.*, 1979). Up to September 1978, 232 patients have been irradiated with neutrons (Table 1). Initially only incurable patients were irradiated to observe reactions of skin and of tumours. In early series 65 to 80 rad $(n+\gamma)$ was applied daily. For about 2 years 12 × 130 rad is given in a three fractions per week schedule. This scheme will allow a comparison with the results of treatments with 7-MeV neutrons used by Dr. Catterall at Hammersmith Hospital, London (Catterall, 1974 a, b, 1976, 1977; Catterall et al., 1975). This schedule also allows more time for physical measurements. During the past few months we have treated a few patients with 4 × 100 rad per week during 4 weeks.

The grade of reactions has not changed evidently since we have recently started to apply 12 × 130 rad neutrons (D_n) only, instead of total dose ($D_{n+\gamma}$). Late reactions are frequently observed of light hyperpigmentation, in single cases some teleangiectasis. Fibrotic indurations of the skin are also rare, up to now only six cases. Necrosis

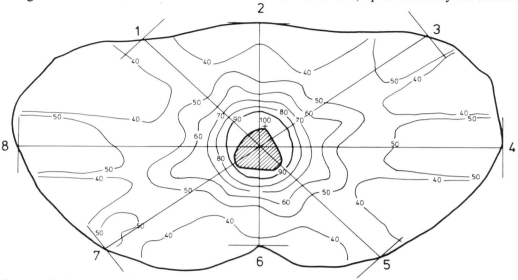

Fig. 1. *Computerized treatment planning with a 14-MeV neutron (DT) beam (total dose) for a prostate cancer: isocentric treatment with eight fields (field size: 4.4 × 10.0 cm, SSD = 80 cm). Application of the same dose at the centre of the tumour by each field* (Franke, 1978).

In the earlier treatments we have applied combined neutron–photon irradiations similar to the schedules employed by Eichhorn (1974, 1976). At present mixed schedules are only employed for glioblastoma, bladder cancer, prostate cancer and rectosigmoid cancer, whereby in general photons are given first followed by neutrons. Commonly we tend to treat tumours with neutrons only.

Table 1. *Types of tumours and number of patients treated with (DT) 14 MeV neutrons between February 1976 to September 1978 in Hamburg–Eppendorf*

1. Tumours of the connective tissue and bone	63
2. Tumours of the skin	21
3. Tumours of the nervous system	25
4. Tumours in the head and neck region	37
5. Tumours of the uro-genital system	26
6. Tumours of the alimentary canal	46
7. Miscellaneous tumours	14
	232

After the application of 12 × 130 rad ($D_{n+\gamma}$), with gamma doses of 5–15% depending on field size and depth of the tumour, given in an overall time of 4 weeks, in most patients only a slight erythema is observed, more intense in large fields or at oblique surfaces (Franke et al., 1978; Franke, in press). Moist skin reactions are rare.

of the skin, muscles or other tissues were not observed. The dose to the spinal cord never exceeded 800 rad. General reactions occur with no higher incidence than in telecobalt therapy. The gut reacts in most cases after 6–8 × 130 rad with heavy diarrhea lasting for a longer time than with X-rays or tele-cobalt γ-rays. Irradiation of open wounds in some cases has not prevented the healing process. Operative procedures after neutron therapy have been performed in six patients: in four without complications, in two with temporary delay of wound healing. Since more than 1 year we try to irradiate more patients with expectation for cure. Controlled clinical trials are running for pre- and/or post-operative irradiation of rectum carcinoma. Such trials are also in preparation for prostate cancer and for glioblastoma.

Results

The analysis of our 232 patients (see Table 1), treated up to now, shows the following results:

1. Tumours of the connective tissues and bone are rare, but in the meantime we have irradiated sixty-three cases in this group (see Table 2): in fibrosarcoma (see Table 3) we can confirm, in accordance with results of Catterall (Catterall et al., 1971, 1974 a, b), that the remission in twelve

Table 2. Tumours of the connective tissue and bone

(a) Connective tissue:	
Fibrosarcoma	18
Xantofibrosarcoma	1
Fibromyxosarcoma	1
Malignant histiocytoma	1
Spindlecellsarcoma	4
Leiomyosarcoma	1
Liposarcoma	11
Rhabdomyosarcoma	6
Synovialsarcoma	1
Hemangiopericytoma	6
Angiosarcoma (Stewart-Treves Syndrome)	1
	51 51
(b) Bone:	
Osteosarcoma	7
Ewing sarcoma	2
Chondrosarcoma	3
	12 12
	63

localized tumours is complete, recidive occurred only in one patient, treated 1 year before the neutron therapy with telecobalt (6000 rad/6 weeks). In spite of the high telecobalt and neutron dose (12 × 130 rad/4 weeks) the excision of the tumour 1 year after neutron

rhabdomyosarcomas (see Table 5) we find only one recidive in a previously irradiated patient, but in three localized spindle cell sarcomas (see Table 6) occurred two recidives, one after "mixed-beam" therapy (mixed photon–neutron schedule). Three patients with osteosarcoma (see Table 7) refused amputation: one child has been free of tumour for nearly 3 years; we combined telecobalt and neutron irradiation of the primary quite near the left shoulder with intensive chemotherapy. One year after irradiation the little girl agreed to plastic surgery of the shoulder joint and the upper humerus, the specimen showed no more tumour cells. The other girl has been free of metastases for more than 6 months, the primary has responded very well, the general condition is excellent; a heavy moist skin reaction occurred at the end of the radiation therapy after combined application of chemotherapy and neutron therapy (1600 rad). The parosseous tumour of an older patient did not respond and amputation of the leg was necessary. Three big chondrosarcomas (see Table 8) were irradiated, but the dose distribution in these huge tumours was not homogeneous; one tumour with low malignancy shows progression, one tumour with intermediate malignancy has been reacting with permanent regression of clinical symptoms since 3 months.

Table 3. Fibrosarcoma

Stage	No.	Dose n Gy (total dose)	Dose distribution homogeneous	Local regression	Recidive	NED[a] since months	Died after months	Remarks
(a) Localized	11	15.60 in 4 weeks	+	+	−	7:>12 4:> 6	−	cytostatica
	1	15.60 in 4 weeks (pre-irrad. before 1 year with Co⁶⁰ 40.00 Gy)	+	+	+ after 1 year	15	−	plastic surgery successful
(b) Not localized	1	about 15.00 15.00 in 5 weeks	−	+	+	−	30	cytostatica
	1	about 16.00 in 5 weeks	−	+	−	17	−	,,

[a]NED = no evidence of disease.

therapy was possible and the plastic surgery successful. The same good result we find for eight localized liposarcomas (see Table 4): no recidive occurred after 1560 rad, one recidive after a lower dose. In four localized

2. *Tumours of the skin.* The experience with metastases and recidives of melanomas was very variable: some tumours have not changed after 1560 rad; after excision, most of these resistant tumours showed no local recidives. In other

Table 4. Liposarcoma

Stage	No.	Dose n Gy (total dose)	Dose distribution homogeneous	Local regression	Recidive	NED since months	Died after months	Remarks
Localized	7	15.60 in 4 weeks	+	+	−	6:>1 y 1:>6 m		
	1	13.65 in 3 weeks	+	+	+ after 1 year and 4 months later	20	−	2 recidives removed by excisions Histology now: partly leiomyosarcoma

Table 5. Rhabdomyosarcoma

Stage	Site	No.	Dose n Gy (total dose)	Dose distribution homogeneous	Regression	Recidive	Living since months NED	Died after months with tumour		Remarks
Localized	larynx		15.60 in 4 weeks	+	+	−	12	−	−	
	foot	1	15.60 in 4 weeks	+	+	−	6	−	−	
	thigh	1	15.60 in 4 weeks	+	+	−	6	−	−	
	lower leg	1	15.60 in 4 weeks (pre-irrad. before 1 y with 60.00 Gy Co^{60}	+	+	+ after 4 months	16	−	−	amputation of the lower leg after recidive

cases we have seen complete remission of the tumour with the same dose.

3. *Tumours of the nervous system.* The expectation of curing glioblastoma with high doses of fast neutrons has failed, the patients did not die due to the growth of the tumour, but presumedly, as a consequence of demyelination of the white matter. These findings of Catterall and her group in London (Catterall, 1975; Lewis, 1975) and Parker et al., (1976) and his group (Laramore et al., 1978) in the USA are in accordance with our own experimental studies in the brain of guinea pigs. Experiments carried out together with Professor Lierse, Director of the Department for Neuroanatomy, University of Hamburg (Franke et al., in press; Lierse and Franke, 1978, in press) showed that doses between 60–900 rad of 14 MeV (DT) neutrons produce 24 hours after irradiation pycnosis of dividing glial cells in the white matter and of not dividing neurons in the cortex of the brain. This process might start demyelination (Franke and Lierse, 1967; Franke, 1979 in press; Franke, Lierse et al., in press; Lierse et al., 1965; Lierse and Franke, 1978; Selle et al., 1968; Wrage et al., 1969). Therefore we irradiate patients with glioblastomas in the following mode: at first the whole brain is irradiated homogeneously with photons up to 4000 rad in 4 weeks, then the tumour region only is irradiated with a boost of 4–6 × 130 rad neutrons. Our first results (see Table 9) show that three of the six patients lived longer than 1 year, two longer than two years in a good condition, one of these with recidive: clinical response occurred very quickly after 720 rad neutrons, whereas the computer tomography often shows no distinct regression of the tumour after these dose levels. But this regression can be seen clinically and in the computer tomography in two of the three chromophobe adenomas of the pituitary after "mixed-beam" therapy (4000 rad with 42 MeV X-rays and 400 rad with fast neutrons): a previously blind patient was able to see persons and big letters of the journals. The computer tomography at 5 weeks after the end of therapy showed increasing necrosis of the tumour (see Fig. 2).

4. *Tumours of the head and neck region.* The experience with large tumours of the lip, the tongue and the floor of the mouth (T3–4,

Table 6. Spindlecellsarcoma

Stage	Site	No.	Dose n Gy (total dose)	Dose distribution homogeneous	Regression	Recidive	Living since months NED	with tumour	Died after months	Remarks
Localized	bottom	1	15.60 in 4 weeks	+	+	−	12	−	−	−
	thigh	1	15.60	+	+	+ after 8 months			20	metastases
	lower leg	1	6.40 in 2 weeks 40.00 10 MeV electrons	+	+	+ after 25 months excision	29	−	−	excision of the recidive

N1,M0) shows local recidives after "mixed-beam therapy". In contrast to this is the experience in a big carcinoma of the tongue (T3,N1,M0): the tongue, irradiated with 1560 rad neutrons (total dose in 4 weeks), is free of tumour for more than one year; the lymph node metastasis, irradiated with "mixed beam" in order to spare the spinal cord, showed no total diminuation; the rest of the metastasis was surgically removed, but local recidive and metastasis occurred some months later, once more treated by excision. The effect on differentiated tumours of the thyroid is excellent (see Table 10). We now treat all patients of this group (without radical surgery or with local recidives) with fast neutrons. The longest time for full remission of a large tumour was 6 months. The small reaction of the normal tissues as skin and larynx is remarkable, we miss edema of the larynx and heavy skin reactions; it is essential that the patient keeps the treatment field strictly dry during and 2 weeks after the end of therapy. The neck is irradiated with moulds of wax and paraffin (50:50%), covered at the surface to the skin with lead of 0.3 mm thickness. In contrast to the excellent reaction of the differentiated tumours of the thyroid two undifferentiated carcinomas (giant cell carcinomas) showed recidives after a short time.

5. *Tumours of the urogenital system.* Tumours of this group are irradiated with mixed photon–neutron schedule: five patients with inoperable bladder cancer (T2–4, NX,M0) are treated, four of these living longer than 1 year without tumour; two of these tumours had invaded the prostate. Five cases with big prostate cancers (T3–4, N1,M0) lived in good condition for more than 1 year and without any sign of local tumour. Now we are beginning to irradiate

Table 7. Osteosarcoma

Stage	Site	No.	Dose n Gy (total dose)	Dose distribution homogeneous	Regression	Recidive	Living since months NED	with tumour	Died after months	Remarks
Localized	upper arm	1	12 × 8.00 in 3 weeks Co^{60}: 22.50 in 3 weeks	+	+	−	32	−	−	cytostatica; plastic surgery after 1 year: histology: no tumour cells!
,,	tibia	1	15.60 in 4 weeks	+	+	−	6	−	−	cytostatica
,,	tibia par-osseous	1	15.60 in 4 weeks	+	−	+ after 4 months: amputation	5	−	−	amputation of the leg

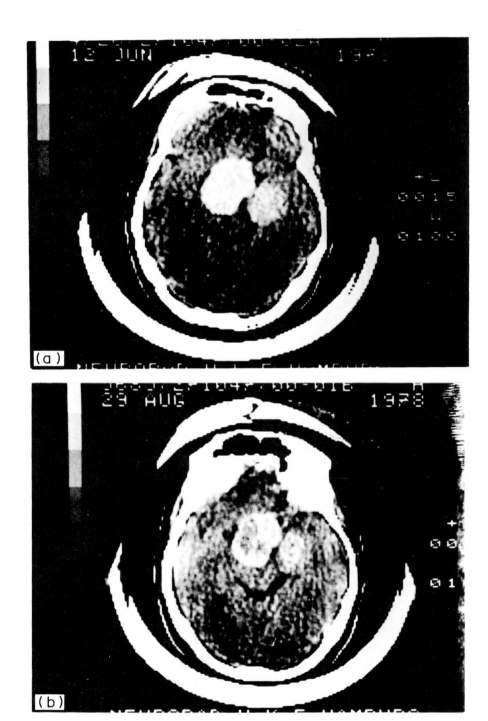

Fig. 2. Computer tomography of a patient (age: 30 years) with an inoperable chromophobe adenoma of the pituitary: (a) just before beginning of the irradiation therapy: (b) 5 weeks after the end of the irradiation therapy (4000 rad with 42 MeV-X-rays in 3.5 weeks and 5 × 80 rad fast neutrons in 9 days): increasing necrosis of the tumour.

Table 8. Chondrosarcoma

Histological grade	Site	Dose n Gy (total dose)	Dose distribution homogeneous	Regression	Progression	NED	Living since months with tumour	Died after months	Remarks
Low malignancy grade I	hip joint pelvis	15.60 in 4 weeks	–	–	+	–	17	–	
Low malignancy grade I	shoulder-joint	13.00 in 3 weeks	–	(+)	–	–	12	–	
Intermediate malignancy grade II	hip joint pelvis	15.60 (neutron dose) in 4 weeks	–	(+)	–	–	2	–	permanent regression of symptoms

tumours of stages B and C (T3,NX,M0) with small fields (see Fig. 1) and increasing neutron dose.

6. *Tumours of the alimentary canal.* In the group of tumours of the colo-recto-sigmoid cancers the control of the radiation effect on the tumour is infrequently so impressively demonstrated with computer tomography as in one of the patients, who had become operable 10 weeks after 9 × 130 rad neutrons in only 3 weeks. In most cases the effect on the inoperable tumours or on recurrences can only be measured in the regression of pains; in others obstruction of the ureter or vaginal infiltration is removed; the regression of pains occurs more often and with arrest for longer time with the mixed photon–neutron schedule than with photons only, the effect increases with neutrons only up to 10–12 Gy (Winkler *et al.*, in press). In general the reactions of the gut after doses of 8–12 Gy (whole pelvis) are at first very intense and last 3–4 weeks, thereafter the general condition improves mostly very well; therefore we have not yet tried to give higher doses of neutrons because we do not yet know the late effects after these heavy early reactions. The therapy of the gut with the mixed photon-neutron schedule is better tolerated than with neutrons only.

Conclusions

Skin reactions of our 14-MeV neutron beam after 12 × 130 rad (total or neutron dose) in 4 weeks are mild, probably as a consequence of the very small portion of low-energy neutrons in the measured neutron energy spectrum. Wound healing after this dose in general is not impaired. Late reactions as teleangiectasies and subcutaneous fibroses are rare, hyperpigmentation is relatively mild. Necrosis of the

Table 9. Malignant glioblastoma

No.	Dose n Gy (total dose)	photons Gy	Effect on clinical symptoms	Effect on computer tomography	Recidive	n Gy (total dose)	Living since months	Died after months
1	4.00 in 2 weeks	40.00 in 4 weeks	+	–	+ after 2.5 years	7.20 in 2.5 weeks	32 in good condition	–
2	4.00 in 2 weeks	40.00 in 4 weeks	+	?	–	–	24 in good condition	–
3	5.00 in 2 weeks	31.00 in 3 weeks	(+)	–	+ after 9 months	5.00 in 2 weeks	–	14
4	4.00 in 2 weeks	30.00 in 3 weeks	–	–	+	–	–	3
5	5.00 in 2 weeks	50.00 in 5 weeks	(+)	?	+	–	–	8
6	5.20 in 2 weeks	40.00 in 4 weeks	(+)	–	+	–	–	4

Table 10. Thyroid carcinoma
with tumour rest
with lymph node metastases
with recidives

No.	Histolog. type	Dose n Gy (total dose)	Telecobalt Gy	NED since months	Recidive	Metastases	Complications skin complaints	Moist skin reaction	Died after months
1	papill.	7.80 in 2 weeks	30.00 in 2.5 weeks	13	–	–	–	–	–
2	papill.	7.80 in 2 weeks	30.00 in 2.5 weeks	14	–	–	–	–	–
3	papill.	15.60 in 4 weeks	–	12	–	–	–	–	–
4	papill.	15.60 in 4 weeks	–	12	–	–	–	–	–
5	trabec. follic.	16.60 in 4 weeks	–	12	–	–	–	–	–
6	follic. papill.	15.60 in 4 weeks	–	6	–	–	–	–	–
7	medull.	15.60 in 4 weeks	–	6	–	–	–	+	–
8	medull.	15.60 in 4 weeks	–	6	–	–	–	–	–

skin or other normal tissues are not yet observed after 12 × 130 rad in 4 weeks. Local control of soft tissue sarcomas as fibrosarcoma and liposarcoma is in accordance with the results of the Hammersmith Hospital after the same total dose and the same type of fractionation. The effect of fast neutron application on the differentiated thyroid cancer is impressive. The therapy of cancer of the bladder, prostate and rectum with mixed photon–neutron schedule shows good results in advanced stages. A higher portion of neutrons in the whole effective dose increases the remission of the tumour. The mixed-beam therapy of glioblastoma seems to give no better results than the therapy with photons only, but this therapy, with different dose levels of neutrons and good concentration on the tumour, should be closely examined, as in the case of the mixed photon-neutron schedule of chromophobe adenomas of the pituitary. The effect of fast neutron therapy on very large, not homogeneously irradiated and very differentiated tumours such as chondrosarcoma of grade I was disappointing.

References

CATTERALL, M. (1974a). The treatment of advanced cancer by fast neutrons from the Medical Research Council's cyclotron at Hammersmith Hospital, London. *Europ. J. Cancer* **10**, 343–347.

CATTERALL, M. (1974b) The results of fast neutron therapy from the Medical Research Council's cyclotron at Hammersmith Hospital, London with particular reference to tumours of the head and neck. *Strahlentherapie,* **148**, 447–450.

CATTERALL, M. (1975) Treatment of Grade III and IV Gliomas; results of treatment with fast neutrons. The Royal Post Graduate Medical School, Medical Res. Council, Fast Neutron Clinic, London, Hammersmith Hospital: Course about the Treatment of Cancer with Fast neutrons, May 27–31, 1975.

CATTERALL, M. (1976) Results of fast neutrons on normal structures and tumours 1–5 years after treatment. In *Proc. Int. Workshop Particle Radiation Therapy,* Americal College of Radiology, pp. 414–422.

CATTERALL, M. (1977) The results of randomized and other clinical trials of fast neutrons from the Medical Research Council cyclotron, London. *Int. J. Radiat. Oncol. Biol. Phys.* **3**, 247–253.

CATTERALL, M., CH. ROGERS, R. H. THOMLINSON and S. B. FIELD (1971) An investigation into the clinical effects of fast neutrons. *Brit. J. Radiol.* **44**, 603–611.

CATTERALL, M., I. SUTHERLAND, and D. K. BEWLEY (1975) First results of a randomized clinical trial of fast neutrons compared with X- or gamma rays in treatment of advanced tumours of the head and neck. *Brit. Med. J.* **II,** 653–656.

EICHHORN, H. J., A. LESSEL and S. MATSCHKE, (1974). Vergleiche zwischen Neutronen- und Telekobalttherapie am Bronchus-, Magen-und Oesophagus-Carcinom. *Strahlentherapie,* **147**, 559–563.

EICHORN, H. J. and A. LESSEL (1976). A comparison between combined neutron- and telecotalt-therapy with telecobalt-therapy alone for cancer of the bronchus. *Brit. J. Radiol.* **49**, 880–882.

FRANKE, H. D. (1974) Aspects of equipment in the new radio-therapeutic department of the University Hospital in Hamburg. In M. KORMANO and F. E. STIEVE (Eds.), *Planning of radiological departments,* G. Thieme, Stuttgart, pp. 347–356.

FRANKE, H. D., A. HESS, E. MAGIERA, and R. SCHMIDT, (1978) The neutron therapy facility (DT, 14 MeV) at the Radiotherapy Department of the University Hospital Hamburg-Eppendorf. *Strahlentherapie* **154**, 225–232.

FRANKE, H. D., W. LIERSE, R. METZNER, and D. TUMBRÄGEL, Histochemische Frühveränderungen im Hirn des Meerschweinchens nach Einzeitbestrahlung mit schnellen Neutronen (DT, 14 MeV). *Strahlentherapie* (in press).

FRANKE, H. D. (1979) Two years experience with fast neutrons (DT, 14 MeV) for clinical tumour therapy in Hamburg-Eppendorf. In: *Proc. of the International Meeting for Radio-Oncology*, Baden/Austria, May 1978, G. Thieme, Stuttgart (in press).

FRANKE, H. D. and W. LIERSE, (1967) Histochemische and ultra-structurelle Veränderungen am Meerschweinchengehirn nach Einwirkung unterschiedlicher Strahlenarten. *Sonderbände zur Strahlentherapie*, Bd. 64, 179–184.

FRANKE, H. D., A. HESS, H. K. KRAUS, E. MAGIERA and B. P. OFFERMANN (1974) Die Neutronentherapie-Anlage in der Strahlentherapie-Abteilung der Radiologischen Universitätsklinik Hamburg-Eppendorf: Aufbau und Dosismessungen. In G. BURGER and H. G. EBERT (Eds.), *Proc. Second Symp. on Neutron Dosimetry in Biology and Medicine*, EUR 5273, Commission of the European Communities, Luxembourg, pp. 955–968.

LARAMORE, G. E., T. W. GRIFFIN, A. J. GERDES and R. G. PARKER (1978) Fast neutron and mixed (neutron/photon) beam teletherapy for grades III and IV astrocytomas. *Cancer* 42, 96–103.

LEWIS, P. (1975) Treatment of grade III and IV gliomas with fast neutrons. Pathology of neutron treated brains. The Royal Post Graduate Medical School, MRC, Fast Neutron Clinic, Hammersmith Hospital, London. Course about "The Treatment of Cancer with Fast Neutrons". May 27–31, 1975.

LIERSE, W., K. GRITZ and H. D. FRANKE (1965) Histochemischer Nachweis von Glykogen und Mukopolysacchariden im Gehirn des Meerschweinchens nach Röntgenbestrahlung. *Fortschritte Röntgenstrahlen* 103, 612–618.

LIERSE, W. and H. D. FRANKE, (1978) Histochemische und elektronen-mikroskopische Reaktion des Meerschweinchenhirns nach Bestrahlung mit schnellen Neutronen (DT, 14 MeV). *Deutscher Röntgenkongresz*, Bonn (in press).

PARKER, R. G., H. C. BERRY, A. J. GERDES, M. D. SORONEN and C. M. SHAW (1976) Fast neutron beam radiotherapy of glioblastoma multiforme. *Amer. J. Roentgenol.* 127, 331–335.

SCHMIDT, R., H. D. FRANKE, E. MAGIERA and W. SCOBEL (1977) Spektroskopische Untersuchungen zur Dosimetrie in gemischten Neutron-Gamma-Feldern. In: *Jahresbericht 1976/1977*, des 1. Instituts für Experimentalphysik der Universität Hamburg, pp. 42–44.

SCHMIDT, R. and E. MAGIERA (1977). Determination of neutron and gamma spectra in a mixed (n, γ)-field. In: G. BURGER and H. G. EBERT (Eds.), *Proc. Third Symp. on Neutron Dosimetry in Biology and Medicine*, EUR 5848, Commission of the European Communities, Luxembourg, pp. 443–454.

SELLE, G., W. LIERSE and H. D. FRANKE (1968) Topochemischer Nachweis von Glykogen und Mukopolysacchariden im Gehirn des Meerschweinchens nach Telekobaltbestrahlung. *Strahlentherapie*, 136, 712–716.

WRAGE, D., W. LIERSE and H. D. FRANKE (1969) Die Aktivierung der alkalischen und sauren Phosphatase in Ganglienzellen nach Bestrahlung des Meerschweinchens mit Röntgenstrahlen (200 kV). *Strahlentherapie* 137, 320–325.

ZYWITZ, F., H. JUNGE, A. HESS and H. D. FRANKE (1979) Response of mouse intestine to 14 MeV neutrons. *Int. J. Radiol. Biol.* 35, 63–72.

Results of clinical applications of negative pions at Los Alamos*

M. M. KLIGERMAN,† S. WILSON, J. SALA, C. VON ESSEN, H. TSUJIL
J. DEMBO AND M. KHAN

Cancer Research and Treatment Center, University of New Mexico, Albuquerque, New Mexico 87131, U.S.A.

†*CRTC/UNM and the Los Alamos Scientific Laboratory, Los Alamos, New Mexico 87545, U.S.A.*

Abstract—Data are presented on forty patients treated under Phase I-II studies at Los Alamos with negative pi mesons (pions), alone or with subsequent conventional radiation, and followed from 6 to 15 months prior to 15 August 1978. Patients had advanced disease (large primary and/or metastatic solid tumors) in or near the thirteen sites planned for randomized Phase III studies, and were selected to provide information on tissue tolerance in those sites. Data on longevity, tumor response, and normal tissue reactions are presented.

Introduction

Phase II trials leading to randomized clinical trials of pion radiotherapy are nearing completion. These studies are being conducted at the Meson Physics Facility of the Los Alamos Scientific Laboratory (LASL) by the University of New Mexico Cancer Research and Treatment Center (UNM/CRTC), Albuquerque. As of 1 September 1978, sixty-seven patients had been treated, all but four of them since June 1976. One patient had two separate primary tumors—histologically distinguishable into a bladder cancer and a prostatic cancer. Counting this patient as two cases, a total of sixty-eight cases have been treated since the start of pion radiotherapy.

Nine of these patients, with skin and subcutaneous metastases randomized to pion and 100 kVp X-radiation, were treated early in the program to establish relative biological effectiveness (RBE) of pions in acute skin reaction (Kligerman and co-workers, 1977). A subsequent paper suggested the possibility of therapeutic gain measured by the time to regrowth of sixteen nodules in one patient participating in that experiment who could be followed for 346 days (Kligerman and co-workers, 1978a). Data are presented here on forty patients treated from 6 to 15 months ago. Illustrative reports on some of these cases have been presented elsewhere (Kligerman and co-workers, 1978b). This paper does not address an additional eighteen patients treated since June 1978, because time available for observation was less than 6 months.

Methods and materials

Miscellaneous locally advanced or recurrent human solid tumors and metastatic lesions, occurring in or near sites selected for the thirteen planned Phase III pion protocols, were selected for Phase I-II study. The Phase III protocols address tumors characterized by little chance for survival by any modern cancer treatment modality or combined modalities in any stage, or those controllable in early stages but not in advanced stages. These include primary tumors of the brain (gliomas), oral cavity, oropharynx, hypopharynx, larynx, esophagus, superior sulcus, pancreas, stomach, urinary bladder, prostate, uterine cervix and rectum. The goal of the Phase III protocols is to seek a two- to three-fold improvement over current long-term survival rates, which in the sites selected range from 1 to 20 per cent survival at 5 years.

---

*These investigations were supported in part by the U.S. Public Health Service Grants No. CA-16127 and CA-14052 from the National Cancer Institute, Division of Research Resources and Centers, and by the U.S. Department of Energy.

Although local/regional control is the goal of the Phase III studies, the relationship between such control and the potential benefit of chemotherapy and perhaps immunotherapy in controlling distant microscopic metastases is important in widening the potential scope of pion radiotherapy.

Of the forty cases reported here, none was lost to follow-up, and all follow-up information was derived from examination by oncologists of the UNM/CRTC or by oncologists at other co-operating institutions. Primary tumors (and nodal metastases, when appropriate) were treated in thirty-one of the forty patients. In nine patients, only metastatic lesions were treated. The reason for treatment of these metastases, and of primaries in some patients who had distant metastases, was to safely obtain information toward determining maximum tolerance of normal tissues in or near those sites planned for study under Phase III trials.

Among the thirty-one patients with primary tumors, fourteen had abdominal or pelvic primary tumors, twelve had head and neck primary tumors, and the remaining five had tumors of the brain (one), lung (two, both superior sulcus), breast (one, a male patient), or skin (one massive recurrent basal cell carcinoma of the anterior chest region).

Fifteen primaries were staged T4. Two of these were staged Nl–M1; three, N3; one, N2; two, N1; two, NX; four, N0; and one (a brain case), T4–G4. Ten cases were staged T3, four of whom were N3, two N2, one N1, and one NX (the latter with a prostatic primary carcinoma measuring 13 cm in cross-sectional diameter). One patient was staged T2–N3, and one patient, a prostate case, T2–NX. Two patients were treated for recurrence after post-surgical local control and two were treated for post-surgical persistence.

Of the nine patients in whom metastases alone were treated, one had metastases to brain (primary breast), two had metastases to lung (primaries melanoma and endometrium), one had metastases to liver (primary rectum) and five had massive lymph node metastases.

In patients who had received chemotherapy prior to pion radiotherapy, the tumor was either actively growing in spite of chemotherapy, or drugs were stopped at least 2 weeks prior to the start of pion radiotherapy. No drugs were given with pion treatment. No patient had received previous conventional radiotherapy to pion-treated portals.

The pion treatment was delivered, generally, in five fractions per week, with daily fractions ranging between 110 and 140 peak pion rads maximum. Total tumor doses ranged from 1000 to 4600 peak pion rads. Only two patients received maximum doses higher than 3300 peak pion rads and nineteen patients received doses under 2700. Of the latter, fifteen received supplemental conventional radiation. Seven of these had conventional radiation of 4000 rad immediately after pion radiotherapy when the target failed in February 1978. Two of the four not receiving conventional radiation were among those in whom pions were being given to a region for the first time, and the dose was kept purposely small. These persons, as well as the remaining two, required other treatment after pions.

All pion patients received static beam treatments, since dynamic scanning (with the potential for tailoring the depth of the peak pion region across the treatment volume) will not be initiated until November 1978. The pion beams ranged in energy from 60 to 110 MeV. The modulated pion peak was spread to dimensions of 5, 6, 8 or 10 cm in depth. When possible, opposing portals were used to optimize uniformity of distribution of both physical dose and high linear-energy-transfer (high LET) particles originating from pion capture. Abutting fields were also applied, as necessary, for treatment of large volumes (i.e. greater than 15 cm in the lateral dimensions).

Results

Longevity and tumor response

Table 1 shows all patients at risk up to 15 months. No patients were lost to follow-up. The majority of deaths occurred within 6 months, an indication of the advanced nature of the disease in most of these patients treated to establish normal tissue tolerance.

Table 1. All pion patients

Months	3	6	9	12	14	15
Patients at risk	40	37	21	14	7	2
Lost to follow-up	0	0	0	0	0	0
Dead	3	10	3	1	1	0
Alive	37	27	18	13	6	2

Table 2 indicates the cause of death at intervals. Four patients had local recurrence as a cause of death and twelve deaths were due to metastases outside the pion treatment volume. No patient died of complications (further, no patient suffered complications due to pion radiotherapy). Two patients died of intercurrent disease (heart attack and pulmonary disease with infection). Both had microscopic tumor deposits at time of death.

Table 2. Pion patients, cause of death

Months	3	6	9	12	14	15
Deceased patients	3	10	3	1	1	0
Cause:						
Local recurrence	1	2	0	0	1	0
Metastasis	2	7	3	0	0	0
Complications	0	0	0	0	0	0
Intercurrent disease	0	1	0	1	0	0

Table 3 shows the status of patients alive at each interval and the status of their treated tumor at each interval for which sufficient follow-up time had elapsed. Of the thirteen patients who exhibited complete regression, ten were treated with pions alone, two of whom were rendered free of disease after post-pion surgery for persistence. Three patients exhibiting complete regression were treated with pions plus X-rays or electrons, two of them found after surgery to have no viable tumor cells in their excised residual mass. All those with complete regression treated with pions alone received not less than 2700 peak pion rad total tumor dose. Two treated with pions plus conventional radiation who exhibited complete regression received 1800-2000 peak pion rad followed by 3900 rad of conventional radiation (photons or electrons), and one received 2700 peak pion rad followed by 2400 rad of photons.

Table 3. Pion patients living, treated tumor status

Months	3	6	9	12	14	15
Alive	37	27	18	13	6	2
Complete regression	11	13[a]	10	7	4	1
Partial regression	15	6	1	0	0	0
No change	11	5	3	2	0	0
Recurrence after complete or partial regression	0	3	4	3	1	0
Regrowth after no change	0	0	1	1	1	1
No evaluation	0	0	0	0	0	0

[a]Two, no disease post-surgery for persistence. All patients seen by oncologist.

There were three recurrences after complete and seven after partial regression, three resulting in death due to local recurrence. Of eleven patients surviving at least 3 months and exhibiting no change in tumor mass, eight were abdominal cases (pancreas, stomach or liver) in whom assessment was ambiguous. Six of these patients have succumbed of disease, exhibiting symptoms of ascites and/or spread to liver or other nearby organs. Two of these patients are still living, one at 12 months and one at 9 months, although the one at 12 months has a biopsy-proven metastasis outside the pion field. One of the non-responders was a patient with a rectal carcinoma, in whom the criterion of at least 25 per cent regression was not met, such that this could be judged a "partial regression", but in whom the lumen did open sufficiently to relieve symptoms of cramps and diarrhea. She later succumbed to local regrowth. Another was a patient with five melanoma metastases to lung (primary skin of thumb), in whom the metastases showed no growth for more than 10 months. She is one of the 15-month survivors. The remaining patient who exhibited no response had a synovial sarcoma of the pelvis and succumbed to liver metastases at 4 months.

Table 4 summarizes results to date with pion radiotherapy, alone and with conventional radiotherapy, for these forty cases as of 15 August 1978, disregarding intervals of occurrence.

Normal tissue reactions

In general, acute reactions of the normal tissues have been remarkably mild. In the head and neck, a severe reaction of the gingival border was observed in two of the earliest of these forty cases. This was subsequently managed by the application of cottonoid soaked with a radioprotectant (acetylcysteine) and held in place during treatment by plastic dental molds, except where tumor involvement was apparent in or contiguous with the gingiva. Other acute oral-pharyngeal reactions were less than would be expected for the degree of tumor response but were related to daily fraction size and overall days. Areas of second-degree reaction did occur, which healed rapidly, often within 4 days after the end of treatment. This reaction usually affected less than half the field.

Patients treated with pelvic portals did not develop diarrhea, mucus, bleeding or spasm. Direct examination of the rectal mucosa failed to reveal any abnormality. An occasional patient had an inconstant minimal increased frequency of urination without burning and without blood. This cleared rapidly soon after the end of pion therapy. Of the thirteen patients who had abdominal irradiation, often with ports over 200 cm², none experienced diarrhea. Only two of these patients exhibited symptoms similar to those of radiation sickness. However, both had extensive non-resectable pancreatic tumor. It is difficult to know if tumor or radiation was causing the symptoms. In one case, the radiotherapists felt that radiation did contribute to mild anorexia and weight loss.

xerostomia occurring at 6 months). This suppression has persisted in all four patients.

The patient with the 13 cm prostatic carcinoma, who received 3300 peak pion rad, 23 fractions, 31 days, developed two stools a day in the second month after treatment. Like the laryngeal dewlap edema noted above, these symptoms occurred at 2 months and lasted only 1 month (in this case, clearing completely).

Table 4. Summary of results with pion radiotherapy

	Total	Pions alone	Pions+conventional
At risk	40	24	16
Alive	22	14	8
Alive, NED	9	6	3
Alive, with disease	13	8	5
Dead	18	10	8
Cause of death:			
Local recurrence	4	2	2
Metastasis	12	8	4
Complications	0	0	0
Intercurrent disease	2	0	2
Tumor response:			
Complete regression	13[a]	10[a]	3
Partial regression	17[b]	7[b]	10
No change	12[c]	9	3[c]
Recurrence after complete regression	3	3	0
Recurrence after partial regression	7	3	4
Regrowth after no change	2	2	0
Lost to follow-up	0	0	0

[a] Includes two with surgery for post-radiation persistence.
[b] Includes the two listed above with partial regression pre-surgery; also two dead prior to 3 months follow-up (both pions only).
[c] Includes one dead prior to 3 months follow-up (pions plus conventional).

Intermediate and long-term reactions recorded to date have also been mild. No patient has exhibited any serious untoward intermediate or long-term effects, including one patient in the previously reported skin metastases series who has now reached 26 months survival. No patients have suffered any serious complications, either during treatment or after treatment.

Two patients treated in the head and neck area developed edema of the larynx and of the subcutaneous tissues of the anterior neck (dewlap) in the second month after the end of pion therapy. This almost completely cleared within 1 month, but a small amount of dewlap and +1 edema of the larynx persists. Four patients have exhibited xerostomia, two nasopharynx patients (one mild, one severe) and one oropharynx patient (mild) treated with pions alone (xerostomia occurring at 11, 9 and 12 months, respectively), and one patient with a larynx carcinoma treated with pions followed by cobalt-60 and iridium implant (severe

Surgeons who have operated on patients after pion radiotherapy have reported that they encountered no difficulty in carrying out their planned operative procedures, and that in some cases normal tissues within the peak pion radiation field responded at surgery as though they had not been previously irradiated.

Conclusions

1. All forty patients tolerated pion therapy well at the doses applied.
2. Complete regressions with pions alone were not noted until doses reached more than 2700 peak pion rad.
3. Conventional radiation and/or surgery was also well tolerated by those patients requiring those treatments after pion therapy.
4. Pions have been least successful to date in treatment of pancreas and stomach, although this appears due to (1) inability to adequately assess regression, (2) inability to adequately

assess tumor extent before pion therapy and (3) metastasis or peripheral spread during or soon after treatment.

5. Although it is too early to assess impact of pion radiotherapy on long-term control or survival, local recurrence after complete regression has been noted in only three patients to date.

Because of the experience gained thus far, including patients treated within the last 6 months, a decision has been reached to establish a *minimum* dose for randomized Phase III clinical trials of pion radiotherapy of 3300 peak pion rad to the target volume, which includes the gross tumor volume and a margin as large as needed to include that volume of tissue beyond which the radiotherapist believes microscopic tumor does not extend. Reductions will be specified for critical structures in the peak pion field. The minimum dose selected has been tolerable to acute reactions in volumes as large as 3300 cm^2. Doses will be delivered 5 days per week, with a daily fraction size of approximately 100 peak pion rad. Split course treatment may be needed to conform to accelerator operating schedules.

References

KLIGERMAN, M. M., A. SMITH, J. M. YUHAS, S. WILSON, C. J. STERNHAGEN, J. A. HELLAND and J. M. SALA, (1977) The relative biological effectiveness of pions in the acute response of human skin. *Int. J. Radiat. Oncol. Biol. Phys.* **3**, 335-339.

KLIGERMAN, M. M., J. M. SALA, S. WILSON and J. M. YUHAS (1978a) Investigation of pion treated human skin nodules for therapeutic gain. *Int. J. Radiat. Oncol. Biol. Phys.* **4**, 263-265.

KLIGERMAN, M. M., C. F. VON ESSEN, M. K. KHAN, A. R. SMITH, C. J. STERNHAGEN and J. M. SALA (1978b) Experience with pion radiotherapy. *Cancer* (in press).

Results of tumor treatments with alpha particles and heavy ions at Lawrence Berkeley Laboratory

J. R. CASTRO,* C. A. TOBIAS,† J. M. QUIVEY,* G. T. Y. CHEN,*
J. T. LYMAN,* T. L. PHILLIPS,‡ E. L. ALPEN§ AND R. P. SINGH*

Abstract—*A clinical trial, starting with 934 MeV helium ions produced at the 184-inch cyclotron, has been underway at Lawrence Berkeley Laboratory since 1975. Eighty-five patients have been treated in a pilot series with the helium ion beam. Treatment techniques have been developed for large field, fractionated, spread out Bragg peak charged particle radiotherapy. A treatment planning computer routine has been developed to compute physical and biological isodoses for charged particle beams taking into account beam perturbations by varying tissue densities as determined by CT scanning in the region of interest. An RBE of 1.2 for fractionated helium ion therapy has been clinically confirmed using daily doses of 200 photon equivalent rad per fraction, four fractions per week to total doses of 6000 rad.*

Pilot studies with heavier ions such as carbon and neon are just beginning at the Bevalac. Patients with locally advanced malignancies not amenable to conventional therapy will be selected for pilot irradiation with carbon and neon ions to determine clinical RBEs, normal tissue reaction and tumor regression. The goal of these pilot studies is a prospective clinical trial utilizing helium ions to evaluate the advantages of improved dose localization, and carbon or neon ions to study the increased biological effect and improved dose localization available with heavy charged particle therapy.

Introduction

At the Lawrence Berkeley Laboratory the utilization of helium ion beams for pituitary irradiation has been underway for many years. The first therapeutic human exposure to high-energy deuterons was performed at the Lawrence Berkeley Laboratory in 1955, and since that time several hundred patients have been treated with small field, high dose per fraction, helium radiation for pituitary tumors, acromegaly and Cushing's disease (Lawrence *et al.*, 1971, 1973a, 1973b). Most of these treatments were accomplished by complex isocentric rotations utilizing the plateau rather than the peak of the beam. A few patients were treated with small field, high dose, single-fraction irradiation of small pulmonary metastatic nodules or brain metastases prior to developing a more systematic clinical trial.

Our efforts to study the use of these beams in humans began in earnest in the spring of 1975, with a detailed effort by physicians, biologists and physicists to carefully plan a clinical trial of heavy charged particle irradiation (Chen *et al.*, 1977).

Among the suggested advantageous sites for heavy charged particle radiotherapy clinical trials were: carcinoma of the esophagus; malignant glioma of the brain; unresectable localized tumors of selected sites in the head and neck such as paranasal sinuses; choroidal melanoma; advanced carcinoma of the cervix uteri and bladder, and localized unresectable carcinoma of the pancreas. Patients with other locally unresectable tumors not likely to be

---

*Radiation Therapy, Lawrence Berkeley Laboratory, Bldg 55, Rm 106, Berkeley, California 94720, U.S.A.
†Radiation Biophysics, Lawrence Berkeley Laboratory, Bldg 10, Rm 202, Berkeley, California 94720, U.S.A.
‡Radiation Therapy, University of California Medical School, M-330, San Francisco, California 94720, U.S.A.
§Donner Laboratory, Lawrence Berkeley Laboratory, Bldg 1, Rm 466, Berkeley, California 94720, U.S.A.

successfully treated with standard modalities and with a low incidence of distant metastases have also been accepted for pilot phase studies. Clinical protocols have been developed for use at Lawrence Berkeley Laboratory and have been submitted to the Radiation Therapy Oncology Group and the Northern California Oncology Group for registration as official protocols. The Bay Area Heavy Ion Association, a group of interested radiotherapists, physicists, biologists, physicians and member radiotherapists of the NCOG and RTOG, will participate in design of protocols, referral of patients, irradiation of control patients and follow-up studies.

With support from the Department of Energy and the National Cancer Institute, a pilot group of eighty-five patients has been irradiated with 934 MeV helium ions produced at the 184-inch cyclotron during the period of July 1975 through October 1978 in order to explore the feasibility of treatment of human cancers with heavy charged particles. The primary tasks of the pilot study have been to develop techniques of large field, spread out Bragg peak multi-fraction radiotherapy including:

1. Modification of the beam line at the 184-inch cyclotron and the Bevalac to provide a large, flat ($\pm$ 2% over 90% of the diameter) field (30 cm circle) of high dose rate (200 rad/min) and deep range in tissue (26·5 cm).
2. Modification of patient positioners at the cyclotron and the Bevalac to accommodate advanced immobilization techniques, bolus and compensators and provide daily reproducibility of patient set-up.
3. Development of necessary radiotherapy clinical facilities.
4. Developing a computerized (PDP 11/34) treatment planning system to display physical and biological isodoses, including perturbations of the particle beam secondary to tissue inhomogeneities in the beam path as determined by CT scan data.
5. Development of a systematic ridge filter design which allows a family of spread out Bragg peaks of different widths to fit a variety of clinical situations and which is adaptable to different ions, providing isosurvival (equal cell killing) across the spread out Bragg peak.
6. Evaluation of clinical RBE for fractionated helium Bragg peak radiotherapy.
7. Clinical radiotheraphy techniques to irradiate tumors as listed above.

Treatment planning

Our computerized treatment planning system for charged particle radiotherapy has been previously described (Chen *et al.*, 1978). The computer program is based on a pixel by pixel range shortening algorithm utilizing detailed quantitative information available from CT scanning. The conversion of the CT data to

Table 1. Heavy ion beams at Bevalac

	Particle	Energy, GeV	Energy per nucleon, MeV/n	Approximate range in tissue, cm	Intensity particles per pulse	Plateau dose rate, rad/min
Carbon	$^{12}_{6}C$	3.7	308	16	10^{10}	500
	$^{12}_{6}C$	4.8	400	26	10^{10}	> 500
						(Note: also available at Bevatron at 5×10^7 particles per pulse)
Neon	$^{20}_{10}Ne$	8.3	415	16	2×10^9	300
	$^{20}_{10}Ne$	11.1	557	26	2×10^9	300
Argon	$^{40}_{18}A$	20	500	12	10^8	40

water equivalent range is performed by a calibration curve using tissue equivalent materials (Chen et al., 1977). We are developing an advanced treatment planning system based on interactive graphic display. The goal is to produce three-dimensional treatment plans which allow the computer to calculate the required compensation for tissue inhomogeneities. The radiotherapist then enters the target volume on each graphically displayed CT slice using a cursor and interactive display terminal. Following selection of the angle for the portals the program calculates the required compensator to stop the beam at the distal edge of the target contour. Lifesize plots of the shape of the compensator are then produced and are used as a template for compensator construction. The program continues to calculate isodose lines for the multiple port treatment. The final step is the production of hard copy isodoses superimposed on the grey scale printout of the CT scan.

compensation of tissue inhomogeneities as determined by CT scanning as well as techniques for *in vivo* verification of dosimetry using diodes placed within body cavities or passing the beam through the patient to stop in a film stack.

We have found the clinical RBE for midpeak helium ion irradiation to be approximately 1.2 as assessed by acute and subacute skin, mucosal and intestinal reactions. There have been only limited opportunities to assess late effects up to 36 months, but we have found no serious variation in this RBE estimate. All patients are irradiated with the spread out Bragg peak placed to cover the target volume.

Dose fraction sizes of 200 cobalt-60 gamma rad equivalent given four times a week to total doses of 6000-6500 CoRE appear well tolerated with helium ion radiotherapy as assessed up to 36 months post-treatment. Patient tolerance has been good with only moderate skin reactions using multi-port therapy. In some instances of irradiation of abdominal tumors, there appears

Table 2

Site of treatment	Number of patients 7/75–10/78		
Intracranial	11		
Head and neck	9		
Thoracic	7		
Abdomen	41 (Ca pancreas—35)		
Retroperitoneal	5		
Pelvis	11		
Skin nodule studies	3		
Ocular melanoma	3		
	90	Helium radiation therapy alone	56
		Photon + helium therapy	29
		Heavier particles	5

Clinical studies with helium

Since July of 1975, we have irradiated eighty-five patients with the helium ion beam for a variety of locally advanced and/or recurrent neoplasms not amenable to control by surgery or standard megavoltage radiation (Table 2). We have concentrated on developing charged particle radiotherapy techniques which permitted daily reproducibility of patient set-up to ± 2 mm for head and neck tumors using individually formed lucite head holders and ± 5 mm elsewhere in the body using light cast or plaster molds. Daily verification films in the treatment position, using X-ray tubes at right angles to and back pointing the beam with the helium port superimposed, have shown these goals to be achievable in most patients. We have also developed techniques for bolus and

to be improved patient tolerance during treatment attributable to more sharply defined dose distributions with sparing of adjacent gut.

Late effects (up to 36 months post-treatment) have shown moderate skin and subcutaneous changes consistent with the stated radiation doses.

Among the patients completing irradiation with helium ions more than 6 months ago have been twenty patients with unresectable carcinoma of the pancreas. Representative treatment portals and biologically effective isodose distributions for irradiation of carcinoma of the pancreas are shown in Figs. 1a, b, c. These patients are treated in the upright position, either seated or standing, because of the fixed horizontal helium beam.

Barium films in the upright-patient position are used to localize the target volume in addition to CT scans (supine) and surgical clipping, when available. Since not all patients have surgical clips placed and we cannot as yet perform CT scans with the patient upright, target volumes have of necessity been larger than might be needed for many patients in order to assure adequate tumor coverage. Doses of 6000-6500 CoRE in 7-8 weeks, 200 CoRE per fraction, 4 fractions per week have generally been delivered except for some early patients treated with initial photon irradiation followed by helium boost.

ten failures, one appeared to have control within the irradiated area but succumbed to distant metastases. One patient expired of intercurrent disease and one of presumed chemotherapy complications at 18 months and 4 months post treatment, respectively. Three patients have progressive disease with inability to complete therapy (Table 3); one refused to continue beyond 3000 rad and is without evidence of disease 14 months post-treatment.

Skin reactions showed only mild erythema

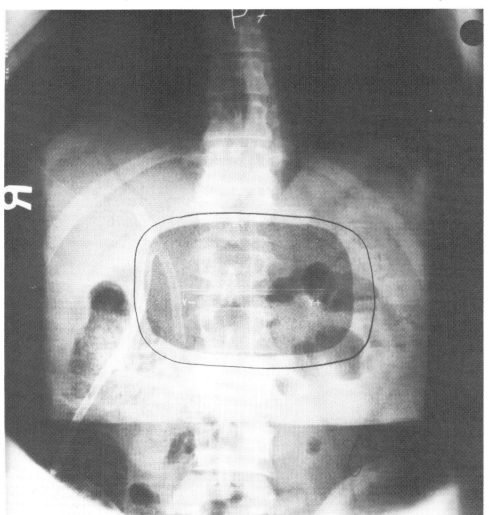

Fig. 1a. AP helium portal radiographs on a 38-year-old patient with localized adenocarcinoma of the pancreas, stage $T_4N_0M_0$. The area included in the heavy pencilled line is actually irradiated with 934 MeV helium ions. The patient was treated in the upright position through anterior, right and left lateral portals (three fields).

Morbidity has been low although in sixteen pancreas patients receiving a minimum of 5000 CoRE, two developed moderately severe upper GI bleeding 4 to 6 months post-therapy. With reduction of the target volume for the final 1000 CoRE of the planned dose, this has not recurred. Four of sixteen patients have no evidence of disease from 8 to 37 months post-therapy; of the

although mild to moderate gastrointestinal reactions also have been encountered probably secondary to unavoidable irradiation of the gastric antrum and duodenum.

Continued experience with pancreatic irradiation is planned with a randomized trial between helium irradiation and megavoltage photon-radiation therapy scheduled to begin shortly. Later it is hoped to extend this trial to include neon or carbon irradiation as a third arm.

We have also irradiated six patients with carcinoma of the esophagus in a pilot series using mixed photon irradiation with helium boost initially and progressing to delivering the entire treatment with helium ions.

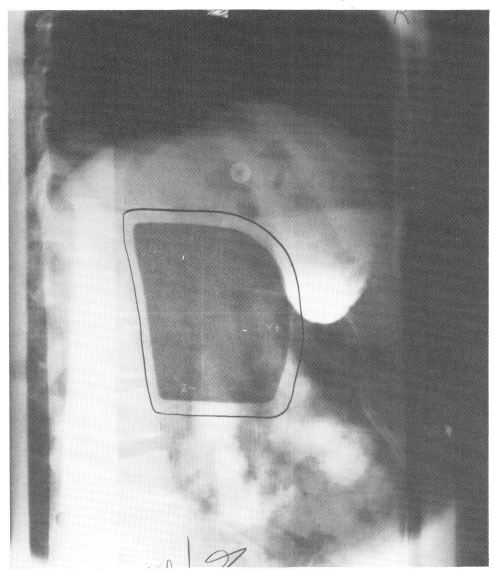

Fig. 1b. Lateral portal for same patient with carcinoma of the pancreas. Heavy line indicates true helium target volume after correction for X-ray magnification.

Treatment planning has been achieved by using CT scan data to determine and compensate for tissue densities encountered by the particle beam traversing soft tissue, bone and air in the lung. The initial target volumes have been generous in order to irradiate a long segment of the esophagus and mediastinum to 4000 CoRE and then have been reduced to carry the primary tumor to 6000 CoRE in 7-8 weeks, 200 CoRE per week, 4 fractions per week. Skin reactions have been moderate although erythema, dry desquamation and patchy moist skin reactions have been encountered because occasionally portions of skin were unavoidably included in the Bragg peak while irradiating the upper esophagus. All healed uneventfully. Dysphagia secondary to treatment was mild. Two of six patients at 4 and 12 months post-treatment have no evidence of disease within the irradiated volume; two patients have local recurrence and two could not complete irradiation because of tumor progression including perforation in one patient. We plan to continue a prospective nonrandomized trial of irradiation for carcinoma of the esophagus utilizing helium and heavier ions such as carbon or neon since they offer biological as well as dose distribution advantages.

Although our clinical impression confirms the

Table 3. Carcinoma of the pancreas (9/75 →10/78)

Not evaluable	4 (incomplete treatment)
Evaluable	16 (>5000 rad tumor dose)
Alive, NED	4/16 (8-37 mos)
Expired, NED, intercurrent disease	1/16 (13 mos)
Expired, persistent tumor in XRT field	9/16 (6 with distant metastases)
Distant metastases	1/16
Unknown	1/16 (expired, decrease in platelets post-chemotherapy)
Complications of radiation therapy	2/16 (upper GI bleeding—both patients NED; pancreatic insufficiency; diabetes)

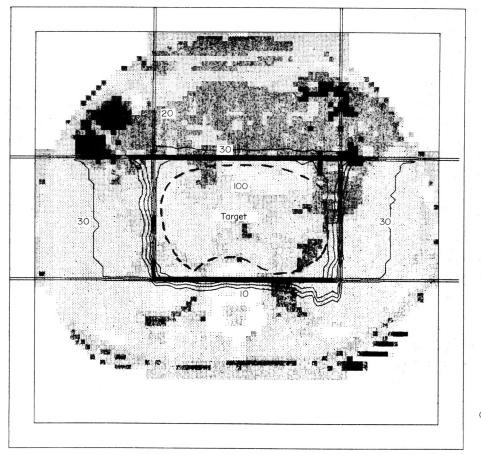

Fig. 1c. The patient received 6250 cobalt^{60} rad equivalent to the 100% isodose which encompasses the target volume, given in 32 treatments over 57 days. Isodoses are depicted on grey level CT scan of patient through the pancreas. Only one cut is shown although isodose calculations are performed at three levels through the target volume.

validity of the CoRE model for helium, it is not possible to draw a definitive assessment of tumor response to helium irradiation since (1) all of these patients had locally advanced or resistant tumors, (2) many of the early patients had combinations of photon and helium ion irradiation and (3) initially total doses were low and a conservative estimate for RBE was used in order to assure patient safety.

Satisfactory treatment techniques have been developed to provide for clinical trials of helium therapy in localized carcinoma of the pancreas, advanced carcinoma of the cervix uteri, carcinoma of the esophagus and continue Phase I–Phase II helium studies for other tumor sites.

Pilot studies with carbon and neon ions

Existing biomedical facilities at the Bevalac are being modified to provide a treatment room for heavy ion patient irradiation, including installation of a Phillips Mark I ram-style treatment couch, X-ray tubes and laser positioning lights. Additional clinical facilities are available in the Research Medicine building which is in close proximity to the Bevalac and which houses the radio-therapy clinical facility including offices, conference room, examination rooms, X-ray simulation room CT scanner suite and treatment planning computer.

Beginning in the fall of 1978, beam time will be made available for carbon, neon and argon patient irradiation 4 days per week starting with 2-week runs and building up to treatment courses of 6 to 8 weeks. Systematic Phase I–Phase II studies with these ions will begin with irradiation of locally advanced, resistant tumors similar to those irradiated in the helium pilot study including carcinoma of the esophagus, carcinoma of the pancreas, malignant glioma of the brain and advanced head and neck tumors.

The goals of the initial pilot studies with carbon, neon and argon ions at the Bevalac are:
1. Evaluation of biologic response of normal tissues including acute skin, mucosal and intestinal reactions.
2. Development of effective treatment techniques building on experience obtained with the helium ion beam at the 184-inch cyclotron.
3. Evaluation of physical and biologically effective dose distributions available with carbon, neon and argon ions correlated with ion energy, range and width of the spread-out Bragg peak.
4. Analysis of potential tumor types of anatomic sites for prospective randomized trials. Such analysis will include a careful review of biological dose distributions available with carbon, neon and argon ions particularly relating biological effects in the entrance region with that of the spread out Bragg peak.

In addition, careful assessment of skin and mucosal reactions will be utilized to confirm preclinical RBE estimates. Observation of other organ and tissue response such as connective tissue fibrosis, CNS effects, pulmonary and renal effects will also be closely monitored both by clinical observation as well as appropriate laboratory and radiological testing.

In order to get an early start on Phase I studies with heavy ions, five patients have been treated to date with carbon and neon ions, using temporary facilities in the Bevalac Biomedical area. They have included patients with the following diseases: Kaposi's sarcoma (carbon ion skin RBE study); metastatic leiomyosarcoma (neon skin RBE study); recurrent carcinoma of floor of mouth and submental area (neon ion boost after helium radiation); melanoma metastatic to brain (carbon and helium radiotherapy) and T_4N_3 squamous carcinoma of the hypopharynx (carbon boost after photon therapy). These patients have been followed only a short time but have had the expected skin and mucosal reactions and toxicity for the predicted RBEs of ~ 2.7 for carbon and ~ 3.4 for neon, assessed at midportion of the spread-out Bragg peak.

The first such patient and the one with the longest follow-up (12 months) was an elderly patient with Kaposi's sarcoma. Six fields were treated with mid-4-cm peak 308 MeV/amu carbon ions at the Bevalac facility. Control irradiations were done with 10 MeV electrons because of their sharp dose fall-off at depth, using the Allis–Chalmers betatron at the Zellerbach Saroni Tumor Institute, San Francisco. Ten (10) equal fractions of 100, 140 or 170 carbon rad were given to 3 cm diameter circular portals and the resultant skin response compared to 10 equal fractions of 250, 300, or 350 rad of electrons using 4 cm diameter circular portals (Fig. 2a).

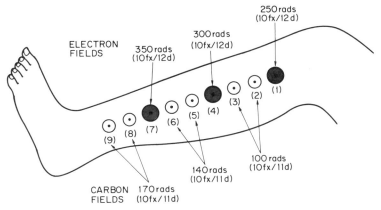

Fig. 2a. Location fields for comparison of electrons and carbon ions respectively.

The skin reactions were scored by two independent observers and analyzed graphically in order to obtain a clinical carbon ion RBE. The graphical analysis showed a similar time course to the peak reaction for both types of radiations suggesting a similar mechanism of epithelial repopulation. Because of the nonlinearity of the skin response scoring system, RBEs were calculated only for average skin response levels that were produced by both control and carbon radiations (Fig. 2b). The calculated RBE for 120 rad per fraction of carbon ions was ≃2.7.

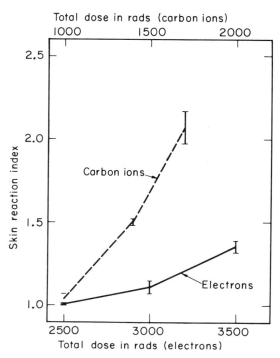

Fig. 2b. Results of analysis of skin reactions after irradiation with electrons and carbon ions respectively.

The patient irradiated with neon ions for metastatic leiomyosarcoma had only one field irradiated with 400 MeV neon ions which was compared to a single adjacent 15-MeV electron field. The resulting skin reactions were relatively similar suggesting the predicted neon RBE of about 3.4 was approximately correct relative to 400 rad applied dose of 15 MeV electrons.

Tumor regression appeared somewhat greater in the neon field (3200 Co^{60} rad equivalent in 8 fractions) than in the 15 MeV electron field.

Further such skin nodule RBE and tumor regression studies are planned as well as careful scoring of mucosal reactions and tumor response of advanced head and neck tumors.

Ultimately, one or two heavy ions, probably either carbon or neon, will be selected for prospective clinical trials in order to compare improved dose distribution and increased biological effect (carbon, neon) with improved dose distribution alone (helium) versus the best available megavoltage low LET radiotherapy.

References

CASTRO, J. R. and J. M. QUIVEY (1977) Clinical experience and expectations with helium and heavy ion irradiation. In: *Proceedings of the International Conference on Particles and Radiation Therapy, Lawrence Berkeley Laboratory, Berkeley, California* (September 1976). Int. J. Radiat. Oncol. Biol. Phys. Part II, **3**, 127-132.

CHEN, G. T. Y., J. T. LYMAN, J. RILEY (1977) Conversion of CT data to relative stopping power for charged particle radiotherapy. *Med. Phys.* **4**, 368.

CHEN, G. T. Y., R. P. SINGH, J. R. CASTRO, J. T. LYMAN and J. M. QUIVEY (1978) Heavy ion radiotherapy treatment planning. Submitted for publication to *Int. J. Radiat. Oncol. Biol. Phys.*

LAWRENCE, J. H., C. A. TOBIAS, J. T. LYMAN and J. A. LINFOOT (1971) Heavy particles in the treatment of acromegaly and cushing's disease and their potential value in other neoplastic disease. In: W. H. BLAND (Ed.), *Nuclear Medicine*, 2nd ed., McGraw Hill, chap. 32.

LAWRENCE, J. H., C. Y. CHONG, J. T. LYMAN, C. A. TOBIAS, J. L. BORN, J. F. GARCIA, MANOUGIAN, J. A. LINFOOT and G. M. CONNELL (1973) Treatment of pituitary tumors with heavy particles. In: P. O. KOHLER and G. T. ROSS (Eds.). *Diagnosis and Treatment of Pituitary Tumors*, Proceedings of a conference sponsored jointly by the National Institute of Child Health and Human Development and the National Cancer Institute, Bethesda, Md., January 15-17, 1973, Excerpta Medica Amsterdam/American Elsevier Publ. Co., New York, pp. 253-262.

LAWRENCE, J. H., C. Y. CHONG, J. L. BORN, J. T. LYMAN, M. D. OKERLUND, J. F. GARCIA, J. A. LINFOOT, C. A. TOBIAS and E. MANOUGIAN (1973) Heavy particles in acromegaly and cushing's disease. *Endocrine and Nonendocrine Hormone-Producing Tumors*, Yearbook Medical Publishers, Inc., pp. 39-61.

LEEMAN, C., J. ALONSO, H. GRUNDER, E. HOYER, G. KALNINS, D. RONDEAU, J. STAPLES and F. VOELKER (1977) A 3-dimensional beam scanning device for biomedical research. LBL-5546, pp. 1-3.

Results of clinical applications of fast neutrons in Japan

H. TSUNEMOTO,* Y. UMEGAKI,* Y. KUTSUTANI,* T. ARAI,†
S. MORITA,† A. KURISU,† K. KAWASHIMA‡ AND T. MARUYAMA‡

Division of Clinical Research, † Division of Hosp., ‡ Division of Physics, National Institute of Radiological Sciences, Chiba City, Chiba, Japan

Abstract—*Clinical trials with fast neutrons were carried out to estimate the indications and schedules for high-LET radiation therapy. Local control rate and incidence of moist desquamation of the patients surviving more than 6 months after completion of therapy was 39.8% and 11% respectively. Radioresistant tumors, such as malignant melanoma, soft-tissue sarcoma or osteosarcoma, and locally advanced tumors have responded well to fast neutrons, whereas late reactions of the normal tissues were slightly more severe after fast neutron therapy than those after low-LET irradiations. It is suggested that improvement of the dose distributions will be indispensable to make the best use of high-LET radiation therapy.*

Introduction

Fast neutron therapy has been advocated because of the radiobiological properties of fast neutron beams, i.e. low oxygen enhancement ratio or low repair capability of the irradiated cells. On the other hand, it was suggested that the late radiation damage of the normal tissue following fast neutron therapy would be more severe than that developed after X-irradiation. The results of clinical trials with fast neutron beams which have been carried out at various radiotherapy centers suggest that the effect for local control of the tumors would be superior in fast neutron therapy than in X-ray therapy, while the difference of the radiation doses associated with local failure and complication might be smaller in the former.

In Japan a clinical trial with fast neutrons has been initiated at the National Institute of Radiological Sciences (NIRS) in November 1975, which was followed by the Institute of Medical Sciences (IMS), Tokyo University, in November 1976. The trials aimed at the assessment of local control of the tumor and late radiation damage of the normal tissues following fast neutron therapy.

Characteristics of the beam

A fast neutron beam, generated by bombarding a thick beryllium target with 30 MeV deuterons, was used at NIRS. The fast neutron therapy has been carried out by applying a vertical beam, obtained by bending deuteron beams through a magnet. The dose rate in tissue equivalent plastic phantom was 42 rad (n, γ)/min per 30 μA for 11.4 × 11.4 cm² field at STD 200 cm. Gamma-ray contamination was estimated to be less than 4% in the beam.

The maximum build up of the fast neutron beam was seen at 6 mm below the surface and the depth dose curves obtained at SSD 175 cm were almost the same as those for Telecobalt γ-rays at SSD 75 cm.

Indications for fast neutron therapy

Radioresistant tumors and locally advanced tumors have been candidated for this trial.

For carcinoma of the uterine cervix, a modified randomization study was adopted. In this case, the patients referred to the hospital, when the machine was in operation, were subjected to fast neutron therapy, whereas the patients referred during the scheduled maintenance service period of the machine were treated with X-rays. For other types of malignancy, the trial was done as a non-randomized study.

Treatment schedule and evaluation of therapy

The relationships between dose and fractionation for treatment schedules of neutron therapy are shown in Table 1.

Table 1. Treatment schedule for fast neutron therapy

I. Fast neutrons only:
  (a) 130 rad × 12 fractions/4 weeks
  (b) 110 rad × 15 fractions/5 weeks
  (c) 90 rad × 18 fractions/6 weeks

II. Mixed schedule:

	Mon.	Tue.	Wed.	Thu.	Fr.	Sat./Sun.
Radiation	N	X	X	X	N	0
Dose	80	170	170	170	80	0

  5 weeks or 6 weeks

III. Fast neutron boost:
  X-rays: 4000–5000 rad/5–6 weeks
  Neutrons: 1500 rad X-ray equivalent dose/1.5–2.0 weeks by shrinking field

These factors have been decided by considering the fact that the RBE value at the tolerance dose (NSD) of normal tissue (skin) for neutrons relative to X-rays would be 1.8. To express the therapeutic results of X-rays and fast neutrons on a same scale of biological effect, a criterion characterized by TDF X-ray equivalent ($TDF_{n(xeq)}$) is introduced.

$$TDF_{n(xeq)} = R_{tol}TDF_n = 3.01 \times 10^{-2} d_n^{1.176} t^{-0.129} \quad (1)$$

$$TDF_x = d_x^{1.538} t^{-0.169} 10^{-3} \quad (2)$$

where TDF_n is the TDF of neutrons, R_{tol} is the ratio of TDF at full tolerance level of X-rays to that of neutrons (=30.1), d_n and d_x are the dose per fraction of neutrons and X-rays in rad respectively, and t is the time interval in days.

For mixed schedules or neutron boost, TDF X-ray equivalent (TDF_{total}) is calculated as follows:

$$TDF_{total} = \sum_{i=1}^{N_n} TDF_{n(xeq)i} + \sum_{j=1}^{N_x} TDF_{xj} \quad (3)$$

where N_n and N_x are the number of fractions of neutrons and X-rays, respectively.

Therapy with fast neutrons only was mainly applied in the treatment of radioresistant tumors. On the other hand, mixed schedule technique or fast neutron boost was used for the patients requiring a relatively large treatment volume to be irradiated, such as carcinoma of the uterine cervix or carcinoma of the lung. Glioblastoma multiforme was treated with mixed schedule or fast neutron boost.

Early or late effects of fast neutron therapy was recorded by a scoring system consisting of five steps. An example of late reactions for tumor or skin is shown in Table 2. For normal tissues, the reactions classified as score 3 mean the disease to be manageable by a conservative treatment, whereas the reactions diagnosed as scores 4 or 5 are diseases requiring a continuous medical care or incurable disease, respectively.

Table 2. Late reaction scores for tumor and skin following radiation therapy

I. Tumor: (1) Complete disappearance
  (2) Incomplete disappearance
  (3) Residual but non growing tumor
  (4) Residual and slowly growing tumor
  (5) Rapidly growing tumor

II. Skin or mucous membrane:
  (1) No change
  (2) Dryness
  (3) Atrophy or teleangiectasis
  (4) Ulceration
  (5) Incurable ulceration

Results

Between November 1975 and July 1978, 352 patients were treated at NIRS and 78 patients at IMS. The largest number of the patients in this series were patients suffering from carcinoma of the uterine cervix. Local control rate and incidence of moist desquamation for the patients surviving over 6 months after completion of the therapy are summarized in Table 3. The higher local control rate for fast neutrons only in comparison to that for other modalities would be explained by the fact that the patients referred to this method were mainly those who were suffering from radioresistant tumors.

On the other hand, higher local control rate obtained in mixed schedule or neutron boost therapy when compared to those treated by fast neutrons only might be a consequence of the improvement of dose distribution within the target volume. Moist desquamation developed in 18% of the patients treated with fast neutrons only. However, ulceration of the skin has not been observed in this series.

Table 3. Local control rate and moist desquamation of the skin following fast neutron therapy (Nov. 1975–Dec. 1977). Evaluated in May 1978 (NIRS)

	Local control		Moist desquamation	
Neutron only	24/76	31.5%	14/76	18%
Mixed schedule	46/101	45.5%	7/101	6%
Neutron boost	24/59	40.7%	5/59	8%
Total cases	94/236	39.8%	26/236	11%

1. Radioresistant tumors

Malignant melanoma. Recently, radiobiological studies have suggested that the radioresistant feature of malignant melanoma cells might be explained by the wide shoulder of the dose survival curve for cultured human malignant melanoma cells, whereas fast neutrons, characterized by the lower repair capability for irradiated cells, would be effective to destroy these radioresistant tumor cells (Dewey, 1971). In this clinical trial, a dose of TDF 100-120 was given to the primary tumor.

As shown in Table 4, the local control rate for the malignant melanoma arising from the head and neck seems to be almost the same as those obtained in the disease of the trunk or extremities, while the regression of the melanoma following fast neutron therapy were more marked in the head and neck than the other sites.

This result might support the suggestion proposed by Veronesi or Mishima of a strong dependence on the site (Veronesi *et al.*, 1971; Mishima, 1967).

Table 4. Results of fast neutron therapy for malignant melanoma at NIRS

	Local control rate			
All cases	Head and neck	Extremity	Trunk	Others
12/23	9/16[a]	2/4	1/2	1/1

[a]Seven patients have received surgical procedures for the residual tumor.

Soft tissue sarcoma. Of twelve patients, seven have been in local control, five of which received fast neutron therapy post-operatively (Table 5).

Complications were seen in four patients. It was our impression that fibrosis developed in the subcutaneous tissue following fast neutron therapy would be slightly more severe than those following X-rays.

Table 5. Results of fast neutron therapy for soft tissue sarcoma at NIRS

No. of cases	Local control	Failure	Complication
12	7	5	4

(June 1978)

Osteosarcoma. Osteogenic sarcoma has been considered to be an incurable disease, because of the radioresistant feature and distant metastases in an early stage. Local infusion of a chemotherapeutic agent, such as Adriamycine, followed by fast neutron therapy was the treatment of choice of this trial. After completion of the radiation therapy, a systemic chemotherapy has been applied at regular intervals.

Table 6 shows local control rates of the patients suffering from osteosarcoma treated with 30 MeV d-Be neutrons, according to the histological examination. There was a difference in radiosensitivity between osteosarcoma and chondrosarcoma. The severe late reactions were seen in the subcutaneous tissue within the treated area, when a dose over TDF 120 had been applied.

Table 6. Local control of osteosarcoma following fast neutron therapy, according to histological examination (NIRS)

	No. of cases	Tumor cell Negative	Tumor cell Positive	Control rate
Osteosarcoma	13	12	1	12/13
Chondrosarcoma	2	1	1	1/2
Total	15	13	2	13/15

(March 1978)

2. Locally advanced tumor

Carcinoma of the uterine cervix. The patients suffering from stage III–B and stage IV–A of carcinoma of the uterine cervix were candidated in this trial. These patients received a whole pelvic irradiation with a dose equivalent to 5000 rad of X-rays in mixed schedule, combined with a high dose rate intracavitary irradiation of 1000–1300 rad at point A. Local failure following fast neutron therapy seems to be lower than that obtained by low-LET radiation therapy, while the complications were observed more frequently in high-LET radiation therapy (Table 7).

In the case of adenocarcinoma of the uterine cervix, the results of fast neutron treatment were disappointing contrary to our expectation. Therefore, reconsideration of the policy for this disease is necessary.

Carcinoma of the esophagus. The patients with disease classified as superficial or serrated type of tumor were included in the study, whereas the patients with the lesion over 10 cm long and with penetrated lesion were excluded.

Of ten patients who received radiation therapy, either mixed schedule or fast neutron boost, 60% were estimated as the local control but with marked complications (Table 8).

In the case of pre-operative irradiation, the tumor was removed surgically after a dose of TDF 50 of fast neutrons.

Table 7. *Results of radiation therapy with fast neutrons or X-rays for squamous cell carcinoma of the uterine cervix, stages III and IV A (NIRS)*

Series	No. of cases	Local failure	Remote metastasis	No. of death or AWD[a]	Complication Rectum	Bladder
Fast neutrons	20	3/20 (15%)	2/20 (10%)	4/20 (20%)	5/20 (25%)	2/20 (10%)
X-rays	32	6/32 (19%)	5/32 (16%)	11/32 (34%)	5/32 (16%)	1/32 (3%)

[a] Alive with disease (June 1978)

Carcinoma of the lung. The effect of fast neutron beam on Pancoast' tumor was appreciable with respect to local control of the tumor and also relief of the pain. The marked effects of fast neutrons were observed in the treatment of adenocarcinoma of the lung. However, the complications, characterized by severe fibrosis of the lung, are considered to be more marked in fast neutron therapy than those produced by X-rays. Therefore, boost therapy applying fast neutrons was recommended as a modality for carcinoma of the lung.

Carcinoma of the urinary bladder and prostate. T_3 carcinoma of the genitourinary organs was treated with mixed schedule or fast neutrons only.

Of five patients suffering from carcinoma of the prostate, three were diagnosed as local control without severe late complication.

Discussion

Clinical trials suggest that the local control rate was improved by applying high LET-

Table 8. *Results of fast neutron therapy for carcinoma of the esophagus at NIRS*

	No. of cases	Radiation therapy alone (ten cases)		Complication
		Local control	Local failure	
Mixed beam	7	4	3	—
Neutron boost	3	2	1	—
	10	6 (60%)	4 (40%)	—

(March 1978)

Malignant glioma. Mixed schedule or boost therapy of fast neutrons were used for the patients suffering from glioblastoma multiforme. This is because of the fact that the patients who received a whole-brain irradiation of fast neutrons did not improve at all and that severe gliosis of the brain tissue developed. A dose equivalent to 5000–6000 rad X-rays was delivered to the tumor tissue in shrinking-field technique. Of seven patients, six survived after completion of the therapy. The longest survivor has been a 16-year-old female patient. She is enjoying her life actively after 2 years following radiation therapy. Fast neutron therapy seems to be a promising modality for the radioresistant malignant glioma, if the indication and the radiotherapy technique is appropriate.

Head and neck tumor. Fast neutron therapy was tried in the cases of carcinoma of the tongue, T_3, and the fixed cervical lymph nodes. However, an interstitial radiotherapy was necessary to manage the residual tumor, even though the large T_3 carcinoma of the tongue had responded well to high-LET radiation.

radiation but that a dose over TDF 120 would bring forth severe late radiation damages.

The tumors which are considered to be difficult to eradicate by radiation therapy, even with fast neutrons, have to be removed surgically after radiation treatment when it is accessible. Therefore, the appropriate selection of the indications as well as close collaboration with other oncologists are necessary to promote fast neutron therapy. On the other hand, it was suggested that improvement of the dose distributions would be indispensable to make the best use of high-LET radiation therapy.

References

Dewey, D. L. (1971) The radiosensitivity of melanoma cells in culture. *Brit. J. Radiol.* **44**, 816.

Mishima, Y. (1967) Melanocytic and neurocytic malignant melanomas, cellular and subcellular differentiation. *Cancer* **20**, 632-649.

Veronesi, U., N. Cascinelli and F. Preda, (1971) Prognosis of malignant melanoma according to regional metastasis. *Amer. J. Roentgenol.* **111**, 301-309.

Five years of clinical experience with a combination of neutrons and photons

H. J. EICHHORN, A. LESSEL AND K. DALLÜGE

Bereich Experimentelle und klinische Strahlentherapie Zentralinstutut für Krebsforschung der Akademie der Wissenchaften der DDR, Berlin-Buch, DDR.

Abstract*—In the Central Institute for Cancer Research in Berlin-Buch (GDR) patients have been treated with fast neutrons since April 1972. The neutrons produced by a Soviet cyclotron have a mean energy of 6.2 MeV with a dose rate of 20 rad/min at a distance of 1 m.

The cyclotron is located in Dresden at a distance of 250 km from the Institute. The neutron beam is only available for clinical use 6 times a year, each time for a 2-week period. For this reason only part of the total tumor dose in each patient was contributed by neutron irradiation, the rest by cobalt-60 treatments.

For the total beam the RBE factor for the daily dose level used can be estimated to be in the range from 2.6 to 3.4.

Up to now 450 patients have been treated this way, comprising 250 cases of inoperable bronchial carcinoma, 44 cases of inoperable esophagus carcinoma and 36 cases of inoperable soft-tissue sarcoma.

The effect of irradiation was established by microscopic examination of autopsy specimen of the complete tumor. If after careful examination of multiple samples no tumor cells could be detected, the result was scored as "tumor negative". Even the finding of a very small number of viable tumor cells made the result "positive". The results obtained in two groups of patients were compared. One group was treated exclusively with telecobalt therapy, the other with a combination of neutrons and telecobalt therapy. The most important clinical model was the inoperable bronchial carcinoma without clinical evidence of distant metastases.

At first a prospective experimental series was started, using a historical control group. The autopsy findings in 149 patients who only received cobalt-60 therapy were compared with 116 patients who had been given part of their total tumor dose (18% and 36%, respectively) in the form of neutron irradiation. The two groups were fairly well comparable with regard to tumor size, degree of differentiation of the tumors, tumor dose and overall treatment time. The results of this first experiment are shown in Table 1.

The difference in favour of the addition of neutrons compared to telecobalt only is statistically significant ($P < 0.01$ for 36% N; $p < 0.05$ for 18% N). A few years ago a controlled randomized trial started, in which again two groups of patients with bronchial carcinoma were compared. The autopsy specimen of thirty-

Table 1. Bronchial carcinoma

	$N_{18\%, 36\%}$ + Co-60 gamma	Co-60 gamma only
Tumor negative	61/116 = 53%	47/149 = 33%
Mean tumor dose	5352 rad ($\times$ RBE)	7044 rad
Mean tumor volume	231 cm^3	212 cm^3
Mean overall treatment time	57 d	52 d

*This abstract was presented at the third meeting on fundamental and practical aspects of fast neutrons and other high-LET particles in clinical radiotherapy in The Hague, The Netherlands, on 15 September 1978 by K. Breur.

Table 2. Bronchial carcinoma (randomized trial)

	$N_{41\%}$ + Co-60 gamma	Co-60 gamma only
Tumor negative	18/46 = 39%	9/39 = 23%
Mean tumor dose	5795 rad ($\times$ RBE)	5544 rad
Mean overall irradiation time	45 d	46 d

nine patients with pure telecobalt gamma therapy and of forty-six patients with combined neutron telecobalt therapy (41% neutrons) are now available for evaluation. The results are shown in Table 2.

Histologically total tumor destruction was achieved in 23% of the cases in the first group compared with 39% of the cases which received also neutron irradiation ($p = 0.10$).

showed to be considerably more effective if the treatment included a proportion of fast neutrons (Tables 3 and 4).

Investigations on the Vascular Connective Tissue

Because of the importance for late normal tissue complications, the late reactions of the

Table 3. Gastric cancer

	$N_{19\%, 39\%}$ + Co-60 gamma	Co-60 gamma only
Tumor negative	9/27 = 33%	0/10
Mean tumor dose	5182 rad ($\times$ RBE)	5200 rad
Mean overall treatment time	55 d	48 d

Similar non-randomized experimental studies, using historical control groups, have been carried out for inoperable carcinomas of the stomach (autopsy findings in twenty-seven patients, neutron proportion of 39%) and in inoperable esophagus carcinomas (autopsy findings in twenty-three patients, 33% neutrons). Also for these two tumor types the irradiation

vascular connective tissues in the subcutis was studied in twenty-six patients who survived the tumor disease for more than $1\frac{1}{2}$ years after radiation therapy. In the course of treatment in these patients, forty-six large fields in the epigastric region, the thorax or the dorsum were irradiated with varying neutron proportions and total dosages. The RBE factor 3.0 established on

Table 4. Oesophageal cancer

	$N_{20\%, 33\%}$ + Co-60 gamma	Co-60 gamma only
Tumor negative	17/23 = 74%	76/136 = 55%
Mean tumor dose	5550 rad ($\times$ RBE)	8200 rad
Mean overall treatment time	65 d	68 d

Table 5.

100% neutron irradiation	3600 rad $\times$ RBE (1200 rad/12 days)	strong fibrosis (6 out of 10 fields)
50% neutron irradiation— proportion (+ Co-60 gamma)	>3800 rad $\times$ RBE	moderate fibrosis (3 out of 10 fields)
	>4900 rad $\times$ RBE	strong fibrosis (7 out of 11 fields)
25–40% neutron irradiation— proportion (+ Co-60 gamma)	>5900 rad $\times$ RBE	strong fibrosis (2 out of 4 fields)
	<5800 rad $\times$ RBE	no fibrosis (11 out of 11 fields)

human skin erythema was used for this study. The results are presented in Table 5.

These investigations seem to suggest that the RBE factor for the delayed reactions of the vascular connective tissue is higher than 3.0.

Reference

EICHHORN, H. J. and A. LESSEL (1977) Four years' experiences with combined neutron-telecobalt therapy, investigations on tumor reaction of lung cancer. *Int. J. Radiat. Oncol. Biol. Phys.* **3**, 277-280

Hyperbaric oxygen and hypoxic cell sensitizers in clinical radiotherapy: present state and prospects

S. DISCHE

Marie Curie Research Wing for Oncology, Regional Radiotherapy Centre, Mount Vernon Hospital, Northwood, Middlesex, England

Abstract—*Experience with hyperbaric oxygen in radiotherapy now extends over 25 years. In randomized controlled clinical trials significant benefit has been shown in the treatment of head and neck tumours and some trials in carcinoma of cervix. Improvement has been shown in one trial in carcinoma of bronchus but not in any of those in carcinoma of bladder. The results of hyperbaric oxygen must now be compared with other methods currently under trial for improving the results of radiotherapy.*

Clinical work with the hypoxic cell sensitizers began in 1973 with metronidazole and in 1974 with the more promising compound, misonidazole. It has been shown that there is a satisfactory penetration of the drug into tumours. Sensitization of hypoxic cells in man has been demonstrated as well as increased effect in tumours when single doses are employed. Neurotoxicity restricts the total dose which may be given. Randomized controlled clinical trials using misonidazole are now underway.

Introduction

Hyperbaric oxygen and the hypoxic cell sensitizers were introduced, like high-LET radiations, to improve the results of radiotherapy. They have a target in common—the radio-resistant hypoxic tumour cell.

When we consider hyperbaric oxygen we must go back to 1953 to Scott's dramatic demonstration of benefit in an animal tumour system (Gray *et al.*, 1953) and to Churchill-Davidson's original work published in 1955 (Churchill-Davidson, Sanger and Thomlinson, 1955). In eight patients half the tumour was treated in oxygen and half in air. In all seven assessable cases there was histologically a greater immediate response in oxygen than in air. This work gave dramatic stimulus to the further use of the hyperbaric chamber, but single treatments and immediate responses are a long way from fractionated radiotherapy with the objective of cure.

What have we to consider when we make an objective assessment of the current position of hyperbaric oxygen? We know that however dramatic and valuable individual case reports and consecutive series of cases may be, in the cold task of assessment only the results of randomized controlled clinical trials can be accepted.

What has been achieved in terms of improved tumour control using hyperbaric oxygen to compare with the results being reported at this meeting? If we look to the radiotherapy of head and neck cancer we find that there have been just six prospective randomized controlled trials (Berry, 1978; Van den Brenk, 1968; Chang *et al.*, 1973; Henk, Kunkler and Smith, 1977; Henk and Smith, 1977; Shigematsu *et al.*, 1973).

In the first three the margin has been in favour of hyperbaric oxygen, but they have been concerned with a limited number of cases and in two short-term follow up. Next are those performed at Cardiff; work initiated by Professor Kunkler and continued by Dr. Henk. The final one at Leeds was similar to the second Cardiff trial and can be added to it. (Table 1).

A highly significant improvement in local results has been achieved in both with improvement in survival in the second when the Leeds cases are included (MRC Report, 1978). We can conclude from this evidence that hyperbaric oxygen improves the result of

Table 1. Medical Research Council: hyperbaric oxygen trials

Head and Neck 1

		Fractions	No. of cases	% local recurrence-free 2 yr	4 yr	p
Cardiff I	Oxygen	10	125	50	37	
	Air	10	151	50	37	
Cardiff II Leeds	Oxygen	10	60	74	62	0.001
	Air	conventional	67	50	24	

Head and Neck 2

		Fractions	No. of cases	% survival 2 yr	4 yr	p
Cardiff I	Oxygen	10	125	57	55	0.0005
	Air	10	151	34	31	
Cardiff II Leeds	Oxygen	10	60	72	69	0.01
	Air	conventional	67	44	41	

treatment of head and neck cancer. There has, however, been much criticism of these results by those who feel that the air-control patients fared rather poorly compared with those managed with other régimes of conventional radiotherapy. Even if it is conceded that only an improvement in result of a less adequate scheme of radiotherapy has been demonstrated, it is still a highly significant biological result and there is the probability that even with the best of conventional radiotherapy a further margin of improvement will be obtained using hyperbaric

Table 2. Medical Research Council: hyperbaric oxygen trials—carcinoma of cervix

Actuarial local recurrence-free rates according to stage (all patients)

Stage	Treatment series	Total patients	Percentages surviving by years since entry to trial					Chance prob. of difference between curves
			1	2	3	4	5	
IIB	Oxygen	12	91	81	61	61	61	
	Air	11	73	73	64	64	64	0.83
III	Oxygen	119	82	77	76	76	71	
	Air	124	67	50	47	47	44	<0.001
IV	Oxygen	30	71	71	71	71	48	
	Air	24	77	53	53	53	53	0.81
All Patients	Oxygen	161	80	76	73	73	67	
	Air	159	68	52	49	49	47	<0.001

Actuarial survival rates according to stage

IIB	Oxygen	12	92	67	58	42	42	
	Air	11	91	73	73	73	73	0.034
III	Oxygen	119	77	57	49	45	37	
	Air	124	70	49	34	30	25	0.012
IV	Oxygen	30	47	30	20	16	16	
	Air	24	38	17	17	17	17	0.76
All Patients	Oxygen	161	73	53	44	40	33	
	Air	159	67	46	34	31	27	0.08

oxygen combined with it. When a conventional technique has been taken to the limits of its possibilities it is no doubt more difficult to show further improvement and an even larger number of cases would be required in any clinical trial.

With bladder cancer no significant improvement in results has been shown in any of the Medical Research Council studies and, as far as I am aware, convincingly, in any other (Cade et al., 1978). On the other hand, in carcinoma of bronchus Miss Cade and Dr. McEwen are showing a good margin in favour of oxygen in their studies at Portsmouth now with a six-fraction technique. (Cade and McEwen, 1978).

Hyperbaric oxygen
Carcinoma of bronchus (squamous ca)
Portsmouth
123 patients   6 treatments over 3 weeks
               3600 rad (max)

		2 yr	4 yr
Survival	Oxygen	25%	16%
	Air	12%	3%

In carcinoma of cervix we now have the most interesting results from the four United Kingdom centres co-operating in the Medical Research Council trials. An overall increase in local tumour control between 20-24% at 2-5 years has been demonstrated—this being highly significant (Watson et al., 1978). The greatest benefit is seen in the stage III cases where there is also a significant improvement in survival (Table 2).

Recently I have heard that in a multi-centre trial carried out by the RTOG in the United States in stages IIb and III carcinoma of cervix the margin of benefit was with oxygen. In other trials, particularly those at Houston performed by Dr. Fletcher and his colleagues and at Capetown by Professor Sealy and his colleagues, each study including over 200 cases, no benefit has been shown.

When we look at the results of these trials we need to find some explanation for the differences encountered. Why, for example, should hyperbaric oxygen contribute in head and neck tumours and not in bladder cancer? We know that hypoxic cells exist in nearly all the solid tumours which have been studied in animals, this despite very varying other biological characteristics of these tumours. In man we have little data to guide us, but it seems unlikely that human tumours are greatly different. Hypoxia is likely to be a cause of radio-resistance in all human cancer including bladder cancer. Perhaps it is such a problem in the bladder that hyperbaric oxygen is not sufficiently able to influence it and improve results. We must, of course, recognize that causes exist for radiation failure other than hypoxia and if other factors dominate then this might account for the findings in carcinoma of bladder (Dische, 1978a).

The variable results in carcinoma of cervix present an even more difficult problem. A further surprising finding in the United Kingdom trials was that the benefit was dramatic in the younger patients below the age of 55, but seemed to disappear in the older ones. A possible explanation is the existence of a number of different types of tumour at this one site. It is conceivable that biological differences may exist between such tumour types so that hyperbaric oxygen may be beneficial in some patients but not in others. A differing distribution of such types from one series to another may account for the variable results (Dische, 1978a).

We have so far been concerned with responses observed in tumours but we are really concerned with an improvement in the therapeutic ratio and so, therefore, must equally concern ourselves with normal tissue effects. We find that with the use of hyperbaric oxygen there is evidence for an increased effect in normal tissues. This is most marked in those known to contain a high percentage of hypoxic cells and in man the cartilage of the larynx is the only site where this is certainly known to apply. The problem was well documented by Dr. Churchill-Davidson in his pioneer work (Churchill-Davidson et al., 1966). In the first Cardiff trial Dr. Henk reported increased cartilage necrosis and for the second trial made a reduction of 10% in dose when the larynx was included in his field of treatment (Henk et al., 1977; Henk and Smith, 1977). He did not, however, find any significant increase in normal tissue damage in other structures in the head and neck region.

In the Medical Research Council trials in carcinoma of cervix using hyperbaric oxygen there was a significant increase in late tissue effects particularly in bowel. The increase in effect which we see under these conditions in tissues which are presumably normally well oxygenated is due to the elevation of oxygen tension to unusually high values. Although the curve associating oxygen tension and radiation effect has largely flattened out over this range it is still, nevertheless, rising and one can expect an increase in radiation effect similar to that which might result from an elevation in dose given in air, by about 3%.

Where there is an increase in normal tissue effects in a clinical trial, as here with hyperbaric oxygen, one has to carefully assess the increase in tumour control against the increase in morbidity. The bulk of the evidence favours the view that with hyperbaric oxygen in head and neck cases and in the cervix benefit outweighs the morbidity and a true improvement in the therapeutic ratio has been achieved (Dische, 1978a).

The benefit seen in the treatment of tumours in the head and neck region and in carcinoma of cervix is a considerable one. Local control improved by over 20% and survival by over 15%. It is strange, therefore, that so few patients are currently being treated with such a beneficial technique. The reason is often given that it is a cumbersome and difficult method. Although there are problems in its application, with sufficient purpose and planning such treatment can readily be given in any radiotherapy centre. Had the trials in head and neck tumours and in carcinoma of cervix been started soon after the pioneer work of Churchill-Davidson had been published, and conducted in more centres so that larger numbers would have been gathered sooner, the results as reported now would have created wide interest and probably a vast sale of hyperbaric chambers. At this time the method is a quarter of a century old and many new and exciting developments have come to interest us, untarnished by any long period of experience.

It is right now that we should explore these new methods and in careful study determine if they can give us the sort of benefit that hyperbaric oxygen has been shown to give. If, with greater ease and acceptance, similar or greater benefits can be achieved then certainly hyperbaric oxygen can be discarded as a method to be used on its own to improve results of treatment though it must be given a very important place in the history of radiotherapy.

The history of development of chemical agents to specifically sensitize hypoxic tumour cells in radiotherapy can be traced back to 1963 when Professor Adams first suggested that chemical agents might replace oxygen in its vital function of determining radiation injury (Adams and Dewey, 1963). Progress was slow in the initial 9 years of study, but finally it was found possible to sensitize hypoxic mouse skin. However, it was not until 1973, when the nitroimidazoles were found to be good sensitizers, that progress towards the clinic really started to be made (Adams et al., 1976).

When, in 1973, metronidazole was first given to patients and then in 1974, Ro 07-0582, now called misonidazole, improved tumour control had been demonstrated in many animal tumours. No other method introduced so far into clinical radiotherapy has been better based upon laboratory study (Adams et al., 1976).

In our early clinical work in 1974 and 1975 using misonidazole we were able to show that the drug is well absorbed and that plasma concentrations reach a peak in 1-2 hours. There is a good plateau period extending to 4 or 5 hours and then the drug concentration falls off with a half-life of about 12 hours. Plasma concentrations rise proportionately with the dose given. Now, in our studies of over 160 patients with some 4000 estimates of plasma concentration, we can show that a higher concentration is usually achieved in women than in men, but that this is compensated for as regards tissue exposure and, therefore possible toxic effects, by a shorter half-life.

Levels in normal tissues and in tumour rise with the plasma concentration and are usually between 60 and 100% of the plasma concentration at the time of sampling. A man presented a secondary node due to a squamous cell carcinoma in the neck. It was 90 mm in diameter and was fluctuant due to central necrosis. Histological study showed much degeneration and evidence for a considerable concentration of hypoxic cells. An identical sample was sent for misonidazole estimation for a large dose of 7.5 g had been given 4 hours prior to biopsy. The concentration was 142 μg/g or 73% of the plasma concentration at that time. The drug certainly seems to reach the hypoxic cells.

We were able to show in our first series of patients given the drug that misonidazole could sensitize hypoxic cells in man as in animals. We irradiated skin under oxic and under hypoxic conditions in order to see whether the drug could restore radiosensitivity. We used a complex system to make the skin hypoxic (Dische and Zanelli, 1976). An Esmarch's bandage was wound around the limb; a sphygmomanometer cuff applied at high pressure and then after uncoiling the Esmarch's bandage, the limb was encircled in a bag of nitrogen. Under hypoxia with this system usually double the dose of radiation is required in order to achieve the same degree of pigmentation after treatment as that achieved when radiation is given under normal oxic conditions. Using hypoxia and the radio-sensitizing drug, however, much enhanced reaction was achieved, in fact in this experiment when 800 rad were given under oxic conditions, we reduced our dose in hypoxia to 1200 rad and even then the response was greater than when 1600 rad were given under these hypoxic conditions, without sensitizer. In this situation, therefore, the radio-sensitivity of artificially hypoxic skin in man has been taken more than half-way back to that of oxic skin by the addition of the radio-sensitizing drug (Dische et al., 1976).

We are, of course, concerned not about normal skin made hypoxic but hypoxic tumour cells. In our original series of eight patients with multiple deposits of tumour we gave a single large dose of misonidazole combined with radiotherapy and compared the result with radiotherapy alone. In three of four assessable cases we were able to demonstrate some increase in radiation response with the use of the sensitizer (Thomlinson et al., 1976).

Unfortunately, misonidazole may give certain toxic effects. The important one which really limits the amount of drug which can be given is neurotoxicity, principally peripheral neuropathy (Dische et al., 1977). However, if a maximum dose of 12 g per square metre of surface area is set when the drug is given over a period of 17 days or greater there is a low and acceptable incidence of peripheral neuropathy (Saunders et al., 1978). We have found that neurotoxicity can be reduced further by monitoring the blood concentrations achieved. In our most recent group of sixty patients the incidence of neurotoxicity was reduced to 11% and nearly all have been of mild severity, of short lasting and really of little clinical significance (Dische et al., 1978a).

It might be that if the course of radiotherapy is a long one and particularly if the dose given in single amounts weekly, that some elevation of the total dose of misonidazole may be allowed. We await further evidence on this subject. A continuing collaboration between centres and a pooling of data should enable us to advance knowledge as to safe tolerance at a rapid rate (Dische et al., 1978b).

How should we administer this dose? With a single treatment as for palliation, the sensitizer would certainly improve the response. When we move to the multi-fraction techniques used in radiotherapy for cure there are many possible patterns of administration. First it can be given in a few large doses, each combined with a high radiation dose. Secondly, it can be given once or twice a week when daily radiotherapy is employed. In this situation when the misonidazole is given the radiation dose is commonly higher than on the other occasions in the week when the radiation is given alone. The third pattern shows a normal daily fractionated course of radiotherapy with misonidazole being given in small doses with each. There is good argument for exploration of all these possibilities.

It is too early to give an estimate as to the likely impact of misonidazole upon clinical radiotherapy. We have our favourable impressions based upon the 160 patients to whom we have given the drug but only randomized controlled clinical trials can give us the answer. With metronidazole we have the important study of Dr. Urtasun in the treatment of glioblastoma (Urtasun et al., 1976). Here a very significant increase in survival is obtained in the patients given metronidazole. The same criticism has been made of this trial as of some of the hyperbaric oxygen trials and perhaps, even more appropriately, for the régime of radiotherapy is not one which is considered the best in most centres of radiotherapy for the palliation of glioblastoma. It is, nevertheless, an important biological result. At Capetown Professor Sealy, in a trial of the drug combined with six fractions of radiotherapy over 3 weeks in advanced oral tumours, is showing a margin of benefit to the sensitizer in the early stages of his study. At Mount Vernon in carcinoma of cervix we showed that the regression observable at the end of a course of radiotherapy directly correlated with the long-term local control. We were able to predict the benefit which was later to be seen in follow up because we noted a greater number of good regressions in the patients treated in hyperbaric oxygen (Dische, 1974). Apparent complete regression has been noted in all eight cases which have now been given misonidazole with radiotherapy for carcinoma of the cervix and so here again we have promise of benefit.

Many clinical trials with misonidazole usually involving the co-operation of multiple centres are now being planned and some are underway. We look forward to seeing the results which will be shown. We also watch with great interest the development of new sensitizers which should bring greater benefit.

I would be most unhappy if hyperbaric oxygen, chemical sensitizers and high-LET radiation were simply regarded as competitors for patients and for financial support. There are important similarities and important differences between these techniques which actually bring them together.

They have as a common target the hypoxic radio-resistant tumour cell. Although such cells are considered to be an important cause for radiation failure we do not know the true extent of the problem. Any success with hyperbaric oxygen or with chemical sensitizers must indicate that hypoxic cells really do exist in tumours and still remain a problem in fractionated radiotherapy. One can, therefore, expect other methods which are pursuing the same target, including high-LET radiation, also likely to show benefit.

We must recognize that none of the three techniques are now, or are likely in the near future, to bring the sensitivity of hypoxic cells to that of oxic ones at every treatment. Benefit may be enough that in a fractionated course of radiotherapy using the technique many more patients will be cured; however with our present limitations with all three approaches, 100% tumour control is not likely to be achieved in every case treated. We have also the problem that when we push these régimes to their limits we run into problems of morbidity. We can see, however, that these régimes are quite different in application and it is logical, therefore, to consider combining them. There is much in the history of medicine as a whole and in oncology in particular to encourage us to do so. Certainly our biological colleagues have shown that, in the laboratory, benefit can be achieved with neutrons and with misonidazole but that a combination of the two does even better (Denekamp et al., 1976).

Human malignant disease presents a vast field with many different tumours each with its own pattern of biological behaviour. It is not unlikely that as we gain further knowledge we will find that different tumours require different conditions of radiotherapy to obtain the best chance of local cure. Combinations of high-LET radiation and chemical sensitizer and even the incorporation of hyperbaric oxygen may in certain instances be the best method of management. We must not shrink from the problems of complexity but intensify our observations upon tumours and of their responses to treatment in order to learn the best way to cure all our patients.

References

ADAMS, G. E. and D. L. DEWEY (1963) Hydrated electrons and radiobiological sensitization. *Biochem. Biophys. Res. Comm.* **12**, 473-477.

ADAMS, G. E., J. F. FOWLER, S. DISCHE and R. H. THOMLINSON (1976) Increased radiation response by chemical sensitization. *Lancet*, **1**, 186-188.

BERRY, G. H. (1978) A clinical appraisal of hyperbaric oxygen in head and neck cancer. *Brit. J. Radiol.* **51**, 150.

VAN DEN BRENK, H. A. S. (1968) Hyperbaric oxygen in radiation therapy. An investigation of dose-effect relationships in tumour response and tissue damage. *Amer. J. Roentgenol.* **102**, 8-26.

CADE, I. S. and J. B. MCEWEN (1978) Clinical trials of radiotherapy in hyperbaric oxygen at Portsmouth (1964-1976). *Clin. Radiol.* **29**, 333-338.

CADE, I. S., J. B. MCEWEN, S. DISCHE, M. I. SAUNDERS, E. R. WATSON, K. E. HALNAN, G. WIERNIK, D. J. D. PERRINS and I. SUTHERLAND (1978) Hyperbaric oxygen and radiotherapy. A Medical Research Council trial in carcinoma of the bladder. *Brit. J. Radiol.* (awaiting publication).

CHANG, C. H., J. J. CONLEY and C. HERBERT (1973) Radiotherapy of advanced carcinoma of the oropharyngeal region under hyperbaric oxygenation. *Amer. J. Roentgenol.* **117**, 509-516.

CHURCHILL-DAVIDSON, I., C. SANGER and R. H. THOMLINSON (1955) High-pressure oxygen and radiotherapy. *Lancet*, **1**, 1091-1095.

CHURCHILL-DAVIDSON, I., C. A. FOSTER, G. WIERNIK, C. D. COLLINS, N. C. D. PIZEY, D. B. L. SKEGGS and P. R. PURSER (1966). The place of oxygen in radiotherapy. *Brit. J. Radiol.* **39**, 321-331.

DENEKAMP, J., S. R. HARRIS, C. MORRIS and S. B. FIELD (1976) The response of a transplantable tumour to fractionated irradiation. II Fast neutrons. *Radiat. Res.* **68**, 93-103.

DISCHE, S. (1974) The hyperbaric oxygen chamber in the radiotherapy of carcinoma of the uterine cervix. *Brit. J. Radiol.* **47**, 99-107.

DISCHE, S. and G. D. ZANELLI (1976) Skin reaction—A quantitative system for measurement of radiosensitisation in man. *Clin. Radiol.* **27**, 145-149.

DISCHE, S., A. J. GRAY and G. D. ZANELLI (1976) Clinical testing of the radiosensitiser Ro 07-0582 II. Radiosensitisation of normal and hypoxic skin. *Clin. Radiol.* **27**, 159, 166.

DISCHE, S., M. I. SAUNDERS, M. E. LEE, G. E. ADAMS and I. R. FLOCKHART (1977) Clinical testing of the radiosensitiser Ro 07-0582: Experience with multiple doses. *Brit. J. Cancer* **35**, 567-579.

DISCHE, S., M. I. SAUNDERS, P. ANDERSON, R. C. URTASUN, K. H. KÄRCHER, H. D. KOGELNIK, N. BLEEHEN, T. L. PHILLIPS and T. H. WASSERMAN (1978b) The neurotoxicity of misonidazole. The pooling of data from five centres. Letter to *Brit. J. Radiol.* (awaiting publication).

DISCHE, S. (1978a) Hyperbaric oxygen. The Medical Research Council trials and their clinical significance. *Brit. J. Radiol.* (awaiting publication).

DISCHE, S., M. I. SAUNDERS, I. R. FLOCKHART, M. E. LEE and P. ANDERSON (1978) Misonidazole. A drug for use in radiotherapy and oncology. (Awaiting publication).

GRAY, L. H., A. O. CONGER, M. EBERT, S. HORNSEY and O. C. A. SCOTT (1953) The concentration of oxygen dissolved in tissues at the time of irradiation as a factor in radiotherapy. *Brit. J. Radiol.* **26**, 638-648.

HENK, J. M., P. B. KUNKLER and C. W. SMITH (1977) Radiotherapy and hyperbaric oxygen in head and neck cancer. *Lancet* **2**, 101-103.

HENK, J. M. and C. W. SMITH (1977) Radiotherapy and hyperbaric oxygen in head and neck cancer. Interim report of second clinical trial. *Lancet* **2**, 104-105.

MEDICAL RESEARCH COUNCIL REPORT (1978) Working party on radiotherapy and hyperbaric oxygen. *Lancet* (awaiting publication).

SAUNDERS, M. I., S. DISCHE, P. ANDERSON and I. R. FLOCKHART (1978) The neurotoxicity of misonidazole and its relationship to dose, half-life and concentration in the serum. *Brit. J. Cancer*, **37**, Suppl. 3, 268.

SHIGEMATSU, Y., H. FUCHIHATA, T. MAKINO and T. INOUE (1973) Radiotherapy with reduced fraction in head and neck cancer, with special reference to hyperbaric oxygen radiotherapy in maxillary sinus carcinoma. (A controlled study). In T. SUGAHARA, L. REVESZ and O. C. A. SCOTT (Eds.), *Fraction Size in Radiobiology and Radiotherapy*, 1st ed. Igaku Shoin Ltd., Tokyo, pp. 180-187.

THOMLINSON, R. H., S. DISCHE, A. J. GRAY and LESLEY M. ERRINGTON (1976) Clinical testing of the radiosensitiser Ro 07-0582. III. Response to tumours. *Clin. Radiol.* 27, 167-174.

URTASUN, R. C., P. R. BAND, J. D. CHAPMAN, A. F. WILSON, B. MARYNOWSKI and E. STARRE (1976) Metronidazole as a radiosensitizer. *New Engl. J. Med.* 295, No. 16, 901.

WATSON, E. R., K. E. HALNAN, S. DISCHE, M. I. SAUNDERS, I. S. CADE, J. B. MCEWEN, G. WIERNIK, D. J. D. PERRINS and I. SUTHERLAND (1978) Hyperbaric oxygen and radiotherapy. A Medical Research Council trial in carcinoma of cervix. *Brit. J. Radiol.* (awaiting publication).

Treatment at low dose rate, by low-LET radiation: present status and prospects

B. PIERQUIN, E. CALITCHI AND R. OWEN

Centre Hospitalo-Universitaire Henri Mondor, 94000 Créteil, France

Introduction

Since January 1970 we have treated 120 patients by telecobalt gamma-radiation at low dose rate; the localization of the tumours was as follows:

Head and neck		83
Soft tissue sarcomas		14
Pelvis—uterus	3	
—rectum	3	10
—bladder	4	
Oesophagus		5
Pancreas		4
Brain (astrocytomas)		2
Bronchus and lung		2
Total		120

This table shows a predominance of head and neck cancers. In fact cancers of the buccal cavity and oropharynx are considered by us to be the most interesting to study, both because they are readily accessible to examination and clinical follow-up and because their prognosis, in relation to their extent, is well known (Calitchi, 1978).

Results of Low-dose-rate Treatments

To begin with, we will say a few words on the results of low-dose-rate treatment at other sites.

Soft-tissue sarcomas

Ten patients have been treated after surgical excision of the tumour and four by telecobalt gamma-radiation only. With the exception of one case no recurrence has been seen following the irradiation (65 Gy), given by a single or by a split course. In those cases, where the tumour was *in situ*, its complete disappearance to clinical examination was obtained after a delay by 2 to 3 months.

The phenomena of muscular and sometimes skin fibrosis became manifest after a delay of 1 year, restricting function particularly in the case of the lower limbs. This fibrosis of muscle, although not equivalent to that following classical fractionated radiotherapy, appears to be considerable.

Pelvis

Full dose (60 to 65 Gy) treatment has been given in only one case of Ca. uterus, one case of Ca. rectum and two cases of Ca. bladder. The results of the uterine and rectal cases have been ambiguous, but satisfactory results were obtained in the two bladder cases. The skin reactions, both immediate and late, have not been observed, provided that the patient was not irradiated through the treatment table. Bowel reactions have been minimal or absent wherever the target volume was less than 11 cm in length or width.

Oesophagus

The five cases of cancer of the oesophagus which were treated by single or split course, have shown excellent short-term results, with the disappearance of the tumour and good restoration of normal function. The late results (at 1 year) have been disappointing, four out of the five patients have developed broncho-mediastinal fistulae. We no longer treat Ca. oesophagus at low dose rate.

Pancreas

The four cases of Ca. pancreas tolerated the irradiation remarkably well for treatment by split-course, with no diarrhoea or vomiting. The target volume was relatively restricted; 10 to 13 cm in length, 6 to 8 cm in width and 5 to 6 cm in thickness. There seemed to be a definite but temporary effect on the tumour.

Astrocytomas

Two astrocytomas have been treated, each by a single course of 50 Gy. The treatment appears to have been well tolerated over the following year. We have not been able to draw any conclusion about long-term effects.

Lung

One pulmonary metastasis was irradiated to 30 Gy; one Ca. bronchus was treated to 60 Gy. These two isolated cases do not permit any conclusions.

Head and neck

It is the eighty-three cases of Ca. oropharynx and buccal cavity (and a few cases of Ca. hypopharynx and larynx) which form the bulk of our experience. It should be noted at the outset that tumours of the larynx and hypopharynx should be excluded because in their infiltrating forms (T3) they are not good indications for low-dose-rate irradiation; cure is very likely achieved but at the price of oedema requiring secondary surgery. In this respect the sequellae of treatment at low dose rate are no different from those of fractionated irradiation. During the first period (until 1972) we treated the cancers of the buccal cavity and oropharynx by a single course of 65 Gy at the rate of 8 Gy per day, with an output of 0.9 to 1.1 Gy per hour. The overall time for the course (including a week-end's rest) was 8 to 10 days. Eighteen patients with advanced tumours (T3 or T4) have been thus treated. The immediate results have been notable for a minimal skin reaction, a maximal mucosal reaction, a spectacular tumour reduction both locally and in lymph nodes and the majority have been apparently sterilized within 3 months following the irradiation. However, there is a high incidence of necrosis (more than 50%) during the following year. Since 1972 we have irradiated our patients by a split course technique; two series of 32.5 Gy separated by an interval of 3 weeks (total dose 65 Gy). This split technique was adopted for three reasons:
  to decrease the mucosal reaction,
  to decrease necrosis,
    to allow re-oxygenation of the tumour remnant during the 3 week interval.

The results at more than a year after the treatment would seem to confirm the three premises:
  the mucosal reactions are markedly diminished, being slight after the first series, moderate after the second and of short duration;
  the skin reactions are virtually non-existent;
  the necroses are less frequent (see Table 1) and are only seen at the site of the primary tumour. They appear to conform in these advanced tumours to the area of ulceration, already biologically necrosed and non-restorable.

As regards the effects on the tumour itself, the lack of randomization between the two groups of patients and the progressive refinement of the criteria for patient selection (less T4 and more T3 tumours) preclude any conclusion. The results would seem to be at least equivalent.

Table 1

Low dose-rate—45 E.N.T. patients
(March 1978)

Analysis of necroses (N), follow-up >1 year. Period 1970 to 1977. 25TO: (a) before 1972, single course; (b) after 1972, split course.

Dose	N/TO	Delay	Site	
			Tumour volume	Normal tissue
65a	7/9 77%	7.6	7	2 (bone)
65b	8/16 50%	8.6	8	0

Current programmes of treatment at low dose rate

At present we are continuing our research along three lines.

1. *A randomized trial between fractionated and low-dose-rate irradiation*, under the control of the EORTC. This trial is limited to one particular site: T3 tumours of the base of the tongue in those patients conforming to strictly defined criteria of local and lymph node extension and general condition. This trial, begun in January 1976, progresses slowly due to the rigidity of these criteria. In the eighteen patients followed for more than 6 months we have noted an enormous difference in the cutaneous effects, identical mucosal effects, half the number of recurrence in those patients treated by low dose rate and as many necroses (proportionately less for T3 tumours, see Table 2).

Table 2. Preliminary data of randomized trial T3 tumours of the base of the tongue

	Low dose rate (L.D.R.)	fractionated (F.T.)
Skin reactions (0 to 3)	0.25	2.3
Mucosal reactions (0 to 3)	2.6	2.5
Tumour = 0 (3 months)	6/8	4/8
Tumour = 0 (9.78)	4/8	2/8
Alive (9.78)	4/8	4/8
Necrosis (9.78)	2/8	2/8
Alive without relapse or necrosis (9.78)	2/8	0/8

2. *A randomized trial on the whole range of ENT cancers* (EORTC trial excepted), where we are comparing irradiation to a basal dose of 45 Gy for a low-dose treatment (a single course of 5 to 7 days) and classical fractionated treatment. The aim of this trial is to determine the effects of this dose on the skin, the mucosa and on tumour reduction 6 weeks after the start of treatment. Additional treatment is then given, by the most applicable technique, without particular randomization. The short-term results (at 3 months) on ten cases confirm the marked difference in cutaneous effects, the more marked mucosal reactions of low dose rate and a more pronounced tumour effect of low dose rate (80% tumour reduction at 6 weeks as opposed to 50% for fractionated treatment, see Tables 3 and 4).

Table 3. Protocol for study of head and neck tumours treated with low-dose-rate irradiation in comparison with conventional fractionated irradiation

Low dose rate—basal dose study—Fractionation

45 Gy in 1.5 weeks
Interval of 5-6 weeks, then additional 25 Gy (technique of choice)
Compared with classical fractionated irradiation

45 Gy, 9 Gy/week (5 fract.), 5 weeks
Interval of 1-2 weeks
Additional 25 Gy (technique of choice)
Length of study: 2 years (October 1977 to October 1979)
Estimated number of patients: 70-80

3. *Research concerning the upper limit of low dose rate to be applied.* A few months ago we began a study of irradiation at double the dose rate previously adopted: from 90 to 100 rad/h, we have changed to 170 to 200 rad/h (1.7 to 2.0 Gy/h). The purpose of this study is to define the upper limit of the "plateau-effect" of low-dose-rate therapy. It is already known that the effectiveness is almost identical within the range of 30 to 120 rad/h (Pierquin *et al.*, 1973). For the moment, it appears that for ENT patients, this plateau extends as far as 200 rad/h (2 Gy/h). If these results are confirmed, the practical implications of this increase in dose rate would be considerable: it would be possible, in fact, to treat twice as many patients and under more comfortable conditions (one session per day of 800 rad in 4 hours). The practicability of this technique thus becomes comparable with that of fractionated high-dose-rate treatment. At present (September 1978) the first patients treated have not shown any appreciable differences; it seems, therefore, that this plateau does extend at least as far as 200 rad/h.

Table 4. Results of study of head and neck tumours treated according to protocol of Table 3

E.N.T. Study 45 Gy—Sept. 1978

5 cases X 2	Low dose rate	Fraction
Skin effects (0 to 3)	0.7	2
Mucosal effects (0 to 3)	2.2	1.6
Tumour effects (0 to 5)	80%	50%
	4.2	3

Conclusions

This study of low-dose-rate irradiation confirms, little by little, an improved differential effect between certain normal tissues (skin and small intestine for example) and certain tumours (squamous carcinomas, fibrosarcomas and myosarcomas for example). It would be desirable if other centres became involved in this clinical research, to confirm the results already obtained and to expedite the conclusions. A comparison with other unorthodox techniques (neutrons and multi-fractionation) is needed to be able to draw firm and precise conclusions. In order to achieve these goals, the principle teams responsible in the world must collaborate in a co-ordinated and comparative study.

References

CALITCHI, E. (1978) Bilan à huit ans, de l'irradiation à faible débit par telecobalt 60. Thesis, Paris.

PIERQUIN, B., D. CHASSAGNE, F. BAILLET and C. H. PAINE (1973) Clinical observations on the time factor in interstitial radiotherapy using iridium-192. *Clin. Radiol.* **24**, 506-509.

Combined chemo/radiotherapy of cancer: present state and prospects for use with high-LET radiotherapy

THEODORE L. PHILLIPS

*Department of Radiation Oncology,
University of California, M-330,
San Francisco, California 94143, U.S.A.*

Abstract—*Cytotoxic chemotherapy added to low-LET radiotherapy has improved survival and in some cases local control. In some instances the combination has yielded increased normal tissue damage. Similar results can be expected, as confirmed by intestinal crypt cell experiments, with high-LET. Less interaction than with low-LET appears to occur with agents blocking sublethal damage repair or causing synchronization.*

Introduction

Soon after the early use of chemotherapy in the treatment of Wilms' tumor, it was noted that increased radiation reactions occurred in normal tissues, particularly skin and lung. Increased tumor response was also attributed to combined radiation and actinomycin. The use of radiation in combination with chemotherapy has become widespread and to a large extent successful in the treatment of a number of pediatric malignancies, including Wilms' tumor, Ewing's sarcoma and rhabdomyosarcoma.

There have been three major clinical rationales for adding chemotherapy to conventional radiotherapy or adding radiotherapy to what was previously conventional chemotherapy in various human neoplasms. In the first situation chemotherapy was added in an attempt to destroy microscopic distant metastases not included in the radiation treatment volume; this was the primary rationale in most of the pediatric solid tumors. The second use of radiation was for the control of tumor cells growing in sanctuaries not reached by very effective systemic chemotherapy. The third situation, which has been widely studied but not yet proven to be particularly efficacious, is the use of systemic or regional chemotherapy to enhance the local control rate by radiotherapy.

Theoretical considerations for combined radiotherapy and chemotherapy

The interactions which occur between chemical modifiers of radiation response and radiation may be classified as enhancement, interference or antagonism (Phillips, 1977). These classifications apply to all chemical modifiers including sensitizers and protectors, but may be used to describe the interactions which occur with cytotoxic cancer chemotherapeutic agents. When an enhanced response is seen, it may be due to synergistic action between the cytotoxic drug and radiation in terms of cell kill in tumor or normal tissue, but may be due only to additivity of cell kill.

Recently Steel and Peckham (1978) have discussed the enhancement of response which may be seen with combined radiation and chemotherapy. They make it clear that enhancement must always be examined critically in order to determine whether this enhancement is additive, subadditive or supra-additive (synergistic or potentiating). They point out that in many cases apparent synergism is in reality additivity, which appears more than additive because of the nature of the survival curves for the individual agents.

More than additive cell kill may be observed when one or more of four basic types of interactions occur. (1) *Physiologic*: radiation or drug treatment may cause a major physiologic change in the tumor or in a critical normal tissue which leads to an increased effect of the other agent when subsequently administered. Examples include tumor shrinkage and reoxygenation caused by drug or radiation-induced tumor shrinkage leading to increased drug access to the tumor. (2) *Cell kinetic*: radiation or drug may induce changes in the cell proliferation or distribution of cells in the various cycle compartments which lead to increased response when the second agent is delivered. (3) *Radiation damage modification*: a drug may change the initial radiation lesion or inhibit repair of the radiation lesion, leading to enhanced effect. (4) *Drug-damage modification*: radiation may interfere with the repair of the drug-induced lesion or with the nature of the initial drug lesion leading to an enhanced effect.

Any or all of these may occur when supra-additivity occurs, and may also occur in what appears to be simple additivity.

Review of current experience with combined radiotherapy and a chemotherapy

A review of the current experience with the clinical use of combined radiation and chemotherapy in terms of (1) observed tumor response and apparent cure rate, (2) normal tissue damage enhancement and (3) the status of current clinical trials sheds some light on the usefulness of adding chemotherapy to conventional radiotherapy. Experience with these combinations using low-LET radiations should also provide insight into those situations in which such combinations may be beneficial with high-LET radiations. This review may only be brief because of space limitations; more detailed information may be obtained elsewhere (Carter and Wasserman, 1975; Muggia *et al.*, 1978; Phillips and Fu, 1976, 1978).

Brain tumors

A wide range of trials has been conducted in gliomas with primary emphasis on nitrosoureas. Trials completed to date show a small benefit in terms of survival when nitrosoureas are combined with high-dose radiotherapy to the whole brain. The nitrosoureas are quite active in treating gliomas when used alone, and it appears likely that the improved life span could also be achieved by sequential administration of radiation, with drug delayed until the time of failure. There has been no evidence of enhanced brain damage in any of the combined trials with nitrosoureas. Methotrexate (MTX), used primarily in the treatment of childhood leukemia, has caused enhanced brain damage in the form of leukoencephalopathy when combined with whole radiation and when administered either intrathecally or systemically in high doses. Current trials are investigating various combinations of nitrosoureas, procarbazine, vincristine and hydroxyurea in gliomas and medulloblastomas.

Head and neck cancer

Attempts to combine radiation and chemotherapy have probably been more extensive in head and neck cancer, with the major goal enhancement of local control, as with brain tumors, and a secondary goal diminution of failure due to distant metastases. Early trials with 5-fluorouracil (5-FU), MTX, hydroxyurea and bromouracil were equivocal. Subsegments of certain randomized trials in terms of institution or site did show enhanced control. Bleomycin (BLEO) has been employed more recently and has appeared to enhance immediate radiation response and short-term control in a few trials. None of these combinations has been widely accepted as standard therapy. Increased acute normal tissue reaction has been observed with all agents which have shown activity in enhancing tumor response, and acute mucositis has been limiting (Fu, 1978). The enhancement of late damage was also reviewed by Fu (1978) and soft tissue and bone necrosis, late wound healing, fistula formation and damage to the eye have been described. Although there is a suggestion that these reactions have been enhanced, it is by no means clear. Current clinical trials include the randomized investigation of BLEO during radiotherapy and maintenance BLEO and MTX, as well as a new trial involving preoperative *cis*-platinum (C-PL) and BLEO, surgery and postoperative irradiation vs. a similar regimen followed by maintenance C-PL, BLEO and MTX.

Carcinoma of the lung

Because of the high incidence of distant metastasis, as well as local failure, following radiotherapy in carcinoma of the lung, combined radiation and chemotherapy have been widely employed. Most such trials have been negative and in some instances the chemotherapy plus radiation patients have had shorter survival times. A few trials have been positive, in particular one trial with high-dose cyclophosphamide (CTX). It can be concluded, however, that there is no proven combination of radiation and chemotherapy of benefit in carcinoma of the lung, with the exception of oat cell carcinoma. In the past few years very aggressive chemotherapy/radiotherapy regimens involving three or four drugs with simultaneous radiotherapy have yielded increased complete regression rates and mean survival times and with the potential for increased long-term cure. In this situation, the chemotherapy regimens have been extremely effective when used alone, inducing a number of complete remissions and apparent cures, and thus the combined effect may be simply additive.

Marked enhancement of normal tissue damage has occurred with combined radiotherapy/chemotherapy in the thoracic region, including an increased incidence of radiation pneumonitis and radiation fibrosis of the lung and a marked increase in the incidence of severe esophagitis. Enhanced radiation pneumonitis has been seen with actinomycin D (ACT-D), ADM and BLEO in particular, but also occurs with CTX. Increased esophageal reaction has been observed with ACT-D, ADM and BLEO. Enhanced cardiac damage has been observed with combined radiotherapy and ADM due to additivity of radiation injury to the capillaries and drug injury to the myocardial cells. Current clinical trials of combined radiation and chemotherapy in lung cancer are concentrated on oat cell carcinoma. A variety of three and four drug regimens with simultaneous or sequential radiotherapy are under investigation. Radiotherapy is also used widely in the prophylactic treatment of the central nervous system.

Gastrointestinal cancer

Combined radiotherapy/chemotherapy has been employed widely in colorectal, gastric and pancreatic carcinoma, and more recently in hepatomas and liver metastases. 5-FU and methyl-CCNU have been the most widely used drugs because of their activity against gastrointestinal carcinomas. Apparently improved survival and local control have occurred in rectal cancer, gastric and pancreatic cancer using combined radiation and 5-FU. A recently completed randomized trial by the G.I. Tumor Study Group in pancreatic cancer has shown best survival with 6000 rad split-course radiotherapy combined with concomitant 5-FU. It is not clear whether the results are due to suppression of metastatic disease or enhancement of local control. Increased acute and late reactions have been seen in the gastrointestinal tract with ACT-D, ADM, 5-FU and in experimental situations BLEO (Phillips et al., 1975). Current clinical trials include 5-FU/methyl-CCNU chemotherapy with radiotherapy in rectal cancer and in carcinoma of the pancreas. The helium trial at Berkeley has introduced combined 5-FU and radiotherapy into both the low-LET control and the helium ion experimental arms.

Genitourinary and gynecologic cancers

There is no clear-cut evidence of enhanced local control or survival with combined radiotherapy and chemotherapy in gynecologic cancer, although there have been a few suggestive trials. Advanced carcinoma of the cervix treated with radiation and hydroxyurea has shown improved survival. The addition of alkylating agents to radiotherapy in carcinoma of the ovary has led to some increase in the local pelvic control achieved with moderate radiation doses. Trials of combined radiotherapy and chemotherapy in testicular carcinoma have shown improved responses of pulmonary metastases, but no significant change in survival vs. drug alone. The recent introduction of C-PL and BLEO into testicular treatment has yielded very successful chemotherapy regimens, and the role of radiotherapy in addition to these is not at all clear. In patients with urogenital and gynecologic tumors, increased radiation effects on normal tissues are primarily due to enhanced damage to the lung and gastrointestinal tract, as previously described. Current clinical trials involve the combination of phenylalanine mustard with radiotherapy in a number of ovarian protocols involving pelvic or pelvic plus para-aortic irradiation and the addition of systemic chemotherapy or hormone therapy to local irradiation for carcinoma of the prostate.

Lymphomas and leukemias

Combined radiotherapy/chemotherapy has been very successful in lymphomas and leukemias. Irradiation of the brain, a sanctuary for chemotherapy, has added significantly to the cure of acute lymphocytic leukemia in children. The addition of multidrug chemotherapy to radiotherapy has led to improved disease-free survival in Hodgkin's disease, although it is not yet clear whether it is superior to sequential radiotherapy and chemotherapy for failure. Enhanced damage has been seen in the form of enhanced leukoencephalopathy with combined radiation and MTX in leukemia and radiation pneumonitis with combined regimens involving alkylating agents and mantle field irradiation. Current trials are evaluating radiation in lower doses as consolidation following systemic chemotherapy for Hodgkin's disease and non-Hodgkin's lymphomas.

Conclusions concerning combined low-LET radiotherapy and chemotherapy

Improvement in cure has been seen in several disease entities due to the sterilization of distant metastases by very effective chemotherapy. Improved survival has been observed with leukemia with radiation treatment of sanctuary sites. Some improvement in local control with radiotherapy has been seen in a few sites where the chemotherapy is very active alone against the tumor under treatment. Synergistic enhanced response is probably not important, since in the few instances in which it has been observed it has caused marked increases in normal tissue damage, requiring a reduction in radiation dosage, yielding similar local control rates. These improved results have been accompanied by a significant price in acute and late normal tissue damage.

Combination of chemotherapy with high-LET radiotherapy

Rationale

The current status of combined low-LET radiotherapy and chemotherapy suggests in certain situations it may be advisable to include chemotherapy with high-LET radiation. It is clear in diseases with a high incidence of distant failure, such as childhood malignancies, oat cell carcinoma of the lung and gastrointestinal cancers, that adding chemotherapy may improve survival and allow a better evaluation of the local control achieved by high-LET radiation. Since high-LET radiotherapy is being tested because of potentially improved local control, it is not logical to add chemotherapy for the purpose of improving local control in the initial attempts to determine the efficacy of high-LET radiotherapy in sites such as brain or head and neck. Because chemotherapy may influence local control, it would be necessary to include it in both arms, i.e. low- and high-LET arms, if it were to be employed for the purpose of sterilizing distant micrometastases. It is unlikely that chemotherapy and high-LET radiation would be combined for the very sensitive hematologic malignancies where radiation therapy is used to treat chemotherapeutic sanctuaries.

Potential differences and similarities of low- and high-LET radiotherapy and chemotherapy

It is likely that the experience with chemotherapy and high-LET radiotherapy in terms of tumor control or survival will be similar to that with low-LET radiotherapy, since most of the improvement has been due to sterilization of micrometastases by chemotherapy or enhanced local control due to additive chemotherapy cell kill. There may be significant differences in normal tissue damage due to the biologic interaction of radiation and chemotherapy when one compares low- and high-LET radiations.

Recapitulating the four methods of potential synergistic interaction of radiation and drugs, it is unlikely that the physiologic mechanisms would be very much changed when chemotherapy is added to high-LET radiotherapy. On the other hand, changes in cell cycle kinetics and synchrony, modification of drug damage by radiation, and modification of radiation damage by drug could be different with high-LET. Reponses to high-LET radiations are known to be less affected by cell age. In situations where drug apparently modifies radiation damage repair, one would expect to see less effect with high-LET radiations. In situations where radiation may modify drug damage or damage repair, the effect is not clear with high-LET radiations.

Experimental data comparing high- and low-LET in intestinal crypt cells

In order to approach some of the questions raised in the preceding paragraph, an experiment has been performed comparing the response of intestinal crypt cells using ^{137}Cs irradiation and carbon ion irradiation in the distal peak region of a beam with 26 cm range and 4 cm spread Bragg peak. Techniques for intestinal crypt cell assay have been described previously (Phillips et al., 1975). Drug was administered intraperitoneally either 3 hr after or 2 hr before irradiation of the whole animal (^{137}Cs X-rays) or the abdomen (carbon ions). Animals were sacrificed 3.5 days following irradiation and the intestine prepared for microscopy; four to six animals and thirty-two sections were examined for each dose point. A fixed radiation dose was used for each drug experiment and the drug dose was varied from 0.25 to 1× the maximum tolerated dose previously established for that particular drug.

The results are shown in Figs. 1 through 7. The agents tested were actinomycin D (ACT-D), adriamycin (ADM), methotrexate (MTX), cis-platinum (C-PL), BCNU, bleomycin (BLEO) and cyclophosphamide (CTX). The per cent MTD is the per cent of the maximum tolerated drug dose. Circles represent low-LET radiation and triangles high-LET radiation, with closed circles or triangles administration of drug prior to irradiation. The intestinal crypt cell survival number with radiation alone is shown by the dotted regions.

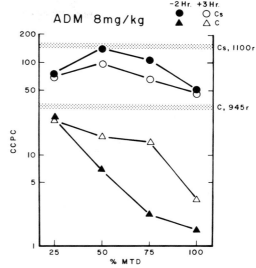

Fig. 2. Survival of crypt cells per circumference (CCPC) with combined ADM and low- or high-LET radiation.

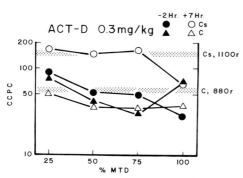

Fig. 1. Survival of crypt cells per circumference (CCPC) with combined ACT-D and low- or high-LET radiation.

In Fig. 1, ACT-D given prior to irradiation is more effective than drug given post irradiation at all drug dose levels with ^{137}Cs. On the other hand, there is no difference between administration of drug before or after with high-LET peak carbon irradiation. Since the dose-response curve we obtained with intestinal crypt cells for carbon shows no apparent shoulder or D_q and since multi-fraction experiments have indicated no repair of sublethal damage in the intestinal crypt cells following carbon peak irradiation, it is likely that this result represents reduction in the shoulder of the dose-response curve and therefore lower survival with low-LET radiation when drug is given before radiation and the absence of this enhancement with high-LET radiation.

In Fig. 2, with ADM there is little difference with low-LET radiation when drug is given before or after irradiation, although there is additive cell kill. The high-LET radiation dose selected yielded more cell kill, and therefore the results may be due to a different level of survival. There is a greater cell kill as a function of drug dose when the drug is given before or after irradiation with high-LET, and pre-radiation drug appears more effective than post-radiation drug. This marked difference could be due to the difference in the high-LET lesions, with changes in drug access to the cell due to permeability changes or it may be due to a different level of injury. Further experiments are required to elucidate the exact mechanism.

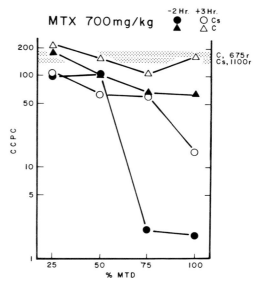

Fig. 3. Survival of crypt cells per circumference (CCPC) with combined MTX and low- or high-LET radiation.

In Fig. 3, with MTX a difference is seen between low- and high-LET radiations. With high-LET radiations higher doses of MTX are much more effective when given before than after irradiation, but this difference is eliminated with high-LET radiations. Our experiments, with time intervals from plus to minus 24 hr, have indicated that MTX in these high doses causes synchronization with wide fluctuations in the total cell kill when radiation is added. Although the synchronization is most certainly caused by MTX when combined with high-LET radiation, decreased dependence of high-LET cell kill on cell age would lead to less enhanced cell kill, which is shown by this experiment.

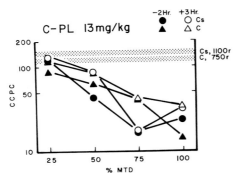

Fig. 4. Survival of crypt cells per circumference (CCPC) with combined C-PL and low- or high-LET radiation.

In Fig. 4, with C-PL the differences between drug administration before or after irradiation with low-LET were small. Split dose studies suggest that C-PL inhibits repair of radiation injury, possibly explaining the slightly greater effect when drug is given before than after irradiation with low-LET at certain dose levels. There is little difference between before or after administration with high-LET. The small differences seen in this experiment are obviously equivocal and repeat experiments are required.

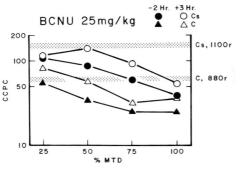

Fig. 5. Survival of crypt cells per circumference (CCPC) with combined BCNU and low- or high-LET radiation.

In Fig. 5, with BCNU the response to low- and high-LET radiation is similar. The drug is more effective when given before than after irradiation. Longer-term timing experiments indicate that this enhanced response occurs from about −12 to 0 hr relative to irradiation. The persistence of the greater effect before than after with high-LET suggests that radiation is interfering with repair of drug injury and that both low- and high-LET radiation interfere to the same degree.

In Fig. 6, with BLEO there is a marked effect when given before irradiation as compared to after irradiation. This effect is similar for low- and high-LET radiations, again suggesting that there is no modification of the radiation injury by drug, but that the radiation enhances drug injury irrespective of the radiation quality.

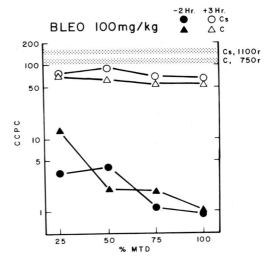

Fig. 6. Survival of crypt cells per circumference (CCPC) with combined BLEO and low- or high-LET radiation.

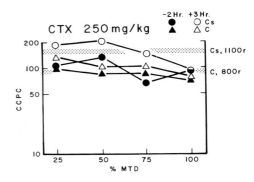

Fig. 7. Survival of crypt cells per circumference (CCPC) with combined CTX and low- or high-LET radiation.

In Fig. 7, CTX has little effect on intestinal crypt cells and there is little difference between before or after administration with low- or high-LET.

Summary and conclusions

The addition of cancer chemotherapy to radiotherapy seems to be of benefit for both low- and high-LET radiation in situations where very effective agents are available when used alone against a particular tumor, where distant metastatic failure is important, and where the agent does not markedly enhance radiation complications in the local and regional treatment volume. There may also be some advantage in the enhancement of local control at certain sites, although this will not be the goal of high-LET trials. Because of this enhancement, it will be necessary to use drug in both high-LET arms if it is elected to employ drug for the control of distant metastases.

In terms of local normal tissue damage enhancement, there may be some differences between low- and high-LET radiations. Theoretical considerations and experimental results suggest that drugs which interfere with sublethal damage repair will interact less with high-LET radiation, that drugs whose injury repair is modified by radiation will behave similarly, and that the interaction with drugs which cause synchrony will be less with high-LET than low-LET.

Work supported by National Cancer Institute Research Grants CA17227 and CA20529.

References

CARTER, S. K. and T. H. WASSERMAN (1975) Interaction of experimental and clinical studies in combined modality treatment. *Cancer Chemother. Rep.* **5**, 235-241.

FU, K. K. (1979) Normal tissue effects of combined radiotherapy and chemotherapy for head and neck cancer. *Frontiers in Radiation Therapy and Oncology* 13: 113–132.

MUGGIA, F. M., H. CORTES-FUNES and T. H. WASSERMAN (1978) Radiotherapy and chemotherapy in combined clinical trials: Problems and promise. *Int. J. Radiat. Oncol. Biol. Phys.* **4**, 161-171.

PHILLIPS, T. L. (1977) Chemical modification of radiation effects. *Cancer* **39**, 987-999.

PHILLIPS, T. L. and K. K. FU (1976) Quantification of combined radiation therapy and chemotherapy effects on critical normal tissues. *Cancer* **37**, 1186-1200.

PHILLIPS, T. L. and K. K. FU (1978) The interaction of drug and radiation effects on normal tissues. *Int. J. Radiat. Oncol. Biol. Phys.* **4**, 59-64.

PHILLIPS, T. L., M. D. WHARAM and L. W. MARGOLIS (1975) Modification of radiation injury to normal tissues by chemotherapeutic agents. *Cancer* **35**, 1678-1684.

STEEL, G. G. and M. J. PECKHAM (1979) Exploitable mechanisms in combined radiotherapy-chemotherapy: The concept of additivity. *Int. J. Radiat. Oncol. Biol. Phys.* 5: 85–92.

SESSIONS III and IV

Physical Aspects and Radiobiology

Review of performance of high-LET radiation sources used in clinical applications

D. K. BEWLEY

MRC Cyclotron Unit, Hammersmith Hospital, Ducane Road, London W12 0HS, U.K.

Abstract—*Sources are reviewed under the following headings: 14 MeV neutrons, accelerator-produced neutrons, heavy charged particles and negative pi mesons. Emphasis is placed on the situation obtaining in 1978.*

Introduction

At the present time these sources can be divided into fast neutrons, negative π-mesons and charged heavy ions. Fast neutrons have been used in the vast majority of the clinical applications. Sources of fast neutrons include the deuterium–tritium (d+T) reaction which needs a relatively low accelerating potential, and reactions between protons and deuterons with targets of beryllium or deuterium.

14-MeV neutron generators

Table 1 lists the performance of four (d+T) generators based on information from the users. Three use sealed tubes with a mixed beam of deuterons and tritons on a mixed target, while the fourth employs a beam of deuterons on a target which initially contains only tritium and titanium. The details are to some extent provisional, particularly for the Haefely generator where the tube in current use is the third one supplied and is being run considerably below the rating specified by the manufacturer. Tube life is clearly very uncertain under these circumstances.

In principle the output of the sealed tubes remains constant. One Philips tube has been reported to give a constant output (with readjustment of parameters) for three-quarters of its life. By contrast, the output from a tritium target falls steadily as the tritium is consumed and diluted with deuterium. For estimating the cost per complete treatment of one patient, taken as 1500 rad, I have assumed that the target of the radiation dynamics machine is changed when the output per mA has fallen to half.

It is perhaps surprising that the cost per rad is so similar. Tritiated targets cost only one tenth as much as the sealed tubes but have to be changed roughly 10 times as frequently. The effective size of the neutron source is considerably smaller in the case of the tritiated target than in the sealed tube.

None of these machines is adequate for modern, effective radiotherapy. To obtain a penetration equal to that of a modern ^{60}Co therapy machine the neutron SSD should be 100 or 120 cm, but the dose-rate would then be too low. An SSD of only 80 cm does not leave enough room for good shielding. Also, close to the opening of the diaphragms there is a large component of degraded neutrons, 27% of the dose according to one set of measurements. This also contributes to the rather poor penetration of the beam. Finally, collimators are heavy and have to be changed mechanically, a process which may take up to 5 minutes.

Neutron generators based on accelerators

Neutron generators based on accelerators vary greatly depending on their design. Deuterons or protons are usually accelerated to energies between 10 and 100 MeV and interact with targets of Be or D_2. The advantage of using a target of Li rather than Be is too small to outweigh the disadvantage of the lower melting point of Li. Other ions such as ^{3}He and ^{4}He give lower neutron yields than ^{1}H and ^{2}H.

Table 1. d+T Neutron sources in medical use

Manufacurer type	Elliott mixed	Philips mixed	Haefely mixed	Radiation dynamics d on T
mA	30	18	150	12
kV	250	250	200	500
n/sec × 10^{12}	1.0	1.0	3	3.5
SSD used, cm	80	80	100	80
Rad/min	6	6	10	20
Target diam., cm	3.5	5	4.5	2.5
Av. tube life, h	150	150	150 (?)	130 mA h (T ½ target)
Cost per tube or target £K	15 (200 h)	?	18 (300 h)	1.5
Cost per 1500 rad, £	300	?	150	170

Figure 1 shows the kerma rate at 1 metre for various ions and targets, as a function of particle energy. The relationship is known most accurately for d on Be. At first sight the lower neutron intensity from protons on Be is a disadvantage, but some modern cyclotrons can accelerate protons to nearly double the energy attainable with deuterons. For example, the CS-30 cyclotron (Cyclotron Corporation) can accelerate deuterons to 15 MeV or protons to 26 MeV. For 100-μA protons on a thick Be target the kerma rates at 1 metre would be about equal at 65 rad/min. But the dissipation of power in the target would be almost doubled using protons.

The neutron spectra from the three principal reactions of Fig. 1 are of very different shapes. With protons on Be there is a large contribution from low-energy neutrons followed by a trough at about 12 MeV and a peak at 20-24 MeV (Johnsen, 1977). The low-energy component is less important with deuterons on Be and there is a peak at about $0.4E_d$. With deuterons on D_2 the spectral shape depends on deuteron energy. The reaction $^2H(^2H,n)^3He$ gives the greatest intensity at about $(E_d + 2)$ MeV, but a lower peak at about $0.4E_d$ becomes more prominent at higher deuteron energies. A deuterium target is therefore most useful for machines accelerating deuterons to relatively low energies, say less than 15 MeV.

These very different spectral shapes make it difficult to compare the mean neutron energies. Instead Fig. 2 compares the penetration of neutrons from d and p on Be (Smith et al., 1974). 15 MeV d and 26 MeV p give neutron beams with rather similar penetration (Goodhead et al., 1978).

Protons are particularly useful at higher energies where target heating is a less severe problem. The penetration of proton-generated beams can be further improved by using a hydrogenous filter to attenuate the lower energy component of the beam (Bewley et al., 1978). With deuterons, a better way of increasing the mean neutron energy is to use a thin target backed by a material of higher atomic number in which neutron production is relatively small.

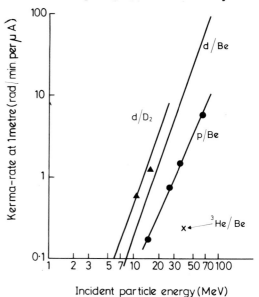

Fig. 1. Neutron intensity in the forward direction from thick targets of Be and ^2H bombarded by p, d and ^3He.

Table 2 shows the equivalence between neutrons from p on Be and X-rays in terms of depth of the 50% contour of neutron dose. To obtain a beam with penetration equivalent to that of a 4–10-MeV linear accelerator one needs 50–90 MeV protons with a filter.

The reliability of modern cyclotrons can be very good, comparable with that of an X-ray generator. Unlike d+T generators the target is not usually the main problem. However, the use of a D_2 target with relatively small accelerators is still in the development stage.

Therapy with heavy charged particles

At the present time the only regular medical application of these ions is the use of sharply defined beams for "neurosurgery", mainly ablation of the pituitary. The radiation is not at high-LET and so this application will not be discussed here.

Heavier ions such as C and Ne are needed to give a significant rise in LET; the effect with He ions is only marginal. Typical currents and energies needed are 50-500 pA at 100-500 MeV per a.m.u. (Bewley, 1974).

These energies are so high that at present the only suitable accelerators are a few in nuclear physics research laboratories. Grunder and Leeman (1977) have discussed the design of a suitable accelerator for medical use; it would have to be a two-stage machine, the main accelerator being a synchotron with an injector in the form of a cyclotron, Van de Graaff or linear accelerator.

accelerate but produce pions less efficiently, by a factor of about 30. Designs for medical pion generators have been discussed by Knapp (1977); these are based on mean currents of 350 μA of electrons at 770 MeV or 30 μA of protons at 650 MeV, to give a dose-rate of 30 rad/min over a tumour 10-cm cube. The current needed depends on the solid angle over which pions are collected and directed at the patient. The designs discussed by Knapp are based on the arrangement to be installed at Stanford which collects pions over one steradian and focuses them into the tumour in a ring geometry.

The only pion generator in medical use at the

Table 2. X-rays to give same penetration as p+Be neutrons

Depth for 50% (12 × 12) (cm)	10	12	13	14	16	18
E_p (MeV) no filter	34	52	60	68	80	—
E_p (MeV) 5-cm filter	—	32	42	52	68	80
E X-ray (MV) 12 × 12 field at 100 cm SSD	1.5	2.2	3.0	3.7	6	9

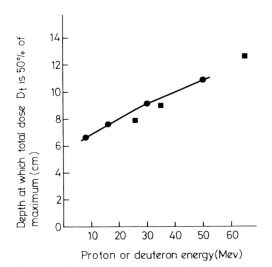

Fig. 2. Depth in water of the 50% depth-dose for neutrons produced by the d+Be (●) and p+Be (■) reaction. Field size 30-60 cm², 100 cm SSD.

Negative pi mesons

Negative pi mesons combine the advantages of neutrons and heavy ions (high-LET, defined range with maximum dose at depth), but also their disadvantages (high current of accelerated particles and high energy). Mesons of 70 MeV penetrate about 15 cm in water, but to generate them one needs protons or electrons of at least 500 MeV. Electrons are more economical to

present time is that at Los Alamos. With 275 μA of 750 MeV protons and a small angle of acceptance, a dose-rate of about 5 rad/min can be achieved over a 10-cm cube (Table 3). The accelerator (SIN) in Switzerland is at an early stage of development; at the present time about 50 μA of 590 MeV protons gives a pion dose rate of a few rad/min over a much smaller volume. An arrangement like that at Stanford will be needed to obtain a useful therapeutic beam. At Vancouver the accelerator (TRIUMF) is intended to accelerate 100 μA of protons to 500 MeV (Henkelman et al., 1977). With these parameters it seems unlikely that a single channel like that at Los Alamos can provide a high enough intensity for practical radiotherapy.

Conclusion

At the present time the only practical form of high-LET therapy for use in a hospital is a neutron beam produced by a cyclotron. Deuterium–tritium generators lack adequate output for satisfactory use. Much larger machines are needed for direct use of heavy charged particles and π-mesons, so that this method can at present only be used in special research institutes.

References

BEWLEY, D. K., B. C. PAGE, J. P. MEULDERS and M. OCTAVE-PRIGNOT (1978) Neutron beams from protons on beryllium. *Poster Session A. Characteristics of Fast Neutron Sources.*

Table 3. π-Mesons

Institute	Design parameters			Operating conditions	
	E (MeV)	I (μA)	Dose rate (rad/min)	I	Dose rate
Los Alamos	750	1000	20*	275	5*
Villigen	590	100	?	50	2†
Vancouver	500	100	?	10(?)	?
Stanford	550	500 (electron)	30*	—	—

*Over a volume 10-cm cube.
†Over a small volume.

BEWLEY, D. K. (1974) Introduction to the discussion on characteristics of installations for radiotherapy with fast neutrons and other types of high-LET radiation. *Europ. J. Cancer* **10**, 201-202.

GOODHEAD, D. T., R. J. BERRY, D. A. BANCE, P. GRAY and B. STEDEFORD (1978) Fast neutron therapy beam produced by 26 MeV protons on beryllium. *Phys. Med. Biol.* **23**, 144-148.

GRUNDER, H. A. and C. W. LEEMAN (1977) Present and future sources of protons and heavy ions. *Int. J. Radiat. Oncol. Biol. Phys.* **3**, 71-80.

HENKELMAN, R. M., L. D. SKARSGARD, K. Y. LAM, R. W. HARRISON and B. PALCIC (1977) Recent developments at the π-Meson Radiotherapy facility at TRIUMF. *Int. J. Radiat. Oncol. Biol. Phys.* **2**, 123-127.

JOHNSEN, S. W. (1977) Proton–beryllium neutron production at 25-55 MeV. *Medical Physics* **4**, 255-258.

KNAPP, E. A. (1977) Accelerators for pion clinical facilities. *Int. J. Radiat. Oncol. Biol. Phys.* **3**, 293-297.

SMITH, A. R., P. R. ALMOND, J. B. SMATHERS and V. A. OTTE (1974) Dosimetric properties of the fast neutron therapy beams at TAMVEC. *Radiology* **113**, 187-193.

Dose distributions of clinical fast neutron beams

B. J. MIJNHEER* AND J. J. BROERSE†

*Antoni van Leeuwenhoek Hospital, Plesmanlaan 121, Amsterdam, The Netherlands
†Radiobiological Institute TNO, Lange Kleiweg 151, Rijswijk, The Netherlands

Abstract—*A review is given of dosimetric properties of sixteen clinical fast neutron beams. The central axis depth dose, gamma-ray contribution, beam profile and dose build-up have been compared for about equal field sizes. The depth in the phantom, at which the central axis total $n+\gamma$ dose has been reduced to 50% of its maximum value, varies between about 8 cm and 14 cm, depending mainly on the neutron-producing reaction and SSD employed. The gamma-ray contribution to the total $n+\gamma$ dose in the phantom shows large discrepancies between the different machines. The average cyclotron values are somewhat lower than the comparable average data, at equal depth, for $d+T$ neutron sources. The penumbral width, measured at 10 cm depth in a phantom, is decreasing with increasing average energy of the neutrons. The edges of the useful beam are less sharp for the $d+T$ neutron generator than for the cyclotron beams. Measurements of the build-up of the dose below the surface show large variations between the different machines, even for comparable neutron energy spectra. Due to the relatively high RBE of the short range particles, the skin-sparing properties of all neutron beams will probably not differ considerably and are more or less comparable to those of electron beams.*

Introduction

A considerable amount of information concerning dose distributions of neutron beams is already available in the literature and is summarized in ICRU report 26 (ICRU, 1977). Part of the information is concerned with experimental machines. Recently a number of centres started patient treatments. It seems, therefore, worthwhile to compare the most recent dose measurements at these clinical fast neutron machines.

The physicists in charge of the dosimetry were asked to give the following latest information on their machine:
1. A percentage depth dose curve (neutron or total $n+\gamma$ absorbed dose) for a field size of about 10 × 10 cm, and the relative gamma-ray contribution at different depths.
2. A beam profile for the same field size at 10 cm depth in a phantom (neutron or total $n+\gamma$ absorbed dose) and relative gamma-ray contribution.
3. A percentage absorbed dose or ionization curve in the build-up region.
4. The types of instruments, phantom size and material, SSD and neutron-producing reaction used for these measurements.

Although not explicitly asked, it has been assumed that patient treatment is based on the provided dose distributions, eventually modified by means of flattening filters. If no new information was available, literature values have been taken, as far as possible. Only macroscopic dose distributions have been compared. Technical characteristics and the radiation quality of the different beams are dealt with at other sessions of this symposium.

Central axis absorbed dose distribution

Collimated fast neutron sources exhibit central axis absorbed dose distributions in a phantom that depend on the energy of the neutrons, the source-to-surface distance (SSD), the field size and gamma-ray contribution to the absorbed dose. The type of the phantom material in which the penetration is determined will also influence the absorbed dose distribution. As can be seen from Table 1, the different therapy centres use different phantom materials which will introduce small variations in dose distributions even when the same neutron source is used. Some typical central axis depth dose curves are presented in Fig. 1.

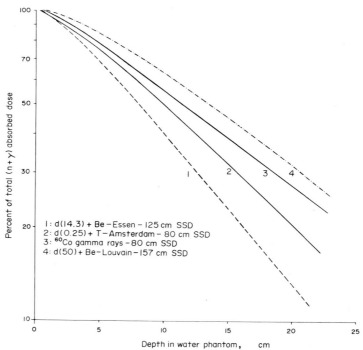

Fig. 1. Central axis depth dose curves for ^{60}Co gamma-rays and three different neutron beams for field sizes of approximately 10×10 cm.

The depth in the phantom at which the total $n + \gamma$ absorbed dose is reduced to half its maximum value, d(50%), was chosen as a criterium for comparison of the depth dose data of the different fast neutron radiotherapy beams. The data are presented in Table 1, together with the field size, SSD and phantom material used for these measurements by the different centres. It

Table 1. Summary of relevant data, including depth dose characteristics and penumbral width, of fast neutron therapy sources

Place	Reaction (E in MeV)	SSD (cm)	Phantom material	d(50%)[a] (cm)	Field size (cm × cm)	x(20%)−x(80%)[b] (cm)
Dresden	d(13.5)+Be	100	TE-sol.[c]	8.0	12 × 15	2.3
Tokyo	d(14.0)+Be	125	TE-sol.[c]	8.1	10 × 10	2.3
Essen	d(14.3)+Be	125	H$_2$O	8.1	10 × 10	2.4
Edinburgh	d(15)+Be	125	H$_2$O	9.0	11 × 11	2.3
London	d(16)+Be	117	TE-sol.[d]	8.7	9.5 × 9.5	2.2
Seattle	d(21)+Be	150	TE-sol.[c]	9.4	10 × 10	1.8
Chiba-shi	d(30)+Be	175	TE-sol.[d]	11.7	10 × 10	1.9
Washington	d(35)+Be	125	TE-sol.[c]	11.1	10 × 10	1.6
Houston	d(50)+Be	140	TE-sol.[c]	13.2	10 × 10	2.7
Louvain	d(50)+Be	157	H$_2$O	13.6	10.5 × 10.5	1.3
Batavia	p(66)+Be	153	TE-sol.[c]	14.6	10 × 10	1.4
Glasgow	d(0.25)+T	80	H$_2$O	9.6	10 × 10	3.8
Hamburg	d(0.5)+T	80	TE-plastic (A-150)	9.7	10.8 × 10.8	2.9
Heidelberg	d(0.25)+T	100	H$_2$O	10.6	11 × 11	3.6
Manchester	d(0.25)+T	80	H$_2$O	9.4	10 × 10	3.9
Amsterdam	d(0.25)+T	80	H$_2$O	9.9	9 × 11	4.6
^{60}Co gamma-rays		80	H$_2$O	11.9	10 × 10	1.6
8 MV X-rays		100	H$_2$O	17.3	10 × 10	0.8

[a] Depth at which the total $n + \gamma$ dose is reduced to half its maximum value.
[b] Distance off-axis between 80% and 20% of the central axis total $n + \gamma$ dose.
[c] Density of 1.07 g cm^{-3}
[d] Corrected to unit density.

may be estimated that the uncertainty in the given values is a few millimetres, which is mainly due to the difficulties in measuring the absorbed dose in the region of its maximum value.

As can be seen from Table 1, the values of d(50%) are for the cyclotrons increasing with deuteron energy. The small irregularities can largely be explained by differences in field size (e.g. between Edinburgh and London) or in SSD (e.g. between Chiba-shi and Washington). A comparison with photon sources shows that a deuteron energy of 30 to 35 MeV is necessary to obtain a penetration comparable to a clinically used ^{60}Co source, whereas the d(50)+Be and p(66)+Be neutron beams have depth dose characteristics comparable to a 4-MV X-ray beam. The values of d(50%) are for the d+T sources lower than for ^{60}Co gamma-rays, although an improvement of about 1 cm can be obtained if the SSD is increased from 80 cm to 100 cm.

The gamma-ray component of the total absorbed dose in a phantom originates from the production of gamma-rays in the target, the target shielding, the walls of the therapy room and the phantom itself. The relative gamma dose increases with field size and depth in the phantom. The influence of SSD on the relative gamma dose in the phantom is probably small as has been demonstrated for a d+T neutron generator (Mijnheer, Visser and Wierberdink, 1978).

The relative gamma dose, expressed as a percentage of local neutron + gamma dose, along the central axis of the beam as measured by the different groups, is listed in Table 2. Also given are the neutron-insensitive devices which have been used to derive the gamma dose. The cyclotron data do not show a tendency to change with deuteron energy. The spread in the measured values is probably due to differences in shield design and uncertainties in the derived gamma dose. The fractional uncertainty contributed to the gamma dose will, for instance, strongly depend on the relative neutron sensitivity, k_U of the instrument which has been used to determine the gamma dose (ICRU, 1977). Consequently, gamma dose determinations in a neutron field with different detectors will show large variations, as has been found during intercomparisons (e.g. Broerse, Burger and Coppola, 1978). This is probably also the reason for the difference between the Manchester and Glasgow data for the same type of d+T machine. From a comparison of the average values for the cyclotrons and the d+T generators one may conclude that the d+T generators have about a 40% higher relative gamma-ray dose component than cyclotrons. This may be due to the thinner shields defining

Table 2. Gamma dose as a percentage of local neutron plus gamma dose at different depths in a phantom along the central axis of fast neutron therapy sources[a]

Place	Reaction (E in MeV)	Depth: 2 cm	Depth: 10 cm	Depth: 20 cm	Neutron-insensitive device
Dresden	d(13.5)+Be	9.3[b]	15.9[b]	28.0[b]	C/CO_2-ion chamber
Tokyo	d(14.0)+Be	3.3	7.5	13.5	Mg/A-ion chamber
Essen	d(14.3)+Be	5.6	10.2	16.9	GM-counter
Edinburgh	d(15)+Be	4.7	7.7	12.9	GM-counter
London	d(16)+Be	5.4	7.6	11.0	GM-counter
Seattle	d(21)+Be	5.5	8.4	11.8	TE-proportional counter
Chiba-shi	d(30)+Be	3.2	4.6	7.8	Teflon+C/CO_2-ion chamber
Washington	d(35)+Be	8.4[c]	10.7[c]	13.5[c]	
Houston	d(50)+Be	5	–	–	C/CO_2-ion chamber
Louvain	d(50)+Be	6.4	7.8	8.6	GM-counter
Average value for cyclotrons		5.3	8.1	12.0	
Glasgow	d(0.25)+T	6.7	8.9	11.7	GM-counter
Hamburg	d(0.5)+T	7.2	9.2	12.5	GM-counter
Heidelberg	d(0.25)+T	–	7.0	–	GM-counter
Manchester	d(0.25)+T	14.5	15.0	19.5	C/CO_2-ion chamber
Amsterdam	d(0.25)+T	8.5	11.6	15.3	GM-counter
Average value for d+T generators		9.2	10.3	14.8	

[a] For the same field sizes, SSD and phantom material as listed in Table 1.
[b] Not included in the average due to the larger field size.
[c] Gamma dose derived by subtracting neutron dose from the total ($n + \gamma$) dose.

the movable d+T beams compared to the better shielded cyclotron beams.

Transverse absorbed dose distribution

The distance off-axis between 80% and 20% of the central axis total $n+\gamma$ dose, $[x(20\%)-x(80\%)]$, at 10 cm depth in a phantom, was chosen as a criterion for the sharpness of the edges of the useful beam. Values of $[x(20\%)-x(80\%)]$, as measured for the different therapy sources, are listed in the last column of Table 1.

The data for the cyclotrons, with exception of the Houston value, show that the penumbral width is decreasing with increasing deuteron energy. This may be explained by the more forward peaking of the neutrons by the higher energetic deuteron reactions and the decreasing scattering cross-section of hydrogen at higher neutron energies.

penumbral width becomes smaller if only the neutron dose is considered.

At larger distances outside the beam the dose levels for neutron beams are higher than those obtained with photon beams (ICRU, 1977). It should be realized, however, that RBE changes outside the beam may occur. The RBE will decrease due to the increase in the relative gamma-ray contribution with increasing distance from the central axis. This is illustrated in Fig. 2. Almost identical curves are observed for d+T sources as for cyclotrons, whereas changes in the deuteron energy do not seem to have a large influence on the shape of this curve, at least in the penumbra region. On the other hand, the RBE may increase due to the lower absorbed dose per fraction and the lower neutron energy outside the beam compared to the centre of the beam. Recently Bewley and Page (1978) showed that when allowance is made for changes in the RBE of the neutron component, even with

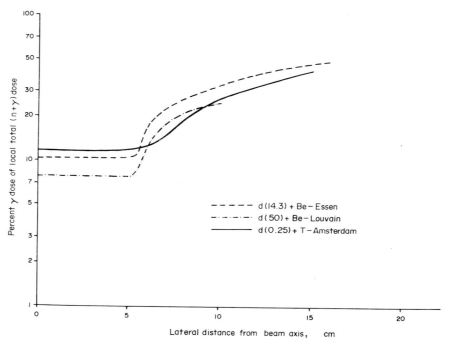

Fig. 2. *Gamma-ray contribution to the total $n+\gamma$ dose as a function of the lateral distance from the beam axis for three different neutron beams for field sizes of approximately 10×10 cm.*

The data for the d+T generators show a much larger penumbra than for the cyclotrons. This is due to the larger target sizes needed to obtain adequate dose rates, the thinner shield and more isotropic emission of the neutrons produced in the d+T reaction. However, the difference in

perfect shielding the biologically effective dose outside the useful beam is much higher for d(16)+Be neutrons than for a ^{60}Co machine. The same statement is probably valid for other fast neutron therapy beams, indicating that in neutron therapy one has to accept a larger hazard to the patient from stray radiation than with megavoltage X-rays.

Charged particle build-up

The build-up of absorbed dose below the surface of a phantom has been measured in all but one centre by means of adding material to a thin-walled ionization chamber. This could be either a chamber with a fixed distance between the electrodes or of the extrapolation type. In Hamburg the increase in optical density of a photographic film behind layers of paraffin was used to study the build-up of the neutron dose (Franke et al., 1978).

in mg cm^{-2}, at which the ionization reaches its maximum value, has been presented in the last column of Table 3. These values are in first approximation comparable to the maximum range of the recoil protons. Electron contribution and low-energy neutron absorption will cause deviations.

The depth at which 90% of the maximum ionization occurs is also given in Table 3. For the broad energy cyclotron and linear accelerator neutron sources this amounts to about one-quarter of the depth of maximum ionization, whereas for the two d+T ionization chamber measurements this depth is about half the depth of maximum ionization. This indicates that the

Table 3. Results of measurements of the build-up of ionization below the surface of a phantom

Place	Reaction (E in MeV)	I_{sur}/I_{max} [a] (%)	$d_{0.9 I_{max}}$ [b] (mg cm^{-2})	$d_{I_{max}}$ [c] (mg cm^{-2})
Tokyo	d(14.0)+Be	31	34	150
Essen	d(14.3)+Be	62	32	170
Edinburgh	d(15)+Be	42	45	200
London	d(16)+Be	29	56	180
Seattle	d(21)+Be	43	96	350
Chiba-shi	d(30)+Be	60	110	450
Washington	d(35)+Be	45	130	450
Houston	d(50)+Be	43	240	900
Louvain	d(50)+Be	57	195	840
Batavia	p(66)+Be	40	430	1660
Manchester	d(0.25)+T	49	110	230
Amsterdam	d(0.25)+T	58	118	250

[a] I_{sur}/I_{max} = ratio of ionization at surface to maximum ionization.
[b] $d_{0.9 I_{max}}$ = depth at which 90% of maximum ionization occurs.
[c] $d_{I_{max}}$ = depth of maximum ionization.

The ratios of the ionization measured at the surface of the phantom to the maximum ionization, which are in first approximation proportional to the relative entrance dose, as measured by the different groups, are listed in Table 3. The data show large scatter, even for comparable neutron energies. Zoetelief, Broerse and Mijnheer (1978) pointed out that build-up curves measured by means of ionization chambers are dependent on several factors like the presence of phantom material behind the ionization chamber, the use of air or tissue-equivalent (TE) gas flushing a TE-chamber and the application of an absorber to remove charged particles from the beam. Recently it has been shown that the surface dose of three different neutron beams, measured under similar experimental conditions, are almost identical (Mijnheer, Zoetelief and Broerse, 1978), therefore the differences in the table may partly be attributed to differences in experimental set-up used by the different groups.

The depth in the phantom material, expressed

build-up curves of the cyclotrons and linear accelerator are steeper due to the presence of more lower energy neutrons.

It should be pointed out that the increase in the measured ionization due to the addition of the TE material does not necessarily mean that a biological skin-sparing effect is present. The alpha-particles, heavy recoils and low-energy protons that are responsible for the dose at the surface, have a much higher RBE than the low-LET particles, fast protons and electrons, that cause the dose build-up (Broerse, Barendsen and Van Kersen, 1968; Bewley, McNally and Page, 1974). Clinical observations have shown that for a d+T source the skin sparing is only marginal (Battermann and Breur, 1978).

Isodose distributions

Central axis depth dose curves of neutron beams in combination with a limited number of beam profiles can be used to generate isodose

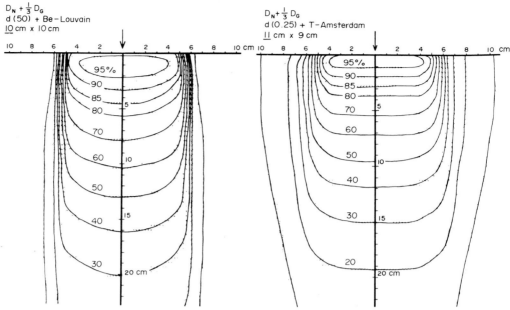

Fig. 3. Isodose distributions for two different neutron beams for the biologically effective dose $D_N + D_G/3$.

distributions in a similar way as for photon beams. Figure 3 is showing isodose charts of the Louvain d(50)+Be cyclotron beam and the Amsterdam d+T generator for the biologically effective dose: $D_N + D_G/3$. Better penetration and a sharper penumbra of the therapy installation with the higher neutron energy are evident. The better depth dose is especially advantageous for pelvic irradiations where a two field irradiation with the d+T source will give a relative large dose just below the surface. This could indicate the need for a multiple-field or rotation-beam therapy. The smaller penumbra of cyclotron neutrons is advantageous during irradiations where critical organs are not far from the target area, e.g. the spinal cord during irradiation in the head and neck region.

Conclusions

The analysis of dose distributions of fast neutron beams for clinical applications has shown the need for consistency in dosimetry procedures, including the choice of phantom material and the introduction of a biologically effective dose which can arbitrarily be taken as $D_N + D_G/3$. The introduction of a single dose value, directly related to the anticipated biological effect, is essential for adequate treatment planning, especially in view of the different photon contributions for different beams. The distributions observed with the higher-energy cyclotron neutron beams are preferable to d+T neutron sources due to a better penetration and a smaller penumbra. However, the latter type of machines has the advantage of a movable beam, compactness of the facility and relatively simple operation and maintenance. It is difficult to predict whether the limited neutron yields of d+T neutron machines will be outweighed by the better dose distributions of high-energy cyclotron beams despite their limitation of a fixed horizontal or vertical beam.

Acknowledgements—The authors wish to express their gratitude to the following physicists in providing them with their most recent, sometimes not yet published data: Drs. H. Abel, S. Matschke and K. Regel (Dresden); Dr. A. Ito (Tokyo); Drs. E. Maier and C. Huedepohl (Essen); Drs. J. R. Williams and D. E. Bonnett (Edinburgh); Drs. D. K. Bewley and J. Parnell (London); Drs. J. Eenmaa and K. Weaver (Seattle); Dr. Kawashima (Chiba-shi); Drs. L. August, R. B. Theus and P. Shapiro (Washington); Drs. A. R. Smith, P. R. Almond, J. B. Smathers and V. A. Otte (Houston); Dr. M. Awschalom (Batavia); Dr. R. C. Lawson (Glasgow); Drs. A. Hesz, E. Magiera and R. Schmidt (Hamburg); Dr. K. H. Höver (Heidelberg); Drs. D. Greene and P. C. Williams (Manchester); Drs. R. van der Laarse, P. A. Visser and Tj. Wieberdink (Amsterdam).

References

BATTERMANN, J. J. and K. BREUR (1978) Results of fast neutron radiotherapy at Amsterdam. In *Proceedings this meeting*, pp. 17–22.

BEWLEY, D. K., N. J. MCNALLY and B. C. PAGE (1974) Effect of the secondary charged particle spectrum on cellular response to fast neutrons. *Rad. Res.* **58**, 111-121.

BEWLEY, D. K. and B. C. PAGE (1978) On the nature and significance of the radiation outside the beam in neutron therapy. *Brit. J. Radiol.* **51**, 375-380.

BROERSE, J. J., G. W. BARENDSEN and G. R. VAN KERSEN (1968) Survival of cultured human cells after irradiation with fast neutrons of different energies in hypoxic and oxygenated conditions. *Int. J. Radiol. Biol.* **13**, 559-572.

BROERSE, J. J., G. BURGER and M. COPPOLA (1978) *A European Neutron Dosimetry Intercomparison Project (ENDIP)*, EUR 6004 (Commission of the European Communities, Luxembourg).

FRANKE, H. D., A. HESZ, E. MAGIERA and R. SCHMIDT (1978) The neutron therapy facility (DT, 14 MeV) at the Radiotherapy Department of the University Hospital Hamburg-Eppendorf. *Strahlentherapie* **154**, 225-232.

ICRU (1977) International Commission on Radiation Units and Measurements, *Neutron Dosimetry for Biology and Medicine*, Report 26.

MIJNHEER, B. J., J. ZOETELIEF and J. J. BROERSE (1978) Build-up and depth dose characteristics of different fast neutron beams relevant for radiotherapy. *Brit. J. Radiol.* **51**, 122-126.

MIJNHEER, B. J., P. A. VISSER and Tj. WIEBERDINK (1978) Clinical neutron dosimetry at the Amsterdam fast neutron therapy facility. In *Proc. Third Symp. on Neutron Dosimetry in Biology and Medicine*, EUR 5848, pp. 203-225 (Commission of the European Communities, Luxembourg).

ZOETELIEF, J., J. J. BROERSE and B. J. MIJNHEER (1978) Characteristics of ionization chambers and GM-counters employed for mixed field dosimetry. In *Proc. Third Symp. on Neutron Dosimetry in Biology and Medicine*, EUR 5848, pp. 565-578 (Commission of the European Communities, Luxembourg).

Dosimetry intercomparisons and protocols for therapeutic applications of fast neutron beams

J. J. BROERSE*, B. J. MIJNHEER†, J. EENMAA‡ AND P. WOOTTON‡

*Radiobiological Institute TNO, Rijswijk, The Netherlands
†Antoni van Leeuwenhoek Hospital, Amsterdam, The Netherlands
‡Dept. of Radiology, University of Washington, Seattle, Washington, U.S.A.

Abstract—*For an adequate evaluation and comparison of clinical results obtained at different centers, the neutron dosimetry procedures should be consistent. A considerable number of neutron dosimetry intercomparisons has been performed during the past 5 years; the general conclusions resulting from these intercomparisons will be summarized. Protocols for neutron dosimetry for radiotherapy have been drafted for the European and American groups involved in fast neutron radiotherapy. The highlights of the protocols will be discussed and the differences between them will be reported. Although adoption of uniform basic parameters is desirable, it seems equally important to standardize the experimental techniques employed for the determination of absorbed dose. The introduction of a secondary standard or a transfer dosimeter would be of great importance for the consistency of neutron dosimetry for biological and medical applications.*

Introduction

A common basis for neutron dosimetry will be one of the essential requirements for the comparison of clinical results obtained at the increasing number of fast neutron radiotherapy facilities. Cooperation between the physicists of clinics involved in fast neutron radiotherapy has been established in Europe by the Fast Particle Therapy Project Group under the sponsorship of the European Organization for Research on Treatment of Cancer, EORTC (Breur, 1974) and in the United States by the Fast Neutron Beam Dosimetry Physics Task Group of the American Association of Physicists in Medicine, AAPM. Both groups quote ICRU report 26 (1977) as the main source for information on methodology and basic physical data for neutron dosimetry. Protocols for neutron dosimetry for radiotherapy have been drafted for the European (Broerse and Mijnheer, 1976) and the American groups (Eenmaa and Wootton, 1978); the main items of these protocols will be summarized. It should be emphasized that efforts are being continued to improve the data base of neutron dosimetry. A few examples of further progress and new experimental results will be given. The international intercomparisons have shown that, in addition to inconsistencies in basic physical parameters, there are also large systematic differences in measurement procedures followed. The possibilities for the introduction of a reference TE ionization chamber will be discussed.

Principles of dosimetry in mixed fields

Neutron fields are always accompanied by gamma-rays originating from the neutron-producing target, from the primary shielding and field-limiting system and from the biological object or phantom being irradiated. The proportion of the total absorbed dose due to the photon component of the mixed field increases markedly with increasing depth of penetration of the incident beam in a phantom.

Because of the differences in biological effectiveness of these two radiation components (which may depend on the specific biological end-point), it is necessary to separately determine the neutron aborbed dose in tissue, D_N, as well as the gamma-ray absorbed dose, D_G, of the radiation field. Two instruments are generally used for the evaluation of the component radiations. One of these will have approximately the same sensitivity to neutrons and photons, while the second instrument, in order to minimize uncertainties in the analysis, will have a reduced neutron sensitivity relative to photons. The dose components in the mixed field can be derived from:

$$R_T = k_T D_N + h_T D_G,$$
$$R_U = k_U D_N + h_U D_G.$$

(In these equations, the subscript T refers to the tissue-equivalent device and the subscript U refers to the neutron insensitive device). R_T and R_U are the quotients of the responses of the two dosimeters in the same mixed beam relative to their sensitivities to the gamma-rays used for the photon calibration. Similarly, k_T and k_U are the sensitivities of each dosimeter to neutrons relative to its sensitivity to the gamma-rays used for calibration and h_T and h_U are the sensitivities of each dosimeter to the photons in the mixed field relative to its sensitivity to the gamma-rays used for calibration.

It is recommended that calibrated tissue-equivalent chambers be preferably used for measuring the tissue kerma in air and the absorbed dose in a phantom. The principle of the use of ionization chambers in neutron dosimetry does not differ from their use in photon dosimetry. The sensitivity, k_T, is determined by the average energy expended to create an ion pair, $\overline{W}$, the dose-conversion factor of chamber TE wall to gas for secondary charged particles created by incident neutrons or photons, S_r, and the ratio of the kerma in tissue to that in the dosimeter material.

Results and discussion of dosimetry intercomparisons

To predict the responses of irradiated biological objects, it is important to determine the energy dissipation with a sufficient degree of precision and accuracy. There are indications that, for biological and clinical applications, the absorbed dose in the biological specimen should be determined with a precision better than ± 2 per cent and an overall uncertainty of less than ± 5 per cent (ICRU, 1976; Broerse, 1977). The most important motive for the performance of dosimetry intercomparisons is to allow an adequate evaluation and comparison of biological and clinical results obtained by different groups. Additional aims of neutron dosimetry intercomparisons are to obtain information on the present adequacy of neutron dosimetry and on the advantages, disadvantages, corrections and systematic errors involved in the various methods. They should also provide participants with the possibility of checking the precision and accuracy of their methods under well-defined and standardized irradiation conditions.

A comprehensive review of all dosimetry intercomparisons performed up to 1976 can be found elsewhere (Almond and Smathers, 1977). The different studies can be divided into two classes: intercomparisons of dosimetry systems performed at specific locations and intercomparisons of the neutron beams actually used for radiotherapy. During the first class of intercomparisons, such as the International Neutron Dosimetry Intercomparison, INDI (ICRU, 1978), and the European Neutron Dosimetry Intercomparison Project, ENDIP (Broerse et al., 1978), all participants brought their systems to central locations (Brookhaven National Laboratory for INDI and Institut für Strahlenschutz GSF and Radiobiological Institute TNO for ENDIP). In the second class of intercomparisons, including those carried out under the auspices of the American Neutron Dosimetry Physics Group, NDPG (Smathers et al., 1976) and of the Japan–U.S. Cooperative Cancer Research Program, CCRP (Ito, 1978), the dosimetry systems were taken to each institution and, where possible, reciprocal visits among institutes were made.

The INDI and ENDIP studies were generally performed with monoenergetic neutron beams produced by the p+T, d+D and d+T reactions at relatively low kerma rates which varied between 5 and 80 rad/h. The quantities to be intercompared were the soft tissue kerma free-in-air and the aborbed dose at three depths in a water phantom for the neutrons and photons, respectively. The results of INDI and ENDIP were basically the same. Although it is recognized that the calculation of a mean value for the participants' results has limited relevance, this procedure was used in order to allow a quantitative comparison of the ENDIP results. In Table 1 the results of the groups participating in the ENDIP sessions at GSF and TNO are grouped for relative differences from the mean between 5 per cent, from 5 to 10 per cent and in excess of 10 per cent. A complete analysis of the ENDIP results has been made (Broerse, et al., 1978); only a few main conclusions will be repeated in the present review:

(a) In general, the variations in total kerma and total dose are smaller than those observed for the neutron kerma and aborbed dose. For the measurements at TNO, the results showed relatively small variations for the free-air condition;

Table 1. Number of evaluated ENDIP results with relative differences, Δx from the mean (Broerse et al., 1978)

Site of inter-comparison	Neutron energy (MeV)	Condition		$\Delta x \leq 5\%$	$5\% < \Delta x \leq 10\%$	$\Delta x > 10\%$
GSF	15.1	free air	K_N	6/12	3/12	3/12
			K_{tot}	8/12	2/12	2/12
GSF	5.25	free air	K_N	7/11	3/11	1/11
			K_{tot}	8/11	3/11	0/11
TNO	15	free air	K_N	11/12	1/12	0/12
			K_{tot}	10/13	3/13	0/13
TNO	15	5 cm depth	D_N	6/12	5/12	1/12
			D_{tot}	10/13	2/13	1/13
TNO	15	10 cm depth	D_N	5/12	6/12	1/12
			D_{tot}	9/13	3/13	1/13
TNO	15	20 cm depth	D_N	3/12	8/12	1/12
			D_{tot}	9/13	3/13	1/13
TNO	5.5	free air	K_N	8/8	0/8	0/8
			K_{tot}	8/9	1/9	0/9
TNO	5.5	5 cm depth	D_N	4/8	4/8	0/8
			D_{tot}	6/9	3/9	0/9
TNO	5.5	10 cm depth	D_N	2/8	5/8	1/8
			D_{tot}	7/9	2/9	0/9
TNO	5.5	20 cm depth	D_N	4/8	2/8	2/8
			D_{tot}	7/9	2/9	0/9

however, larger variations are observed for measurements in the phantom, especially for the neutron-absorbed dose.

(b) For the in-phantom conditions, the deviation of the results from the average value for neutron kerma or aborbed dose show standard deviations in the order of 7 to 8 per cent. These variations seem to be in accordance with the relatively large systematic uncertainties quoted by the participants. Only for a few specific situations were maximum differences up to 20 per cent observed.

(c) The values reported for the gamma-ray kerma and absorbed dose showed large variations up to 100 per cent from the mean value. These variations are not acceptable for the measurements in the phantom where relatively large photon contributions have been measured (up to 25 per cent of the total aborbed dose).

(d) In the ENDIP and the INDI studies, the participants employed a variety of dosimetry systems and they applied divergent basic physical parameters characterizing the detector response for identical experimental conditions. Analysis of the responses of homogeneous tissue-equivalent ionization chambers showed that, in addition to inconsistencies in basic parameters, there are considerable systematic differences in measurement procedures.

The NDPG and CCRP measurements have been made primarily on cyclotron-produced neutrons with fairly high dose rates for tissue

Table 2. Intercomparison results: ratios of measurements of total dose or kerma with respect to Tamvec measurements during the autumn of 1975 and February 1976 (Smathers et al., 1976)

Participants	CHHRI TAMVEC	MRC TAMVEC	NRL TAMVEC	Louvain TAMVEC	Chiba TAMVEC	Chiba TAMVEC
Location	CHHRI	MRC	NRL	Louvain	TAMVEC	TAMVEC
Beam	d+T	d(16)+Be	d(35)+Be	d(50)+Be	d(30)+Be	d(16)+Be
Tissue kerma in air	0.95[a]	1.01	1.00		1.01	1.01
Dose at depth		0.97 (1 cm)		1.02 (2 cm)	0.97 (5 cm)	
		0.96 (5 cm)		1.02 (10 cm)	0.96 (10 cm)	
		0.97 (10 cm)			0.97 (15 cm)	
					0.96 (20 cm)	
Photon calibration			1.00 (^{137}Cs)	1.00 (^{60}Co)	0.99[b] (^{60}Co)	

[a] Choice of parameters account for 4% of the 5% difference.
[b] Ration of the calibration made in Japan to calibration made at M. D. Anderson Hospital.

Table 3. Summary of dosimetry intercomparisons between U.S. institutions (Wootton and Eenmaa, 1978)

	(1974-1975)				(1977-1978)	
Participants	U.W. NRL MDAH–TAMU	NRL U.W. FNL	FNL GLANTA	GLANTA NRL FNL MDAH–TAMU	MDAH–TAMU GLANTA	S/K U.W.
Location	U.W. NRL MDAH–TAMU	FNL	GLANTA	FNL	GLANTA	U.W.
Beam	d(21.5)+Be d(35)+Be d(50)+Be	p(66)+Be	d(25)+Be	p(66)+Be	d(25)+Be	d(21.5)+Be
Detectors	TE ion chambers	TE I.C.	TE I.C.	TE I.C.	TE I.C.	TE I.C. TE calorimeter
Tissue kerma in air	99%	>99.5%	99.5%	>99.5%	>99.9%	—
Dose at depth in phantom	99% at 2 cm 98% at 10 cm	—	—	—	—	>98% at 5 cm
^{60}Co calibration	99%	>99.5%	99%	>99.5%	>99%	99.5%

kerma in air and aborbed dose at depth. The tissue equivalent liquid as suggested by Frigerio et al. (1972) with a density of 1.07 is used as phantom solution for these studies. An example of the measurements of total dose and kerma performed by the NDPG is given in Table 2. The intercomparisons of neutron dosimetry at the institutes cooperating within the NDPG in the United States showed an even better result: all of these groups agree within 2 per cent in their total dose measurements (see Table 3). It should be emphasized, however, that the American groups involved in neutron radiotherapy all used a common set of commercial TE ionization chambers. Typically, the NDPG uses a 1.0 cm³ spherical chamber as the primary calibration instrument for measurements of neutron tissue kerma in air and aborbed dose in a phantom and a 0.1 cm³ cylindrical chamber for spatial dose-distribution measurements.

Dosimetry protocols

The cooperation between the physicists of the EORTC-sponsored Fast Particle Therapy Project Group in Europe and within the NDPG in the United States was established to ensure compatibility of the physical dosimetry among therapy centers engaged in coordinated trials of fast neutron beam therapy and to assist the cancer therapy effort in general by the development of a fast neutron beam dosimetry protocol. Both groups have, in a more or less independent manner, drafted a number of recommendations concerning basic physical parameters for neutron dosimetry, the nature of phantoms to be used in depth-dose measurements and determination of displacement correction factors. The protocols produced to date (Broerse and Mijnheer, 1976; Eenmaa and Wootton, 1978) have not yet received the final approval of the respective European and American participants. This is partly due to insufficient knowledge of quantitative values of the physical parameters involved. New experimental results have recently become available on, for example, the average energy to produce an ion pair (Bichsel and Rubach, 1978; Goodman, 1978), the correction for displacement for in-phantom measurements with ionization chambers (Zoetelief et al., 1979) and the relative neutron sensitivity, k_U, for Geiger–Müller counters (Mijnheer et al., 1979) and nonhydrogenous ionization chambers. A number of authors, including Ito (1978) and Mijnheer et al. (1975), accounted for the displacement by taking the effective measuring-point at approximately three-quarters of the radius of the gas cavity of the chamber in front of the geometrical centre in direct accordance with the results for electrons and photons. If the displacement correction is confined to only one-quarter of the radius, this will imply that the absorbed dose quoted at a reference point has to be increased (for measurements with a TE ion chamber of 8 mm internal radius, a correction of approximately 5 per cent should be applied for d+T neutron beams in phantom). Measurements and calculations of k_U for nonhydrogenous ion chambers, including the studies of Makarewicz and Pszona (1978), indicate a continuously increasing relative neutron sensitivity with increasing neutron energy, contrary to previous results obtained by others (see Fig. 1). These examples are only included to stress the need for continuous evaluation of neutron dosimetry data

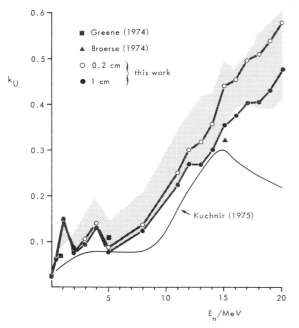

Fig. 1. *Relative neutron sensitivity of a C–CO_2 ionization chamber as calculated by Makarewicz and Pszona (1978) for two spacing distances between the electrodes.*

and photons are compared in Table 4. One of the essential differences between the European and the U.S. neutron dosimetry protocols is that the European group recommended that water be used as the standard phantom material for the following reasons: (a) its neutron attenuation properties are rather similar to those of most biological tissues; (b) its composition is constant and it is available in all centers; (c) it is the medium on which a considerable body of information on the spatial distribution of absorbed dose in photon and electron therapy is based. The U.S. groups, on the other hand, prefer to use tissue equivalent liquid as the standard phantom material with the intent of correcting for the presence of fat and other tissues of different density and, even more important, of different elemental composition, on the absorbed dose delivered to tissues beyond the inhomogeneity.

The European and American dosimetry physics groups are in close contact and maintain a continuous exchange of viewpoints. In order to facilitate the comparison of biological and clinical results, both groups agreed to include the following information in physical protocols describing the neutron dosimetry methods and irradiation conditions:

1. *Beam quality.* The energy of the accelerated particle, target characteristics

and for a substantial extension of our knowledge especially for the higher neutron energies.

A number of important items included in the present-day dosimetry protocols for neutrons

Table 4. *Some main items of dosimetry protocols for fast neutrons (EORTC and AAPM) and photons (HPA)*

Item	Fast Particle Therapy Project Group, EORTC, Europe	Fast Neutron Beam Dosimetry Physics Task Group, AAPM, United States	Code of practice, X- and gamma-rays, HPA, England
Description of absorbed dose	dose in ICRU muscle tissue, total dose with neutron dose in brackets	dose in ICRU muscle tissue, total dose with photon component in brackets	dose in water
Phantom material	water	TE liquid according to Frigerio *et al.* (1972)	water
Kerma ratios	from ICRU 26	from ICRU 26	—
Correction for attenuation and scattering in the phantom	different procedures applied	multiplication factor according to Shapiro *et al.* (1976)	included in G_λ
$\overline{W}_C/\overline{W}_N$	0.95 for TE gas	0.94 for TE gas, 0.93 for TE/air	—
Dose conversion from gas to wall	mass stopping power ration of 1 with homogeneous chamber	dose conversion factor S_r, function of gas, wall, chamber size and neutron energy (for 1-cm³ TE chamber 0.98)	—
Neutron-insensitive detector	preferably $k_U < 0.15$	tabulated values for C/CO_2 and Al/Ar ion chamber and Geiger–Müller counter	—

and neutron energy should be specified. Attempts should be made to obtain information on changes in radiation quality with depth.
2. *Dose specification.* The neutron and photon absorbed dose should be specified at a reference point, e.g. site of the tumor, and the material for which the absorbed dose has been quoted should be mentioned.
3. *Beam geometry.* Information on the geometrical conditions of the irradiation should be provided, e.g. source to surface distance, field size, phantom dimensions and material, etc.
4. *Penumbra and collimation.* The design and material of the collimators may affect the dose distribution at the edge of the beam and this effect should be investigated. Where relevant, the absorbed dose to shielded sites should be specified.
5. *Field intensity distribution.* The degree of field uniformity should be established. Scattering material may be introduced into the beam path to obtain more flattened isodose curves. The effect of beam flatteners on radiation quality should be determined.

Discussion and conclusions

The dosimetry intercomparisons performed among the European institutes involved in fast neutron therapy have shown larger variations among the participants than the results obtained among the American groups. In Europe, the applications of fast neutrons in clinical radiotherapy increased gradually as compared with the United States, where three institutes had to take instantaneous action. For this reason, the American institutes purchased their dosimeters from the same commercial firm and adopted one common ionization chamber. In Europe, the institutes developed and constructed their own dosimeters; these unavoidably show some mutually different characteristics. It has to be concluded, however, that, unless neutron dosimetry groups adopt one reference ion-chamber system, considerable efforts will have to be expended in performing intercomparisons and worrying about the correction factors to be applied for different chambers. Although each institute should be free to develop instruments of special design, it has been recommended (Broerse and Mijnheer, 1976) that all European groups use the same type of secondary standard to check their other dosimeters. It is preferable that an international body be involved in the supervision of the production and use of such a common dosimeter system. At the initiative of the Commission of the European Communities, a committee which is involved in the collection and evaluation of neutron dosimetry data (CENDOS) has been formed (Broerse, 1978). One of the future tasks of the CENDOS committee will be to introduce a set of reference TE ion chambers for neutron dosimetry in biology and medicine. In connection with this program, information is presently being collected on the characteristics of eight different types of tissue-equivalent ionization chambers (Broerse, 1979). In addition to the introduction of a secondary standard or transfer dosimeter, a consistent set of basic physical parameters should be used and agreement should be reached on the procedures to convert instrument response into absorbed dose values.

For an evaluation of fast neutron radiotherapy results obtained at different centers, it will be essential to compare the absolute neutron-dosimetry methods, preferably by direct on-site measurements. Possible differences in radiation quality can be demonstrated by comparing the response of a biological dosimeter in the different beams. If a standard biological system is being taken to the different institutes, a uniform dosimetry system should accompany the radiobiology. The dosimetry procedures followed for the irradiation of the biological dosimeter should also be standardized; irradiation of cells through the medium or through the polystyrene bottom of the culture dish will result in considerable differences in energy deposited in the cell (Broerse and Zoetelief, 1977). If the dosimetry procedures are consistent, variations in the response of the biological dosimeter should reveal true radiobiological differences in the beams. It should be emphasized, however, that, for any specific neutron beam, the appropriateness of the various dose-fractionation schedules will be finally judged by the clinical observations of tumor response and degree of radiation damage to normal tissues.

References

ALMOND, P. R., and J. B. SMATHERS (1977) Physical intercomparisons for neutron radiation therapy. *Int. J. Radiat. Oncol.* **3**, 169-176.

BICHSEL, H., and A. RUBACH (1978) Uncertainty of the determination of absolute neutron dose with ionization chambers. In G. BURGER and H. G. EBERT (Eds.), *Proc. Third Symp. Neutron Dosimetry in Biology and Medicine*, EUR 5848, Commission of the European Communities, Luxembourg, pp. 549-563.

BREUR, K. (1974) International cooperation with regard to clinical trials of fast neutron radiotherapy. *Europ. J. Cancer* **10**, 385-386.

BROERSE, J. J. (1977) Standards in medical neutron dosimetry. In C. D. BOWMAN, A. D. CARLSON, H. O. LISKIEN and L. STEWART (Eds.), *Proc. Symp. Neutron Standards and Applications*, Special publication 493, National Bureau of Standards, Washington, pp. 106-114.

BROERSE, J. J. (1978) CENDOS—A coordinated program on collection and evaluation of neutron dosimetry data. In G. BURGER and H. G. EBERT (Eds.), *Proc. Third Symp. Neutron Dosimetry in Biology and Medicine*, EUR 5848, Commission of the European Communities, Luxembourg, pp. 107-112.

BROERSE, J. J. (1979) Compilation of characteristics of tissue equivalent ionization chambers, CENDOS report, in preparation.

BROERSE, J. J., G. BURGER and M. COPPOLA (1978) *A A European Neutron Dosimetry Intercomparison Project (ENDIP), Results and Evaluation*, EUR 6004, Commission of the European Communities, Luxembourg.

BROERSE, J. J. and B. J. MIJNHEER (1976) Second draft protocol for neutron dosimetry. In J. J. BROERSE (Ed.), *Basic Physical Data for Neutron Dosimetry*, EUR 5629, Commission of the European Communities, Luxembourg, pp. 311-321.

BROERSE, J. J. and J. ZOETELIEF (1978) Dosimetric aspects of fast neutron irradiations of cells cultured in monolayer. *Int. J. Radiat. Biol.*, **33**, 383-385.

EENMAA, J. and P. WOOTTON (1979) Protocol for Neutron Beam Dosimetry (in preparation).

FRIGERIO, N. A., R. F. COLEY and M. J. SAMPSON (1972) Depth dose determinations. I. Tissue-equivalent liquids for standard man and muscle. *Phys. Med. Biol.* **17**, 792-802.

GOODMAN, L. J. (1978) $\bar{W}_N$ computed from recent measurements of W for charged particles. In G. BURGER and H. G. EBERT (Eds.), *Proc. Third Symp. Neutron Dosimetry in Biology and Medicine*, EUR 5848, Commission of the European Communities, Luxembourg, pp. 61-74.

HPA (1969) A code of practice for the dosimetry of 2 to 35 MV X-rays and caesium-137 and cobalt-60 gamma-ray beams. *Phys. Med. Biol.* **14**, 1-8.

ICRU (1976) *Determination of Absorbed Dose in a Patient Irradiated by Beams of X- or gamma-rays in Radiotherapy Procedures*, ICRU report 24, International Commission on Radiation Units and Measurements, Washington.

ICRU (1977) *Neutron Dosimetry for Biology and Medicine*, ICRU report 26, International Commission on Radiation Units and Measurements, Washington.

ICRU (1978) *An International Neutron Dosimetry Intercomparison*, ICRU report 27, International Commission on Radiation Units and Measurements, Washington.

ITO, A. (1978) Neutron dosimetry intercomparisons between Japan (University of Tokyo) and U.S.A. In G. BURGER and H. G. EBERT (Eds.), *Proc. Third Symp. Neutron Dosimetry in Biology and Medicine*, EUR 5848, Commission of the European Communities, Luxembourg, pp. 113-124.

MAKAREWICZ, M. and S. PSZONA (1978) Theoretical characteristics of a graphite ionization chamber filled with carbon dioxide. *Nucl. Instrum. Meth.* **153**, 423-428.

MIJNHEER, B. J., J. E. BROERS-CHALLISS and J. J. BROERSE (1975) Measurements of radiation components in a phantom for a collimated d–T neutron beam and H. G. In G. BURGER, H. SCHRAUBE EBERT (Eds.), *Proc. Second Symp. Neutron Dosimetry in Biology and Medicine*, EUR 5273, Commission of the European Communities, Luxembourg, pp. 423-443.

MIJNHEER, B. J., P. A. VISSER, V. E. LEWIS, G. GULDBAKKE, H. LESIECKI, J. ZOETELIEF and J. J. BROERSE (1979) The relative neutron sensitivity of Geiger-Müller counters. These proceedings, pp. 162–163.

SHAPIRO, P., F. H. ATTIX, L. S. AUGUST, R. B. THEUS and C. C. ROGERS (1976). Displacement correction factor for fast-neutron dosimetry in a tissue-equivalent phantom. *Med. Phys*, **3**, 87-90.

SMATHERS, J. B., P. R. ALMOND, V. A. OTTE and W. H. GRANT (1976). Dosimetry intercomparisons between fast neutron radiotherapy centers in the United States and Europe. In J. J. BROERSE (Ed.), *Basic Physical Data for Neutron Dosimetry*, EUR 5629, Commission of the European Communities, Luxembourg, pp. 267-270.

WOOTTON, P. and J. EENMAA (1978) private communication.

ZOETELIEF, J., A. C. ENGELS, J. J. BROERSE, B. J. MIJNHEER and P. A. VISSER (1979) Effective measuring point for in-phantom measurements with ion chambers of different sizes. These proceedings, pp. 169-170.

Characteristics of fast neutron sources

J. B. SMATHERS

Bioengineering Division, Texas A & M University, College Station, Texas 77843, U.S.A.

Introduction

The opening address of the conference by Mr. Van Spiegel, Director General for Science Policy of The Netherlands Government, posed a question which continued to resurface during the discussion of the Poster Session on the Characteristics of Fast Neutron Sources. Succinctly put, the question is, "Can governments afford to spend increasingly higher percentages of the national income on medical care?" With this question posed to the attendants, the search for the elusive answer as to what is the optimum neutron treatment source for neutron therapy, machine and reaction, was found to require the inclusion of a much broader spectrum of considerations than simply depth dose, dose rate and neutron/gamma dose ratio.

The economic aspects of this consideration might be based on a cost per rad analysis which includes total costs associated with the acquisition, operation, maintenance and disposal of the machine at the end of its useful life. Surprisingly enough, no one at the conference knew of such an analysis for the neutron sources in use yet alone for sources proposed for use.

The session on neutron sources can be subdivided into three catagories: deuterium–tritium, D–T reaction sources, cyclotron-based sources using either protons or deuterons impinging on targets of either beryllium or lithium, and other sources, such as reactors or linear accelerators.

D–T sources

Of the sources presently in clinical use, those using the D–T reaction have resulted in facilities which have been less expensive to construct than other sources and for this reason receive continued attention. The greatest limitation of the D–T source at this time is the low neutron yield allowed by present accelerator–target designs. A number of papers at the conference described the present status of existing designs which will be considered in the succeeding paragraphs.

The mass analyzed, D^+ ion only, beam on a tritium metal hydride target design is typified by the new materials testing accelerator being designed by Lawrence Livermore Laboratory, U.S.A. (Davis). In this design a 150mA D^+ ion beam with a 1-cm diameter spot size incident on a water-cooled metal hydride target containing 5000 curies of tritium and rotating at 5000 rpm is expected to yield 4×10^{13} n/s. This is equivalent to 1 Gy/min (100 rad/min) at 1.25 m TSD. As the new system is a scale up of existing D–T accelerator sources of similar design, no problems were anticipated in reaching the design specifications (Booth and Barschall, 1972; Booth *et al.*, 1977). This will be the most intense D–T source in operation when construction is completed and the useful lifetime of the target of 50 to 200 hours could be greatly extended if a 2-4-cm diameter spot size were used, because of the reduced local heating. Surprisingly enough, little interest by either government or industry has been shown in developing this system into a clinical neutron therapy source.

The power dissipation limitations of occluded metal hydride targets are minimized in the differentially pumped tritium gas target being developed at the University of Wisconsin, U.S.A. by DeLuca and colleagues. The design allows an order of magnitude reduction in beam current from that used on metal hydride targets for an equivalent neutron output. However, the contamination of the tritium target by the accelerated deuterium results in a target (neutron output) half-life of only 2 hours, with little hope of improvement. The gas target can be remotely changed in about only a minute though, so the half-life is of no consequence were it not for the tritium-disposal problem. At this time it is not economical to isotopically separate the waste target gases and recycle the tritium. Source strengths of 3×10^{12} n/s have been achieved and a maxium design output of 4.8×10^{12} n/s is anticipated for the 0.61 kPa m target.

The sealed-tube system accelerating both T^+ and D^+ ions is a concept which has been readily accepted by the medical community for hospital utilization because of the ease of changing targets and the avoidance of breaking the tritium containment system during target change. A neutron yield of 1×10^{12} n/s or approximately 0.05 Gy/min (5 rad/min) at 0.8 m TSD was reported for the Hiletron system by Greene, Gill and Jones. Tube lifetime is still to be determined but the six tubes in use have an average usage of 160 hours and all are still operational. To achieve adequate dose distributions for abdominal tumor volumes, the groups at Amsterdam and at Manchester are both using six-field "box" techniques which coupled with the low dose rate of the sealed tubes places a severe constraint on the number of patients that can be treated per day.

Several companies are pursuing the development of mixed beam D–T sources which incorporate a differential pumping system to recycle the gases back from the target to the source and maintain proper system pressure. This technique allows for greater neutron yields but at some increase in complexity of the system. The Cyclotron Corporation was reported to have achieved a yield of 5.6×10^{12} n/s using a dual source system on a single, two-sided target (Bloch, 1978). It was anticipated that a yield of 8×10^{12} n/s with satisfactory target lifetime will be achieved shortly using a four source–four target system which is in the final stages of construction.

Other research which will have some bearing on the final design of D–T systems are studies of such basic questions as what is the optimum metal hydride to use in targets. Research in progress at the Sandia Laboratories should yield answers to this question and the results will hopefully be reflected in increased target yields due to higher acceptable power loadings (Bacon *et al.*, 1977).

One method present systems use to increase the D–T yield or target lifetime is to spread the ion beam over a greater area on the metal hydride target and thus reduce the power/cm² dissipated in the target. Much has been said about the effect this spreading might have on the neutron beam penumbra but little has been said on what effect it might have on neutron energy flux at the treatment TSD. The paper presented by Gileadi addressed this point and the calculated results indicate that more than penumbra may be effected by increasing the effective target area. For the spherical geometry of a 15-cm void and a collimator composed of 30 cm each of iron and polyethylene, the calculations showed that the incident neutron energy per cm² at 95 cm TSD decreased by a factor of 2 when the target area was changed from a point source to a 3-cm radius disc source (28.3 cm²). The effective target area of 28 cm² far exceeds that proposed by any designer to date. However, the implications the calculations have for smaller target areas are yet to be determined.

For those systems in which a nonanalyzed deuterium ion beam is accelerated, a method to increase the target lifetime by a factor of 3 with no decrease in neutron yield was reported by Hess and Franke. While an analyzed D^+ beam is optimum, a partially analyzed beam separating the D^+, D_2^+ and D_3^+ ions into separate spots on the target results in a factor of 3 improvement in target half-life compared to a nonanalyzed beam on the same target. Using this technique each ion produces neutrons in proportion to its cross-section for doing so but there is no poisoning of the target by one ion and thus there is no interference with the production of neutrons by the other ions. This modification was a retrofit to the existing accelerator and should not be confused with the optimum situation of a D^+ ion analyzed beam which results in maximum neutron yield and target life for a given target power loading per unit area.

Cyclotron sources

The newer cyclotron facilities are reflecting in their design the desire on the part of radiotherapists to eliminate patient set-up problems previously encountered in the use of horizontal beams. The new facility at Louvain-la-Neuve has a fixed vertical beam with a grate-covered pit immediately below the table to minimize transmitted neutrons scattering within the room and floor activation (Wambersie, Meulders and Winant). Certainly the pit cannot but help to reduce room activation and the degree to which this is achieved will no doubt have a strong influence on future facility designs.

Three facilities reported on the use of isocentric units in conjunction with cyclotrons to give a neutron source with a clinical flexibility similar to that of low-LET sources. The Edinburgh isocentric unit, 15 MeV d^+, has had some problems with undesired activation near the target in the treatment head and field modification by the manufacturer resulted in an average reduction of activation levels by a factor of three (Bonnett and Williams). Induced activation of the treatment head still results in whole-body gamma-ray doses to the radiography staff of 2×10^{-6} Gy/treatment field (0.2 mrad/treatment field). This staff dose is presently limiting the number of patients that can be treated per day.

The isocentric unit at Essen, 14 MeV d^+, was constructed by the same manufacturer as that at Edinburgh. This unit routinely operates with 100 μA beam current and is reported to have target heating problems due to the small beam spot size on target (Maier and Huedepohl).

Initial tests on the isocentric unit to be installed at The King Faisal Specialist Hospital in Saudi Arabia have been completed for both 26 MeV p^+ and 15 MeV d^+ ion beams (Tom, Kuo and Hendry). Maximum induced head activation approximated by thirty 1- Gy (100 rad) treatments spaced over 8 hours yielded the following radiation levels 5 minutes after the end of the last irradiation at the position 20 cm in front of the treatment head; p^+–15 mR/hr, d^+-8 mR/hr.

The dose rate limited, lower energy, charged particle reactions force the designers to optimize the treatment head shielding to the point where indeed induced head activation may be a serious problem. The use of higher-energy charged particles with the accompanying higher dose rates and thus larger acceptable TSD distances might be the only way to allow for adequate shielding to be incorporated into the design. The compromise between dose rate and cost (charged particle energy) again confronts us.

In the design of cyclotrons the energy of the particles which can be accelerated is generally limited either by the maximum magnetic field strength or the maximum accelerating frequency of the RF system. For machines presently using the Be(d,n) reaction some interest has been shown in conversion to the Be(p,n) reaction because, providing the RF system has sufficient frequency range, protons with almost twice the energy are possible. Bewley and colleagues have compared 50 MeV d^+ on Be with 75 MeV p^+ on Be filtered by 5 cm polyethylene and found that the use of protons results in an improved depth dose, ~ 17.5 vs ~ 15 cm, and a dose rate/μA reduced by a factor of 3.3. For the proton energy range considered, 30-75 MeV, the use of the 5 cm of polyethylene filtration is equivalent to an increase in proton energy of approximately 15 MeV as reflected in the measured depth-dose distribution.

Lower energy neutrons, which the polyethylene filters are designed to selectively remove, are undesirable because of the adverse effect they have on the neutron-dose distribution. Lone and co-workers reported that theoretical considerations confirm experimental results that these neutrons are unavoidable as they are created by inelastic scatter and excitation within the target material, Be or Li. Experimental data indicating that the contribution of the lower energy tail diminishes with increasing charged particle energy is summarized in the Table 1 (Lone, Ferguson and Robertson).

Table 1. *Relative yield of neutrons below 2 MeV in energy*

Charged particle Energy	Deuterons	Protons
10 MeV	33%	70%
25 MeV	15%	49%

Other sources

Plans to use a nuclear reactor as a neutron source for the treatment of superficial tumors was described by Koester and co-workers. A tissue depth of 5.5 cm for the 50% depth dose was reported and the system allows great flexibility in varying the neutron/gamma dose ratio by use of selective transmission filters. For those long active in the field, this project brings back memories of the work at Brookhaven National Laboratory in the late 1950s.

The desire to develop a cost-effective hospital-based pion source has led to a research effort at Los Alamos referred to as the Pigmi Program. The conceptional design for a 650-MeV, 100-μA proton accelerator has been completed (Knapp). The accelerator would be 100 m long and cost an estimated $12 million U.S. dollars. Dose rates of 0.1 Gy/liter/min (10 rad/liter/min) in the conventional channel and 0.5 Gy/liter/min (50 rad/liter/min) in an enhanced solid angle channel are anticipated. A fall out of this effort may be a 40 MeV, 100 μA proton accelerator composed of the first two stages of the 650-MeV machine (Knapp). The 40-MeV proton accelerator would be but 19 m in length, including an isocentric unit, and yield a dose rate of 1 Gy/min (100 rad/min) at 1.25 m TSD. Cost is estimated at between 2 and 3 million U.S. dollars.

Summary

From the posters and the discussion which ensued, it was obvious that there are a number of different neutron sources being used and developed for use in neutron therapy. The basic question as to what minimum clinical physics criteria any source used in neutron therapy should meet was a subject of discussion but a consensus was not reached among those in attendance. One therapist expressed the opinion that neutrons must be at least cobalt equivalent. It was then pointed out that 4-6 MeV electron accelerators, at twice the cost of a cobalt unit, are replacing cobalt units throughout the United States because of the improved dose distribution they offer. Another expressed the opinion that an acceptable cost for a neutron source would be three to six times that of a high-energy electron therapy accelerator.

As cost is strongly related to the achievable physics characteristics of the neutron sources, it is difficult to separate the two. However, no neutron source will achieve the desirable physics characteristics of 4-6 MeV electrons and only the more expensive ones will achieve a depth dose similar to cobalt.

From the radiobiological studies to date, there is little data to support the selection of one energy cyclotron over another. If anything, the most promising clinical results are reported from a cyclotron at the low end of the energy scale.

With increasing amounts of money one buys higher dose rates and/or improved depth dose capabilities and thus the ability to treat deeper-seated tumors. Three million dollars was recently committed to build a hospital-based cyclotron neutron source (42 MeV p^+ on Be) in the U.S. Thus one can quote this figure as a presently acceptable cost for a neutron therapy source in the United States. How high this figure will go is anyone's guess.

References

BACON, F. M., A. A. RIEDEL, D. F. COWGILL, R. W. BICKES and J. E. BOERS (1977). *"Intense Neutron Source Target Test Facility: Second Semi-Annual Progress Report"*. SAND-1326.

BLOCH, P. (1978). Private Communication. University of Pennsylvania, Philadelphia, Pennsylvania.

BOOTH, R. J. C. DAVIS, C. L. HANSEN, J. L. HELD, C. M. LOGAN, J. E. OSCHER, R. A. NICHENSEN, D. A. TOHL and D. J. SHUMACHER (1977) *Nucl. Instrum. Meth.* **145**, 25-39.

BOOTH, R. and H. H. BARSCHALL (1972) *Nucl. Instrum. Meth.* **99**, 1-4

POSTER SESSION A

Contributions on characteristics of fast neutron sources

D. K. BEWLEY, B. PAGE (London, England), J. P. MEULDERS and M. OCTAVE-PRIGNOT (Louvain-la-Neuve, Belgium): *Neutron beams from protons on beryllium*

Modern cyclotrons can produce protons with nearly double the energy of deuterons. Neutron beams produced by protons on beryllium are therefore of great interest for radiotherapy. Neutron spectra from p and d on Be have very different shapes; with protons there is a much larger fluence of low-energy neutrons coupled with a maximum at around 20 MeV. Hydrogenous filters can therefore be more effective with p than with d on Be.

We have studied the penetration in water of neutrons produced by p on Be at the cyclotron at Louvain-la-Neuve. The field size was 12×12 cm² at 218 cm SSD. The results were converted to 125 cm SSD. The depth of the 50% isodose (D_n) is shown in Fig. 1 as a function of proton energy and thickness of polythene filter. A filter 5 cm thick increases the penetration about as much as an increase in proton energy of 15 MeV. With the filter the beam is more penetrating than that produced by deuterons of the same energy. In fact neutrons from 50 MeV d on Be have about the same penetration as neutrons from 60 MeV p on Be with no filter or 45 MeV p on Be with 5-cm filtration, and all are about equivalent to X-rays from a linear accelerator running at 3 MV (at 100 cm SSD).

An apparant disadvantage of protons is the lower dose rate per μA on the target, about a factor of 5 at the same particle energy. But as the cyclotron can produce protons at nearly double the energy of deuterons, the dose rate per μA at the maximum energy of each from the cyclotron is about the same. The power dissipated in the target is, however, greater for protons. With a 5-cm polythene filter the neutron intensity is reduced by a factor of 1.4 to 2 depending on energy. Use of a filter also produces a small increase in the partial dose of γ-radiation.

Finally, the higher neutron energy from p on Be also results in a more favourable build-up curve below the surface, with the maximum at a greater depth.

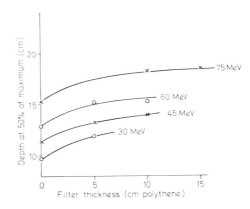

Fig. 1. *Depth of the 50% neutron isodose at 125 cm SSD as a function of proton energy and thickness of filter (Bewley et al.).*

D. E. BONNETT and J. R. WILLIAMS (Edinburgh, Scotland): *The isocentric fast neutron facility at Edinburgh*

Radiotherapy using cyclotron-produced fast neutrons was started in Edinburgh in March 1977 with a fixed horizontal beam. This facility has been described in detail by Williams et al., (1978). Recently (May 1978) treatments were begun in a second treatment room which houses an isocentric neutron therapy unit. Both the cyclotron (type CS30) and the isocentric unit were manufactured by The Cyclotron Corporation, Berkeley, California.

After extraction from the cyclotron, the deuteron beam (energy 15 MeV) enters the isocentric system and passes through a 45° bending magnet followed by a 135° bending magnet before hitting a thick beryllium target. Both bending magnets and the target are located on a gantry which can rotate around the patient with a target–axis distance of 125 cm. Around the target area the shielding consists of Benelex, a high-density compressed wood (Masonite Corporation) with some lead added to the outside in a few strategic positions.

At 125 cm TSD a maximum dose rate of 43 rad/min can be achieved with a deuteron beam current of 100 µA, although to increase target life a dose rate of 26 rad/min is used at present. Interchangeable collimators made of Benelex are used to define field size, and polythene wedges, with nominal angles of 25°, 35° and 45°, are also available. The isodose and depth dose data for the unit are the same as measured for the fixed horizontal beam by Williams et al. (1978).

Radiation leakage through the shielding, and reaching the patient, was assessed with the collimator port blocked with Benelex. Measurements were made along a 3-m line passing through the isocentre and perpendicular to the plane of rotation of the target. The neutron leakage was found to be less than 0.2% of the total dose at the isocentre using an 11 × 11-cm² field size and the gamma dose was less than 0.7% of the total dose.

Prior to the commencement of therapy an extensive programme of measurements and modifications was carried out, with the active co-operation of the manufacturers, in order to reduce the levels of neutron-induced radioactivity in the isocentric unit. The major source of activity was in materials in the immediate vicinity of the beryllium target and this was reduced in three ways, viz.

1. the installation of boron-loaded rubber around the inner walls of the target area to absorb thermal neutrons produced by the Benelex;
2. the replacement of materials with large thermal and fast neutron cross-sections;
3. the installation of additional shielding where necessary.

The reduction in radiation levels varied with position around the head with the greatest reduction being by a factor of 120. In most cases, however, an improvement factor of approximately 3 was achieved.

Since the start of therapy the doses received by radiography staff have been closely monitored. At present between four and eight patients are treated daily on the isocentric unit, with an average of 2.9 fields per patient. After an initial period of familiarization with the unit the whole-body gamma-ray doses to radiography staff were found to be approximately 0.2 mrad/treatment field. It should be noted that the radiographers wait 1 minute before entering the room after the completion of an irradiation to allow the shorter-lived radioactive products to decay.

During this initial period of use, the possibility of build-up of long-lived isotopes in the isocentric unit is being studied.

Reference

Williams, J. R., Bonnett, D. E. and Parnell, C. J. (1978) The fixed horizontal neutron therapy beam at Edinburgh: *Dosimetry and Radiation Protection, Brit. J. Radiol.* **52,** 197–208.

J. C. Davis (Livermore, U.S.A.): *A high-intensity rotating target d+T neutron source**

The Lawrence Livermore Laboratory is completing construction of a facility (Booth et al., 1977) containing two intense 14-MeV neutron sources. Each source consists of an air-insulated 400-kV accelerator and a 5000 rpm rotating target. A source strength of 4×10^{13} n/s will be produced by focusing a 150-mA beam of atomic deuterons (D^+) on the 50-cm rotating target that contains 5 kCi of tritium in a titanium–tritide layer. Based upon experience with the existing rotating target source (Booth and Barschall, 1972) at Livermore, useful lifetime of the target is expected to be between 50 and 200 hours with the 1-cm beam spot desired for materials applications; with the larger beam spots acceptable for clinical applications (i.e. 2-4 cm) target lifetime would be considerably longer. In test operation with H_2^+ beams, currents in excess of 40 mA have been accelerated and transported to a rotating target. Initial operation as a neutron source will be at the 1×10^{13} n/s source strength produced at this current; the design goal of 4×10^{13} n/s is expected to be reached by January 1979. A view of one of the two sources is shown in Fig. 2.

Although these sources were designed for materials research, and not for clinical applications, many of the subsystems required for clinical use, e.g. tritium scrubbers and remote-handling equipment for target and collimator changes, have been developed for this facility. The source diameter is adjustable from 1 to 4 cm, making collimation to produce treatment fields of various sizes relatively simple. At the design source strength of 4×10^{13} n/s, a dose rate of 80 rad/min can be produced for a source-to-skin distance of 125 cm. Cost of one of these sources reconfigured for clinical use is estimated to be about a million U.S. $ exclusive of shielding.

---

*Work performed under the auspices of the U.S. Department of Energy by the Lawrence Livermore Laboratory under contract number W-7405-ENG-48.

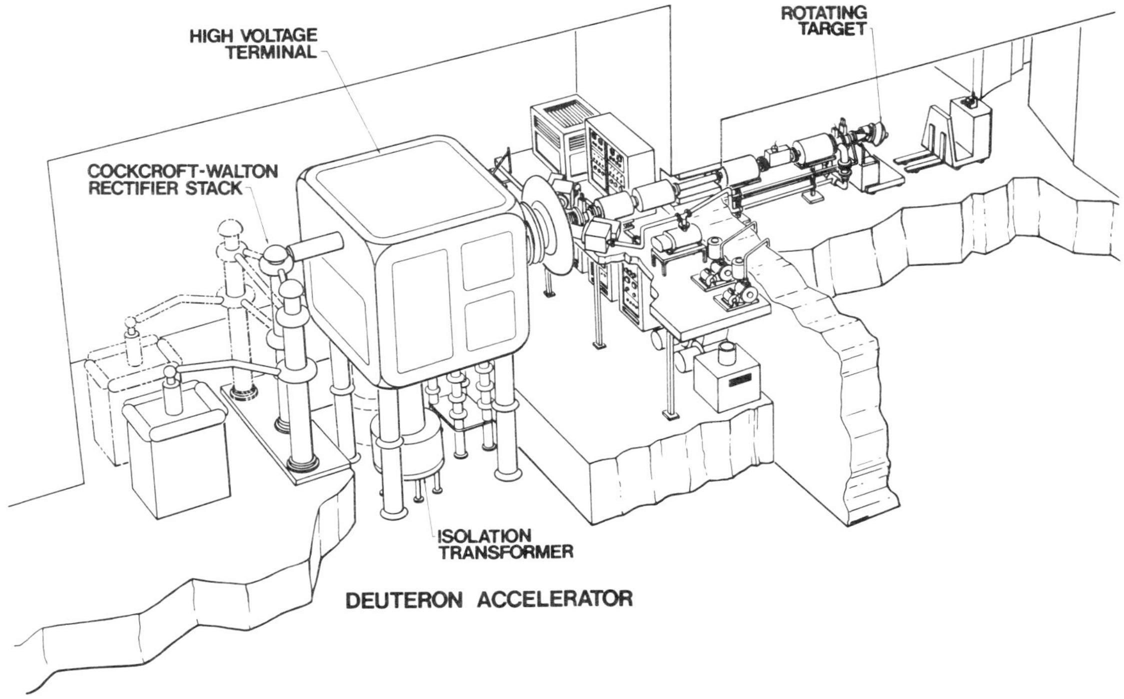

Fig. 2. Major components of RTNS-II source (Davis).

References

BOOTH, R. and H. H. BARSCHALL (1972) Tritium target for intense neutron source. *Nucl. Instrum. Meth.* **99**, 1-4.

BOOTH, R., J. C. DAVIS, C. L. HANSON, J. L. HELD, C. M. LOGAN, J. E. OSHER, R. A. NICKERSON, B. A. POHL and B. J. SCHUMACHER (1977) Rotating target neutron generators. *Nucl. Instrum. Meth.* **145**, 25-39.

P. M. DELUCA, R. P. TORTI, G. M. CHENEVERT and M. E. BRANDAN (Madison, U.S.A.): *A high-intensity gas target d+T neutron source**

The therapeutic value of high-LET beams of fast neutrons is becoming increasingly clear. Suitable sources of fast neutrons can be formed by bombarding low atomic number targets with deuterons and protons of ~30 MeV. To provide a compact and efficient source of comparable radiative properties, the ^{3}H(d,n) reaction (d+T) is a potential alternate. Although beam energies of several hundred keV are adequate, large beam currents are required. Complications introduced by the 20-40 kW of power dissipated by occluded metal targets have limited source strengths and lifetimes. The University of Wisconsin gas target neutron source (GTNS) minimizes these effects by employing an elemental gaseous hydrogen target. Despite an order of magnitude reduction in beam current and target power for equivalent source strengths, the GTNS required development of a windowless differentially pumped target. Reasonable target pressures restrict the aperture sizes, spacing and pumping requirements. The target acceptance is < 10 cm. mrad which requires a well-focused, low-emittance beam. Moreover, the beam–gas interaction generates a new set of difficulties, principally hydrogen-exchange effects, tritium-handling problems and the unusual radiation field associated with an axially extended target.

After several years of development, a suitable neutron source was constructed. Briefly, a deuteron beam is extracted, accelerated and focused into a target chamber. Beam emittance and target acceptance restricted transmission through the pumping apertures to 50%. Ultimately, a 1.33-kPa tritium target was bombarded by a 210-keV, 4-mA deuteron beam. The source strength and target half-life were 1.7 × 10^{12} s^{-1} and 120 min, respectively. Total dose rates of 0.05 Gy min^{-1} were measured 0.6 m from the target end. Specifics of the target

*Supported by Public Health Service Grant 5R01 CA 13469 from the National Cancer Institute.

performance and radiative properties have recently been reported (Chenevert et al., 1977; DeLuca et al., 1978).

To alleviate the established beam–target limitations, a source redesign was initiated. To improve the beam current and energy limits, the accelerating structure was reassembled with new ceramic insulators and the final accelerating stages were redesigned to provide electric field gradients consistent with a space-charge limited beam. A new differentially pumped target with 6 cm mrad acceptance was constructed. Initial tests have been most encouraging. Beams of 247 keV and 16 mA have been accelerated. The beam current and energy were limited by available power supplies. Beam transmission into the target was 77%, i.e. 12.3 mA. Power density in the smallest target aperture was 16 kW cm^{-2}.

Neutron measurements are presently underway. Source strength, neutron-production profiles, dose rates and depth doses are being studied as a function of beam power and energy as well as target thickness. A 0.75-m-long collimator, producing a geometric field 10.5 × 10.5 cm^2 at 0.9 m from the target end, will be employed for many of these studies. Thus far, a 0.87-kPa tritium target was bombarded with a 230-keV, 8-mA deuteron beam. The measured source strength was $3.0 \pm 0.3 \times 10^{12}$ s^{-1} while the dose rate 1.0 m from the target end was 0.06 ± 0.01 Gymin^{-1}. For these conditions, the 0.48 kPa. m target thickness was somewhat less than the 0.61 kPa. m required for maximal neutron production. The measured strength was in agreement with the calculated value of 3.4×10^{12} s^{-1} for a 0.48 kPa. m target (the thick target yield would be 4.8×10^{12} s^{-1}). Although still being investigated, there is clear evidence of "thermal blockage" of the gas in the pumping apertures. Target gas leakage into the accelerator vacuum system is reduced while the ultimate target pressure attainable in the target is increased.

References

CHENEVERT, G. M., P. M. DELUCA, C. A. KELSEY and R. P. TORTI (1977) A tritium gas target as an intense source of 14 MeV neutrons. *Nucl. Instrum. Meth.* **145**, 149-155.

DELUCA, P. M., R. P. TORTI, G. M. CHENEVERT, N. A. DETORIE, J. R. TESMER and C. A. KELSEY (1978) Performance of a gas target neutron source for radiotherapy. *Phys. Med. Biol.* **23**, 876-887.

H. J. GOMBERG, A. E. GILEADI and I. LAMPE (Ann Arbor, U.S.A.): *Computed shielding, collimation and spectral characteristics for a proposed radiation therapy unit*

Recent clinical experience indicates that fast neutrons may be efficient in treating certain types of cancer. Considering its unique space, time and energy characteristics, it appears that a neutron source generated by laser fusion could be used in a therapeutic facility with considerable potential advantages. Conceptual design of a therapeutic facility of this kind is presently being considered at KMS Fusion, Inc. As part of this effort, a powerful and versatile design tool, TBEAM, has been developed which enables the user to determine physical characteristics of the fast neutron beam generated in the facility under consideration, once certain relevant design parameters of the facility are specified.

TBEAM uses the method of statistical sampling (Monte Carlo) to solve the space, time and energy-dependent neutron-transport equation describing the conceptual design specified by the user. The code traces the individual source neutrons and determines the energy and the position of each neutron incident (a) on the field of irradiation, (b) on the adjacent penumbral areas, and (c) on a circular annulus with an outer radius of 100 cm surrounding the field of irradiation and the penumbral regions. This area is large enough to monitor the flux due to leakage and scattered neutrons to which the total body may be unintentionally, but unavoidably, exposed.

This paper presents a sample case in which TBEAM is used to compare the physical characteristics of incident fast neutron beams generated in two facilities identical in all their design parameters with the exception of source geometry. Both conceptual designs use spherical, shell-shaped, layered shields with cone-shaped collimator apertures to define the therapeutically useful beam. Dimensions and compositions are given in Fig. 3. As indicated in the figure, Case A uses a point source while Case B uses a circular plate source with uniformly distributed source intensity.

Comparison of Cases A and B reveals significant differences in the physical characteristics of the two neutron beams

Discussion

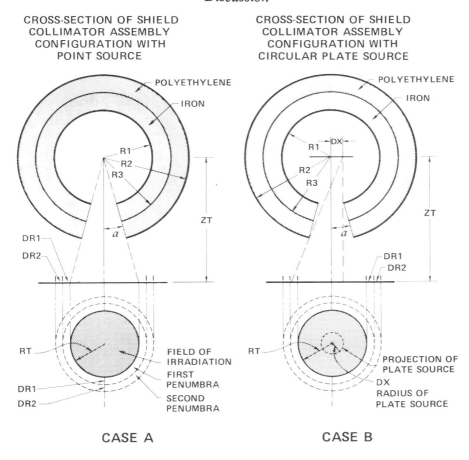

Fig. 3. Cross-section of shield collimator assembly configuration with point source, case A, and circular plate source, case B (Gomberg et al.)

Table 1. Comparison of physical characteristics of fast neutron beam A (point source) vs. beam B (plate source)

	Case A point source	Case B plate source
Relative energy flux incident on field of irradiation	192	100
Fraction of 14.1 MeV neutrons in total number flux incident on field of irradiation	85:100	72:100
Ratio of energy flux incident on first penumbra vs. that on field of irradiation	1.7:100	10:100
Ratio of energy flux incident on second penumbra vs. that on field of irradiation	0.17:100	4:100
Ratio of energy flux incident on the total body area vs. that incident on field of irradiation	0.07:100	0.83:100

generated by the facilities that are identical except for the geometry of the source, as indicated by the entries in Table 1.

The results in Table 1 suggest that, all other design characteristics being identical, the facility with the point source generates a beam of

(a) higher incident energy flux on the field of irradiation;
(b) a spectrum containing more 14.1 MeV neutrons (85% vs. 72%);
(c) better collimation (lower penumbra/umbra ratio); and
(d) better shielding (i.e. lower ratio of scattered and leakage energy flux to energy flux on the field of irradiation) than does the beam emitted under otherwise identical circumstances by the distributed source.

Based on the above indications, it is to be expected that a fast neutron beam generated in laser fusion—being the closest experimental realization of a point monoenergetic source—will have better penetration properties and better collimation (lower penumbra/umbra ratio) than a fast neutron beam generated in an otherwise identical facility, but emitted by a distributed (plate) source.

D. GREENE (Manchester, England) C. H. GILL and R. E. JONES, (Borehamwood, England): *Operating experience with the Hiletron fast neutron therapy machine*

The Hiletron neutron therapy machine has now been in clinical service for just over a year, giving its specified neutron output of 1×10^{12} ns^{-1} at an SSD of 80 cm.

Most of the patients treated had bladder cancer, and for this large body section a five- or six-field isocentric treatment plan is required as shown in Fig. 4.

The exposure times are rather long, but have been well tolerated by the patients. However, the long treatment times limit the use of the machine to about four patients per day. So far no unexpected reactions have occurred on the basis of an RBE of 3. The future intention is to operate a small-scale clinical trial for bladder patients.

The neutron source is a sealed tube using the well-known d+T reaction and operates at a power level of 250 kV and 30 mA, i.e. 7.5 kW.

Six tubes have been made to the current design and they are all still operational. To date they have an average life of 162 hours, with the longest at 336 hours. Within the measuring accuracy the neutron output remains constant. It is anticipated that lives in excess of 400 hours will be achieved.

The tube is housed in a protective canister which is contained within the radiation shield in the treatment head. The shield is designed to reduce the radiation intensity outside the main beam to less than 1% of the level within the beam

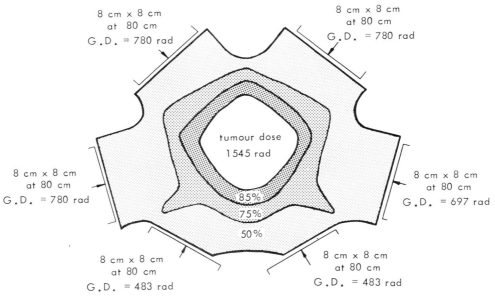

Fig. 4. Treatment plan for a bladder cancer using six fields of 14 MeV neutrons. The term G.D. refers to the given dose, or maximum dose given for each field. (Greene et al.).

at the same distance from the target of the tube. The movement of the treatment table in conjunction with the head provide full isocentric positioning facilities, with an isocentric stability within a 4-mm sphere for all positions of the head.

A series of removable collimators allow twelve different field sizes to be obtained. The collimators can be easily interchanged with the aid of a portable trolley.

The equipment has been designed to reduce the effect of residual activity to a minimum by careful choice of materials. For example, the maximum measured dose rate of the residual activity (from all causes) 100 cm from the collimator and immediately after the neutron beam is switched off is 10 mR hr^{-1}. This level is reduced to 1 mR hr^{-1} after 15 minutes. The maximum measured activity from a collimator only is 2 mR hr^{-1} at a distance of 100 cm. The collimators do not therefore require involved handling or shield facilities.

A. Hesz and H. D. Franke (Hamburg, Federal Republic of Germany): *The effect of an analysed deuterium ion beam on the lifetime of TiT targets used at the fast neutron therapy facility, Hamburg–Eppendorf*

The fast neutron therapy facility (DT, 14 MeV) at the University Hospital Hamburg-Eppendorf (Franke et al., 1978) is used for clinical tumour therapy since February 1976. The fast neutrons are generated by bombarding a rotating TiT target with accelerated deuterium ions (d+T reaction). The deuterium beam is produced by a duoplasmatron type of ion source as a mixture of atomic and molecular deuterium ions. Operation conditions of 8 to 12 mA total beam current and 500 kV accelerating voltage are usual. The neutron yield at 12 mA is about 3.5×10^{12} n s^{-1} giving a dose rate of more than 20 rad/min for a field size of 20×20 cm^2 at the nominal treatment distance of 80 cm SSD. The economy of this unit depends essentially on the lifetime and cost of the tritium targets. For therapeutic applications the targets are generally used for the "initial" half-life only.

The neutron yield produced by a mixed deuterium ion beam incident on a rotating tritium target of about 160 cm^2 active area decreases with a half-life of (140 ± 20) mAh. Several authors (Booth and Barschall, 1972; Stengle et al., 1975; Stengle and Vonach, 1977), however, have reported about a significant increase of the lifetime of TiT targets, when separating the atomic and molecular ions of a mixed deuterium ion beam and bombarding the active target area with the atomic deuterons only. In this case the neutron yield is reduced according to the loss of molecular deuterium ions. Therefore, we studied the effect of separating the total ion beam into atomic and molecular ions and bombarding the target area with all components at different beam spots. Using the analysed beam in this way, the increase in lifetime is lower than using the atomic deuterons only, but the initial neutron yield is not reduced in the same order as for a mixed total beam.

Figure 5 shows the different decays of neutron yield irradiating a stationary TiT target with a mixed or an analysed deuterium ion beam. In the case of an analysed beam three different decay lines can be distinguished. The neutron yield data calculated from these lines are presented in Table 2. The atomic deuterons D_1^+ produce 50% of the initial neutron yield and the molecular deuterium ions D_2^+ and D_3^+ about 20% and 30%, respectively. With these data and the assumption of a total break-up of the molecular deuterium ions at the surface of the target (Ormrod, 1974), the primary beam current is calculated to consist of 70% of atomic ions and 15% of each kind of

Table 2. Neutron yield data for stationary TiT targets (12 Ci/in.^{2}) irradiated with an analysed deuterium ion beam for a total beam current of 1 mA and an accelerating voltage of 500 kV (Hesz and Franke).

Deuterium ion	Initial neutron yield (%)	Relative half-life (mAh)	Beam spot size (cm^2)	Primary beam current[a] (%)	Target current density (mA/cm^2)	Specific half-life (mAh/cm^2)
D_1^+	50	4.60	0.3	70	2.3	10.7
D_2^+	20	1.45	0.3	15	0.5	0.7
D_3^+	30	0.75	0.3	15	0.5	0.4

[a] Assumed break up at the target surface of: D_2(500 keV) → $2 \times D_1$ (250 keV), D_3(500 keV) → $3 \times D_1$ (167 keV).

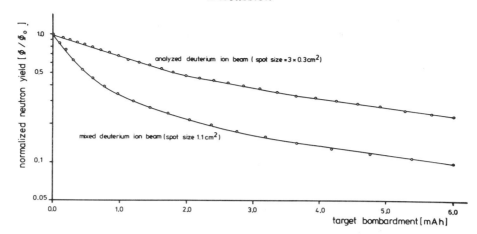

Fig. 5. Neutron yields for stationary TiT targets (12 Ci/in.²) irradiated with different deuterium ion beams for a total beam current of 1 mA and an accelerating voltage of 500 kV (Hesz and Franke).

molecular ions. For a duoplasmatron ion source similar values have been reported (Ormrod, 1974; Hepburn et al., 1973). With the obtained values of relative half-lifes, beam spot sizes and fractions of primary beam current, the specific half-life for the used TiT target irradiated with atomic deuterons is 10.7 mAh/cm². This value is lower than the values reported by Stengl and Vonach (1977). However, it has to be realized that our value is obtained for the higher target current density of about 2.3 mA/cm² and that the estimated beam spot size is an upper limit only. The real beam spot size is less than 0.3 cm², by which an increased specific half-life is calculated. The same considerations are valid for the specific half-lifes in the case of molecular deuterium ions.

The loss in neutron yield is about 50% using the atomic ions of an analysed deuterium ion beam only. To avoid this disadvantage we irradiated TiT targets with all separated components of an analysed deuterium ion beam. Figure 5 shows that during the "initial" half-life the decays of the initial yields of a mixed and an analysed deuterium ion beam follow straight lines with different slopes. The comparison of the two lines indicates a difference in half-life by more than a factor of 3. The performed experiments have shown that analysing a deuterium ion beam and bombarding the active target area with all separated ions, TiT targets are to be used more effectively without any loss in initial neutron yield. These facts are most important for the economy of the clinical application of this type of neutron therapy facilities.

References

BOOTH, R. and H. H. BARSCHALL (1972) Tritium target for intense neutron source. *Nucl. Instrum. Meth.* **99**, 1-4.

FRANKE, H. D., A. HESZ, E. MAGIERA and R. SCHMIDT (1978) The neutron therapy facility (DT, 14 MeV) at the Radiotherapy Department of the University Hospital Hamburg–Eppendorf. *Strahlentherapie* **154**, 225-232.

HEPBURN, J. D., J. H. ORMROD and B. G. CHIDLEY (1973) Design of an intense fast neutron source. *IEEE Trans. Nucl. Sci. N.S.* **22**, 1809-1812.

ORMROD, J. H. (1974) Decay of neutron yield from titanium–tritide targets. *Can. J. Phys.* **52**, 1971-1980.

STENGL, G., H. VONACH and H. FABIAN (1975) Production of DT neutrons by means of TiT targets and analyzed atomic deuteron beams. *Nucl. Instrum. Meth.* **126**, 235-239.

STENGL, G. and H. VONACH (1977) Production of DT neutrons by means of TiT targets and analyzed atomic deuteron beams. *Nucl. Instrum. Meth.* **140**, 197.

E. A. KNAPP (Los Alamos, New Mexico, U.S.A.): *PIGMI—A pion generator for medical irradiation*

The PIGMI Project is an accelerator research and development program directed toward ascertaining the feasibility of manufacturing a low-cost pion source for hospital siting. Several innovations in accelerator technology have been proposed and are being investigated in this program which reduce the cost of such a facility enormously compared to previous practice. At

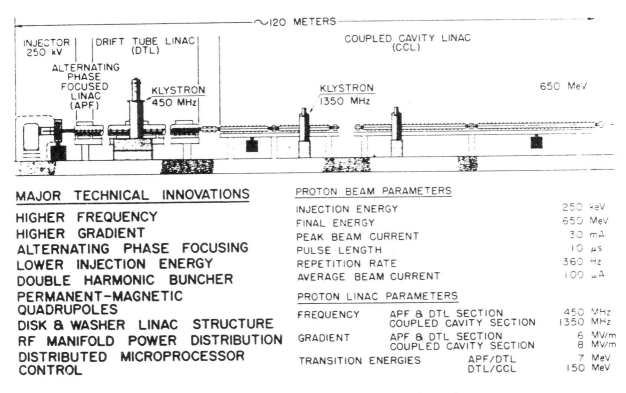

Fig. 6. PIGMI, pion generator for medical irradiation (Knapp).

this time, 2 years into a 3-year program, significant success has been demonstrated in a number of the technologies being studied in this program.

1. Accelerator gradient. A hospital-based accelerator must be compact. The PIGMI program has demonstrated a particle acceleration gradient of 8 MeV/meter, significantly higher (4×) than previous experience in drift tube accelerators.
2. Particle focusing. Permanent magnet focusing systems have been prototyped which would dramatically reduce the cost of particle focusing in both the accelerator and in the treatment head.
3. Higher-frequency operation. Prototype accelerator sections have been operated at significantly higher frequency than common in current practice, leading to reduced size and cost for the final accelerator systems.
4. Low-energy injection system. Low-voltage injection leads to a low cost, compact accelerator system. An ion source and accelerating column has been completed and preliminary tests are continuing.
5. A very efficient accelerator structure for high-velocity proton acceleration (200 MeV > E > 800 MeV) has been developed and prototype design is beginning.

The parameters for a PIGMI accelerator (see Fig. 6) now appear (in 1978 $) to be as follows:

Proton energy:	650 MeV.
Average proton current	100 μA
Pion dose rate	10 rad/min/liter conventional channel, ~ 50 rad/min/liter enhanced solid angle channel.
Length:	< 100 m.
Cost:	~ $12M with two channels (approximate).

No technical obstacles have been encountered which would compromise this approach to economical pion generation for medical application. A working prototype of the first 7 MeV of a PIGMI accelerator is nearing completion and will be operating in the fall of 1978.

E. A. KNAPP (Los Alamos, New Mexico, U.S.A.): *A proton linear accelerator neutron source for radiation therapy*

Sources of fast neutrons for medical use should be reliable, relatively inexpensive, easy to operate and maintain, and compact. Proton linear accelerators appear to offer exceptional capabilities to the medical field for neutron generation, thanks to recent technical advances in accelerator science developed in the PIGMI pion generator program. It now appears possible to build a neutron generator, based on a proton linear accelerator, with performance specifications and size ideally suited to hospital siting. Some typical parameters might be:

Proton energy: 40 MeV.
Proton current: 100 μA
Dose rate (p+Be): >100 rad/min at 1·25 m.
Accelerator length: <6 m.
Switchable beam to two or more treatment rooms.
One klystron tube.
Permanent magnet transport.
Accelerator room size: $10 \times 3 \times 3 m^3$.
Cost: Accelerator ~ \$1.5M,
   Isocentric system ~ \$0.5M each.

Linear accelerators have many characteristics which make them ideally suited to routine use. Clearly the linear electron accelerator has dominated the radiation therapy equipment market for X-ray production because it has proven to be reliable, easy to operate and maintain, and inexpensive to produce. Many of the same characteristics are available in proton linacs, and we should expect the same advantages to accrue.

A low-voltage injector (< 200 keV) injects a proton beam into a drift tube linac, which consists of an alternating phase-focused capture section followed by a conventional quadrupole-focused drift tube accelerator. Permanent magnet focusing in the drift tubes keeps the beam centered on the axis of the accelerator. A single r.f. drive tube provides reliable operation at modest cost. The isocentric delivery system provides achromatic beam transport from accelerator to production target, and the excellent linac beam quality allows a minimum cost bending-magnet system to be designed in the C-arm system. We expect all of the technological features of the accelerator to be demonstrated in a working prototype in the fall of 1978.

L. KÖSTER, A. BREIT, G. BURGER, H. KNESCHAUREK and H. SCHRAUBE (Munich, Federal Republic of Germany): *RENT—a reactor neutron therapy project*

Fast neutron radiotherapy is generally performed by means of d+T generators or cyclotrons. The use of these high energetic neutrons is recommended for the therapy of deep-seated tumors. There are, however, numerous applications where a limited depth dose does not seem to be a severe handicap. About 30% of all cases treated up to now with neutrons in the U.S.A. are carcinomas of the head and neck region. In all these cases moving-field treatments or at least the application of two opposing fields are possible. In this way sufficient homogeneous depth dose distributions are achieved at the tumor site, while restricting the dose at the skin and the spinal cord. A beam of fission neutrons produced by means of a converter facility at a reactor could also be considered as a valuable tool for a fast neutron therapy pilot study.

The device is being installed at the swimming pool research reactor FRM of the Technical University in Munich (Fig. 7). It consists of a horizontal beam line passing the reactor core tangentially. A uranium-235 converter plate is placed at the core edge, resulting in a fission rate in excess of 10^{12} cm^{-2} s^{-1} and a source strength of 4 to 5×10^{14} s^{-1}. The FSD is about 5 m, and the free-in-air kerma rates are $\dot{K}_n = 33$ rad/min and $\dot{K}_\gamma = 25$ rad/min. The depth doses calculated for water and measured in a polyethylene phantom are shown in Fig. 8. Filtering the beam with lead reduces the free-in-air kerma rates as shown in Table 3. This allows the easy application of mixed beams with different and independently variable-depth dose distributions for the gamma and neutron component. The technique is especially interesting with respect to the assumed advantages of mixed radiation schedules. The installation will be built up within the next 2 years. It is planned to start radiobiological experiments and probably the first clinical studies in 1981.

Table 3. Estimated neutron and gamma kerma rates free in air behind lead filters of different thickness, d (Köster et al.)

d (cm)	0	2.5	5	7.5
$\dot{K}_n$ (rad/min)	33	25	18	14
$\dot{K}_\gamma / \dot{K}_n$	0.77	0.30	0.14	0.06

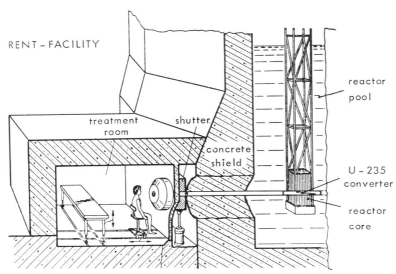

Fig. 7. Schematic layout of the RENT facility at the swimming pool reactor FRM (Köster et al.).

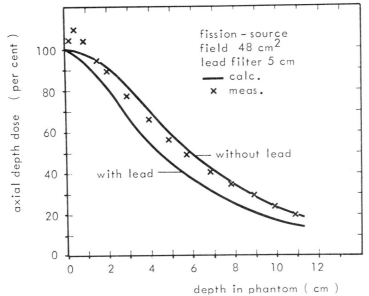

Fig. 8. Neutron depth–dose curves calculated for water and measured in a polyethylene phantom (Köster et al.).

M. O. LEACH and J. H. FREMLIN (Birmingham, England): *A proposed neutron therapy facility at Birmingham*

The Nuffield cyclotron, in the Department of Physics, is a classical fixed-frequency 60-inch cyclotron, presently used to a large extent for medical studies and for isotope production. It is capable of delivering beams of p^+, d^+, $^3He^{++}$ and α^{++} at energies of about 10 MeV per nucleon. External beam currents of ~ 50 μA of deuterons are readily available, maximum internal beam currents being of the order of 1 mA. The Physics Department, which has considerable experience in accelerator design, medical measurements, neutron spectrometry and neutron transport calculations, is situated close to the Queen Elizabeth Medical Centre, which provides a radiotherapy service for a local population of 2 million in a region of total population 5.5 million (the West Midlands).

The cyclotron has already two irradiation areas used for *in vivo* activation analysis measurements, using neutrons produced by proton or ^{3}He bombardment of light elements. The deuteron beam line is currently being extended into one of these areas, in which collimation to provide a facility similar to those in use elsewhere for neutron therapy will be installed, together with the provision of facilities for mammalian irradiations with several neutron beams. Preliminary studies, using this facility, are planned to include the study of the variation of neutron spectra with depth in water and in phantoms containing tissue-equivalent solutions for various incident spectra, together with the effect of changes in collimation and field size on neutron spectra. Further work will extend the method of liquid scintillation neutron spectroscopy to a direct measure of biological effect, by measuring the biological effect of individual charged particles. This will be carried out by calculating the charged particle spectra expected for a given neutron spectrum, and measuring directly the biological effect of given charged particles and heavy ions obtained from the Nuffield cyclotron.

Long-term plans are being prepared for the establishment of a patient irradiation facility at Birmingham. Different types of facility are presently being considered, all involving the construction of suitable irradiation rooms, patient support facilities and staff areas, probably in the form of an out-patient department.

M. A. LONE, A. J. FERGUSON (Chalk River, Canada) and B. C. ROBERTSON (Kingston, Canada): *Low-energy neutron emission from the $d+{}^9Be$ reaction*

The low-energy neutron emission from the d+Be reaction has not been thoroughly investigated. Early results on the neutron spectral distributions (see Grand and Goland, 1977) implied essentially gaussian shapes with very little low-energy neutron component. More recent measurements by the CRNL group (Lone, 1977) and the ORNL group (Saltmarsh *et al.*, 1977) show two component spectral distributions; a broad high-energy peak due to deuteron stripping and an exponentially falling low-energy tail.

In order to understand the original low-energy component, the neutron spectral distributions for $E_n > 0.4$ MeV were measured at $E_d = 3$ to 23 MeV with thick Be targets at the CRNL MP Tandem. The neutron intensity and average neutron energy $\bar{E}_n$ for two neutron thresholds are given in Fig. 9. Data for $E_d > 23$ MeV is taken from other publications (see Lone, 1977). The relative magnitude of low-energy neutrons ($E_n < 2$ MeV) at $E = 10, 15, 20$ and 25 MeV is 33, 28, 21 and 15% for deuterons, and 70, 62, 52 and 49% for protons, respectively. (The proton data have been taken from Lone, 1977).

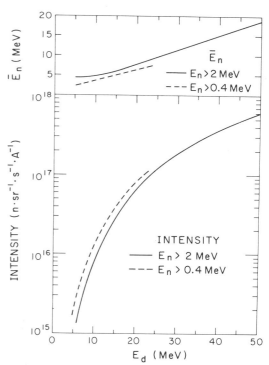

Fig. 9. Neutrons from the d+Be reaction on thick targets at zero degrees (Lone et al.).

Several nuclear reaction mechanisms, e.g. compound nucleus decay and multi-body break-up reactions, have been discussed as possible sources of the low-energy neutron component. The systematic studies carried out by us indicate that the neutron decay of lower states in ^{9}Be excited by direct reactions (inelastic scattering) can also play a significant role in producing low-energy neutrons, particularly with proton beams. To estimate the contribution to low-energy neutrons from inelastic scattering, the expected yield at a beam energy of 15 MeV was calculated (Robertson et al., 1978) using cross-section information available in the literature and was compared with the observed yield. This latter quantity was inferred from Be-Li low-energy yield differences (Lone, 1977). For protons the calculated contribution is 27% of the total yield and the observed contribution is 31%; for deuterons the corresponding values are 3% and 5%, respectively. These results show that the inelastic scattering process contributes significantly less to the low-energy neutron yield from the d+Be reaction than the p+Be reaction.

References

GRAND, P. and GOLAND, A. N. (1977) An intense neutron source based upon the deuteron-stripping reaction. *Nucl. Instrum. Meth.* **145**, 49-76.

LONE, M. A. (1977) Intense fast neutron sources. In M. R. BHATT and S. PEARLSTEIN (Eds.), *Proc. Symp. on Neutron Cross Sections for 10-40 MeV*, BNL-NES-50681 Report, pp. 79-116.

SALTMARSH, M. J., C. A. LUDEMAN, C. B. FULMER, and R. C. STYLES (1977) Characteristics of an intense neutron source based on the d+Be reaction. *Nucl. Instrum. Meth.* **145**, 81-90.

ROBERTSON, B. C., LONE, M. A., FERGUSON, A. J. and EARLE, E. D. (1978) To be published.

E. MAIER and G. HÜDEPOHL (Essen, Federal Republic of Germany): *Dosimetric results for d(14)+Be neutron beams of cyclotron isocentric facility C I R C E in Essen*

In January 1978 the first patients were irradiated with fast neutrons from the CIRCE cyclotron (Rassow et al., 1978a). Since then thirty-seven patients (July 1978) were treated 4 times a week to a total tumor dose of 16 Gy. Most of the treated tumors (26) were in the head and neck region.

The neutrons are generated by bombarding a thick Be-target with deuterons of 14 MeV. With twelve collimator inserts the field sizes can be varied from 5×5 cm^2 to 20×20 cm^2 at the target–isocentre distance of 125 cm.

The dosimetry was performed with tissue equivalent ionization chambers and a Geiger–Müller counter, all manufactured by EG&G, U.S.A. The tissue equivalent ionization chambers were flushed with tissue equivalent gas. The depth dose curves were measured with a 0.1-cm^3 chamber, the absolute dose values were determined with a 1-cm^3 chamber and in the build-up region an extrapolation chamber was used. The relative neutron sensitivities of the tissue equivalent chambers were assumed to be 1. The Geiger–Müller counter has a ^{6}LiF shield to reduce the sensitivity to thermal neutrons and for the relative neutron sensitivity a value of 0.002 was adopted.

All doses were normalized to the total absorbed dose for a field size of 10×10 cm^2/125 cm and a depth of 5 cm in the phantom (Rassow et al., 1978b). The dose rate at this reference point with a deuteron current of 100 μA is 0.286 Gy min^{-1}.

With a modified radiation field analyzer (RFA-1) from Scanditronix, Sweden, the isodose distributions were measured and Fig. 10 shows the total and the gamma-ray isodose curves for a field size of 10×10 cm^2/125 cm. The maximum

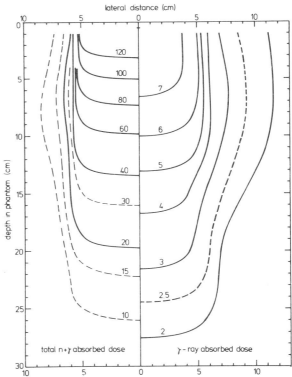

Fig. 10. Total and gamma-ray isodose distributions for a field size of 10×10 cm^2/125 cm and a target-surface distance of 125 cm, measured in a water-filled lucite phantom of $40 \times 40 \times 40$ cm^3 (Maier et al., 1978).

of the total absorbed dose of 145% relative to the reference point at a depth of 5 cm is reached at a depth of 0.2 to 0.5 cm. At a depth of 8 cm 50% of the maximum dose is delivered (Maier et al., 1978).

The lateral decrease of the total absorbed dose occurs at a depth of 5 cm within 1.2 cm from 80% to 20% of the dose at the beam center.

The relative gamma-ray absorbed dose D_γ/D_t at the dose maximum is 5% and rises up to 17% at a depth of 20 cm. Laterally the relative gamma-ray absorbed dose increases rather steep and reaches, for example, at a depth of 10 cm and 10 cm outside the beam axis a value of 33%.

References

Maier, E., G. Hüdepohl, P. Meissner and J. Rassow (1978) Erste Dosimetrie- und Strahlenschutzmeszergebnisse an dem isozentrischen Zyklotron-Neutronentherapiegrät CIRCE. In *Medizinische Physik*, Dr. Alfred Hüthig Verlag, Heidelberg, p. 297.

Rassow, J., G. Hüdepohl, E. Maier and P. Meissner (1978a) Initial measurements at CIRCE (Cyclotron Isocentric Therapy Facility, Radiation Physics Essen). *Proc. Third Symp. on Neutron Dosimetry in Biology and Medicine*, EUR 5848, pp. 327-337.

Rassow, J., E. Maier and P. Meissner (1978b) Proposal for a practical dose calculation scheme for neutron irradiations. *Strahlentherapie*, **155** (in press).

J. L. Tom, T. Y. T. Kuo and G. O. Hendry (Berkeley, U.S.A.): *Dosimetric and radiation measurements made on the Isocentric Neutron Therapy Unit for the King Faisal Specialist Hospital*

An isocentric neutron therapy unit which produces fast neutrons using a model CS-30 cyclotron has been built for installation at the King Faisal Specialist Hospital in Riyadh, Saudi Arabia. This unit is similar to the isocentric neutron therapy installations at Edinburgh, Scotland, and Essen, Federal Republic of Germany. Fast neutrons produced by both the d(15)+Be and the p(26)+Be reactions have been measured and the results are summarized elsewhere (Tom, 1978). The dosimetric properties of the neutron beams have also been measured by Goodhead et al. (1978) and are found to be quite similar.

The characteristics of neutron beams produced by different proton energies on beryllium and lithium were measured by Quam et al. (1978). During these measurements, the effects of filtering the p+Be beam with 6 cm of polyethylene was also investigated and the penetration of the neutrons was found to be significantly enhanced by the use of the filter. Neutron spectra measurements by Johnsen (1978) and Graves et al. (1978) also show that neutron beams produced by the p+Be reaction can be hardened by use of polyethylene filters in the neutron beam. By using a polyethylene filter on the neutron beam of the p(26)+Be reaction, the depth for 50% dose was found to be increased by 0.8, 1.6 and 2.0 cm, respectively, when a 2-, 4- and 6-cm filter is placed at the upstream end of a collimator for a 10 × 10 cm² field at 125 SSD. The tissue kerma rate was correspondingly reduced by 27, 44 and 56%. The isodose curve at 125 cm SSD for the p(26)+ Be neutrons with a 6 cm filter was comparable to the isodose curve for a similar field with cobalt-60 gamma-rays at 80 cm SSD.

The radiation leakage and residual radiation around the treatment head from the deuteron and proton reactions were measured. The total dose $(n+\gamma)$ and the gamma component measured 5 cm outside of a 10 × 10 cm² field at 125 cm SSD was found to be 1.6% and 0.8%, respectively, with deuterons and 2.8% and 1.4%, respectively, with protons. A gamma spectrum analysis of the residual radiation revealed gamma peaks from fast neutron activation of copper and steel and thermal neutron activation of copper and manganese in the steel. Residual radiation from build-up of long-lived isotopes was measured for the deuteron and proton reactions. The residual radiation from build-up approaches a maximum after approximately thirty 100-rad fractions equally spaced over an 8-hour period. The maximum activity, measured 5 minutes after irradiation, 20 cm from the front of the treatment head, approaches approximately 8 mR/hr with the deuteron and 15 mR/hr with the proton-produced neutrons. A significant portion of the residual radiation appeared to come from the treatment room.

References

Goodhead, D. T., R. J. Berry, D. A. Bance, P. Gray and B. Stedeford (1978) Fast neutron therapy beam produced by 26 MeV protons on beryllium. *Phys. Med. Biol.* **23**, 144-148.

Graves, R. G., J. B. Smathers, V. A. Otte, P. R. Almond and R. Grant (1978) Neutron energy spectra of radiation fields produced by the Be(d,n) and Be(p,n) reactions, to be published in *Medical Physics*.

Johnsen, S. W. (1978) Polyethylene filtration of 30 and 40 MeV p-Be neutron beams. *Phys. Med. Biol.* **23**, 499-502.

Quam, W. M., S. W. Johnsen, G. O. Hendry, J. L. Tom, P. H. Heintz and R. B. Theus (1978) Dosimetry measurements of 26, 35 and 45 MeV p-Be and p-Li neutron beams. *Phys. Med. Biol.* **23**, 47-54.

Tom, J. L. (1978) Preliminary results on the dosimetric measurements of the model CS-30 cyclotron and neutron therapy system. *The Cyclotron Corporation Report*, 6014.

A. WAMBERSIE,* J. P. MEULDERS,† FRANÇOISE RICHARD* and M. WINANT* (* Brussels, Belgium, † Louvaine-la-Neuve, Belgium): *The fast neutron therapy facility at CYCLONE, Louvain-la-Neuve*

The neutron therapy program at Louvain-la-Neuve is carried out with the cyclotron "CYCLONE" of the Catholic University. This isochronous, variable-energy cyclotron produced by CSF (Corbeville, France) accelerates different types of charged particles; in particular, deuterons can be accelerated at energies ranging from 13 to 50 MeV.

The treatment room, and related medical facilities, are located one level below the main level of the cyclotron. This permits the use of a vertical therapeutic neutron beam by bending the deuteron beam at 90° (see Fig. 11). This vertical beam appears to be more adequate for positioning the majority of the patients.

The collimation system consists of a fixed shielding and a series of interchangeable inserts. The shielding itself consists of a proximal part (precollimator) and distal part in which the different inserts can be fitted and rotated along their vertical axis. This distal part is conical and consists of a steel mould filled with a mixture (50%) of epoxy and borax. The precollimator, 50 cm steel, which determines the largest available field (25 × 25 cm²), contains two independent transmission ionization chambers used as monitors and located about 25 cm below the target. The interchangeable inserts are cylindrical in shape; their height is 80 cm and their external diameter 40 cm. The proximal part

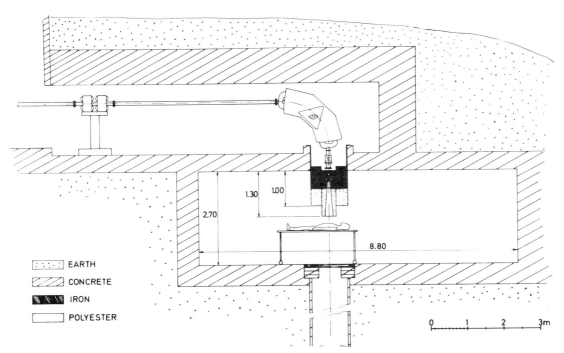

Fig. 11. Schematic view, in vertical projection, of the irradiation facility. The last part of the beam transport with the last quadrupole pair, the 90° bending magnet and the beryllium target are indicated. A pit (neutron trap) is introduced along the beam axis in order to reduce the room activation (Wambersie et al.).

Neutrons are produced by bombarding a thick (10 mm) water-cooled beryllium target with 50 MeV deuterons. This maximum available deuteron energy has been used, up to now, for the clinical applications. A beam current of about 4 μA is used for patient treatments with a dose rate of about 0.4 Gy/min at a target–skin distance (TSD) of 157 cm. The dose rates can be increased in excess of 1 Gy/min. High dose rates allow increasing treatment distance; this in turn permits the construction of relatively long collimators which improves the beam profiles.

(50 cm) is a mixture of iron and epoxy, and the distal part (30 cm) is a mixture of borax and epoxy. A set of twenty inserts are at present available, the field sizes will range from 5 × 5 cm² to 25 × 25 cm² (at a TSD of 157 cm). Due to their weight, positioning of the inserts requires an electrohydraulic device.

The first clinical applications were performed in December 1976 with the horizontal radiobiological beam. Routine clinical applications were started in March 1978 in the permanent medical area with the vertical therapeutic beam. In a first stage, the cyclotron was available 3 times a week for therapy. Under these conditions, mixed irradiations were performed combining three neutron sessions (0.65 Gy each) and two photon sessions (2.00 Gy) per week. A "clinical" RBE of about 3 was adopted. The possibility of using p(70)+Be neutrons is explored. Pretherapeutic dosimetric and radiobiological experiments are in progress.

The efficient collaboration of Y. Jongen, C. Ryckewaert and of the cyclotron staff, at the different stages of the neutron therapy program, was highly appreciated. The collimation system is produced in a joint program with the Centre d'Etudes de l'Energie Nucléaire (CEN) in Mol.

The neutron therapy program is supported in part by the Fonds de la Recherche Scientifique Médicale (FRSM, Convention 3.4525.76); operation of the Cyclotron is supported by the Institut Interuniversitaire des Sciences Nucléaires (IISN).

Neutron dosimetry, radiation quality and biological dosimetry

J. BOOZ

Institut für Medizin, Kernforschungsanlage, Jülich, Federal Republic of Germany

---

THE contributions in this session were concerned with radiation quality, RBE measurements, dosimetry and neutron spectra. Half of the poster contributions dealt with the problem of RBE and radiation quality in a homogeneous phantom: collimated fast neutrons that enter the human body experience elastic and inelastic scattering. As a consequence, with increasing depth and increasing lateral distance from the beam, the mean neutron energy decreases and hence the biological effectiveness increases. On the other hand, also the number of gamma quanta increases which will tend to decrease the average biological effectiveness. These two processes are interdependent in a complex way and counteract with respect to the relative biological effectiveness. Therefore the eventual local variation in RBE is a difficult problem which has been with us for many years.

Most of the posters, discussing this problem, use microdosimetric parameters or spectra to characterize radiation quality. A very useful parameter is y^*, the dose-averaged lineal energy, corrected for the saturation effect, because it is roughly proportional to neutron RBE for 100% survival. The qualification "roughly" means that RBE rises a little faster with LET than y^*, and at maximum a difference of a factor of 2 is possible. Therefore, Günther from Berlin-Buch has developed a new microdosimetric formalism which he presents in his poster. He explains the relation between his system and the established microdosimetric spectra and presents a collection of useful microdosimetric parameters for fast neutrons between 0.1 and 20 MeV.

After this excursion to theory, let us return to the practical problem of neutron-radiation quality and RBE in a homogeneous phantom. Booz and Fidorra from Jülich investigated the problem of the possible change of RBE with depth. For neutrons from the $p(15)+$Be reaction they found that the y^*_n, y^* of the fast neutron component did not change with depth. The same was found by Burger and Morhart from Neuherberg for 15 MeV neutrons and by several other authors for other neutron irradiations (Menzel et al., 1976; Burger et al., 1978). Therefore any change in RBE can only be due to the increasing gamma-component. Booz and Fidorra concluded that RBE as a function of depth can be calculated if the survival curves for both, the neutron and the gamma-components are known. They did the measurements, checked the calculation with the biological data of Railton et al. (1974), found good agreement, and finally evaluated the RBE as a function of depth for both, 14 MeV neutrons and 6 MeV neutrons from d on Be, using the inhibition of colony formation of CHO-cells and T-cells as the biological end points. The result shows that the change in RBE is very small. For 80% survival a maximum decrease of about 10% was calculated between the surface and 20-cm depth.

Such changes, of course, are hard to measure. In fact, Hogeweg, Zoetelief and Broerse from Rijswijk did not find any significant change in RBE with depth for both, 6.5 and 15 MeV neutrons, when measuring survival of mammalian cells in culture. Mijnheer, Haringa, Gorter and Deys from Amsterdam used the impairment of the clonogenic capacity of RUC-2 cells as the biological end point. For 14 MeV neutrons they found a difference in RBE of less than 10% between the surface and 10-cm depth.

The situation, however, seems to be different for smaller neutron energies, smaller collimators, smaller source skin distance and large depths. Mountford from Birmingham measured neutron energy spectra from a Pu–Be source with a liquid scintillator and evaluated the mean neutron energy as an indicator for radiation quality. He found a slight increase in neutron energy with depth, which was more pronounced for the smaller collimators of 1 and 4.5 cm. An increase in radiation quality with depth was also observed by Booz and Fidorra for 6 MeV neutrons, when the depth became larger than 20 cm, and by Burger and Morhart for 15 MeV neutrons, when the source–skin distance was as small as 30 cm.

The problem of the change of radiation quality and RBE with lateral displacement from the beam axis is more complex. There is not such a nice agreement between physical and biological data. All authors who investigated the microdosimetric radiation quality as a function of the lateral displacement agree unanimously that there is a distinct difference in radiation quality between the central beam and the penumbra region. Schumacher and Menzel from Homburg evaluated the contribution of the different radiation components to the lateral dose profile. They found for both, 14 MeV neutrons and neutrons from the Heidelberg cyclotron, that the fast neutron component decreased more quickly than the slower neutron component, and that the gamma-ray component became dominating with increasing distance. The same was found by Hogeweg, Zoetelief and Broerse for 6.5 and 14 MeV neutrons. Mijnheer, Haringa, Gorter and Deys evaluated the neutron energy spectra of 14 MeV neutrons with threshold detectors. They also found an increase in the number of low-energy neutrons with distance from the center. Burger and Morhart from Neuherberg performed neutron-transport calculations and evaluated y^* for 15 MeV neutrons and different field sizes, taking into consideration experimental gamma dose fractions. They obtained a rise in radiation quality, of about 10%, at the beam edge, followed by a continuous decrease for larger lateral distances. However, the authors also discovered that the increase in radiation quality at the beam edge depends on the type of collimator. For the collimator at TNO, Rijswijk, they did not find any increase in y^*. This result could be due to the small source–skin distance of only 45 cm which is used at TNO. Booz and Fidorra, using a commercial collimator and a source–skin distance of 125 cm, also report on an increase of y^* at the beam edge of about 10%, measured for 6 MeV neutrons.

According to the Dual Radiation Action hypothesis the parameter y^* is roughly proportional to the initial slope of survival curves. Hence one should also expect the neutron RBE to increase at the beam edge. However, this has not been observed experimentally. Hogeweg, Zoetelief and Broerse, employing the survival of mammalian cells, did not observe a significant change in RBE at a lateral distance of 4 cm from the beam edge. This is in agreement with the calculated values of y^* for this collimator, mentioned earlier. Hendry, Thomas and Greene from Manchester measured the lateral distribution of RBE for 14.7 MeV neutrons employing the survival of colony formers in mouse bone marrow and of mouse spermatogonia as the biological end points. At a dose level of 43 rad the authors did not observe a significant change in RBE for both systems up to a lateral distance of 48 cm. Therefore the authors conclude that for these two systems the iso-dose curve is a good approximation for the iso-effect curve. This result is further confirmed by the Amsterdam group who did not observe a significant change in the RBE of RUC-2 carcinoma cells between the beam center and the penumbra region.

Only for the TNO-facility we have both, data on RBE and on y^*, and here they are in agreement. For the other collimators and neutron energies we have either RBE or y^* as a function of the lateral distance, Thus there is no reciprocal confirmation, but also no inconsistency, between the physical and the biological results. Obviously the decrease in neutron energy is nearly compensated by the increase in the gamma dose component, and the remaining change in RBE of about 10% or less is perhaps too small to be detected experimentally. Summing up the eight posters on radiation quality and RBE it can be concluded that the local changes of the total dose RBE in a homogeneous phantom for both, the depth in the phantom and the lateral distance from the beam axis, are small, and thus are possibly less important than the change in RBE for the different dose levels and for the different cellular systems in the human body.

Several specific problems of neutron dosimetry are discussed in the following six posters. McDonald, Chang Ma and Laughlin from New York presented a portable, tissue-equivalent calorimeter. The absolute accuracy of aborbed dose measurements with this instrument is $\pm 2.5\%$. Most of that uncertainty is due to the lack of data on the thermal defect of the Shonka plastic. The authors hope to be able to reduce this uncertainty to less than $\pm 1.5\%$.

Mijnheer and Visser from Amsterdam, Lewis from Teddington, Guldbakke and Lesiecki from Braunschweig, and Zoetelief and Broerse from Rijswijk have prepared a joint poster on the neutron sensitivity of Geiger–Müller counters. The poster reviews old data and reports on new measurements performed with counters of different types and with different shields in the standard neutron fields at Teddington and Braunschweig. The typical k_u-value of 0.5% for neutron energies below 4.2 MeV increased with increasing neutron energy, reaching values of 1.6% to 2.8% for different types of counters at 15 MeV.

The poster by Scarpa, Caffarelli, Massari and Moscati from Casaccia reported on the usefulness of different TLD-detectors and films for mixed field dosimetry. For a degraded fission spectrum of 0.4 MeV mean energy, contaminated with 7 to 12% gamma-rays, the authors tested LiF, BeO, CaF_2 and $CaSO_4$. For the last material they found a high sensitivity of gamma-rays and a very low relative sensitivity for fast neutrons, if the measurements were performed in free air. However, if the detector was irradiated inside a mouse phantom, the k_u-value indicating the neutron sensitivity, raised from 4% to about 140%. Also the k_u-values of the other detectors were considerably increased inside the mouse phantom, in general by a factor of 4. The best material in this respect was BeO. It had the smallest k_u-value when irradiated inside the mouse phantom.

Lewis from Teddington reported on the neutron dosimetry facility of NPL. It is based on a 3-MeV Van de Graaff and two 150-kV generators and utilizes the usual target reactions. Dosimeters may be intercompared in collimated beams or in a low scatter environment.

Zoetelief, Engels and Broerse from Rijswijk and Mijnheer and Visser from Amsterdam presented new data on the displacement correction of cavity ionization chambers in neutron fields. For 14 MeV neutrons and spherical tissue equivalent chambers the authors found a displacement of the measuring point of only one-quarter of the radius of the chamber which is considerably smaller than reported before. In the poster, the authors discuss the reason for this discrepancy.

Rassow and Meissner from Essen proposed a dose-calculation scheme for practical dosimetry in neutron therapy. Their scheme makes use of the total absorbed dose, the gamma-ray-absorbed dose, the effective absorbed dose which takes the neutron RBE into account, the RBE of neutrons and gammas and the usual irradiation and geometrical parameters. All dose calculations are performed with reference to one absolute total absorbed dose reading which is the only one to be calibrated periodically by an external dosimeter. The reference point is at a depth of 5 cm in a water phantom, and all absorbed doses are expressed as depth dose ratios with reference to that point and to the standard geometry. A similar system is proposed for isocentric multiple and moving-field irradiations. The authors base their proposal on the argument that the calibration of a reference point at 5 cm depth is much more precise and reliable than the measurement at the point of maximum dose, which often is inconveniently close to the surface.

Smathers and Graves from Texas A&M University and Almond, Grant and Otte from M. D. Anderson Hospital performed a critical comparison of the neutron yields and spectra of the $p(41)+$Be and $d(49)+$Be reactions. The reaction employing protons produced much more neutrons in the low-energy region than the reaction with deuterons. The spectrum could be hardened significantly by 6-cm polyethylene; however, this filtration reduced the integral neutron yield to 48% of the unfiltered beam and still contained slightly more low-energy neutrons than the spectrum from $d(49)+$Be. However, considering the lower price of the proton machine, the authors believe that the $p(41)+$Be reaction is a serious alternative, if the lower neutron yield can be accepted.

The last poster from Octave-Prignot, van Dam, Meulders and Wambersie from Brussels and Louvain-la-Neuve presented the new therapeutic neutron facility at Louvain-la-Neuve. Fast neutrons from the $d(50)+$Be reaction, a dose rate of 40 rad/min at a target-skin-distance of 157 cm, and a maximum field size of 25×25 cm^2 characterize the therapeutic possibilities. The group intends, however, to use the $p(75)+$Be neutrons in the future.

The discussions of the posters was centered on the problem of a biologically and clinically significant description of the radiation quality of fast neutron and high-LET radiations. In particular, the usefulness of y^*, the saturation corrected, dose average lineal energy, was discussed. During this debate two opposing opinions could be distinguished.

The first group criticized y^* because it did not take into account all radiobiological mechanisms. It was said that

- y^* did not consider the dose dependence of RBE correctly;
- y^* did not consider radiation mechanisms at the molecular level and therefore did not show the correct radiation quality dependence;
- the saturation correction was incorrect because it was based on the results of track segment experiments which reflect a much smaller scale;
- different biological end points show different RBE-values for the same radiation, whereas y^* is constant;
- microdosimetric spectra in general and y^* in particular cannot be used in a predictive

way, but only *a posteriori*, for the explanation of unexpected biological results.

The speakers from the other group expressed a more practical opinion. Whilst they appreciated a good part of the arguments on the fundamental incompleteness of y^*, nevertheless they considered y^* an empirical parameter of practical importance for the needs of clinical radiation therapy. With regard to this practical application they said that:

- y^* is a physical quantity of the radiation and must not be misunderstood as a measure of biological effectiveness for particular end points;
- y^* can be used to predict relative changes of RBE in the phantom, but cannot predict absolute RBE-values.

within the range of neutron energies mainly used in neutron therapy, i.e. between 3 and 15 MeV, and for a given mammalian cell end point, y^* is roughly proportional to the initial slope of the survival curves. Therefore y^* is a fair single parameter description of radiation quality for the purpose of neutron therapy, and could be used to estimate the normal tissue damage produced by different radiations;

- y^* is based on the linear quadratic dose dependence and therefore describes correctly the dose dependence of RBE for those biological end points which follow this law;

for new fast neutron and high-LET radiations of unknown radiation quality the comparison of their microdosimetric spectra with the spectra of known radiations is a useful qualitative tool for estimating the RBE-values to be expected.

It can be concluded that most of the speakers agreed that microdosimetric models need to be refined with regard to a more precise description of radiobiological mechanisms. However, the problem of the usefulness, in clinical radiation therapy, of a physical quantity describing radiation quality, and the future requirements of such a description, remained open.

References

BURGER, G., E. MAIER and A. MORHART (1978) Radiation quality and its relevancy in neutron radiotherapy. In J. BOOZ and H. G. EBERT (Eds.), *Proc. 6th Symp. Microdosimetry*, EUR 6064, Harwood Academic Publishers, London, pp. 451-468.

MENZEL, H. G., A. J. WAKER and G. HARTMAN (1976) Radiation quality studies of a fast neutron therapy beam. In J. BOOZ, H. G. EBERT and B. G. R. SMITH, (Eds.), *Proc. 5th Symp. Microdosimetry*, EUR 5452, Commission of the European Communities, Luxembourg, pp. 591-608.

RAILTON, R., D. PORTER, R. C. LAWSON and W. J. HANNAN (1974) The oxygen enhancement ratio and the relative biological effectiveness for combined irradiations of Chinese hamster cells by neutrons and gamma-rays. *Int. J. Radiat. Biol.* **25**, 121-127.

POSTER SESSION B

Contributions on neutron dosimetry, radiation quality and biological dosimetry

J. BOOZ and J. FIDORRA (Jülich, Federal Republic of Germany): *Application of microdosimetry to radiation therapy with special reference to collimated fast neutrons from 15 MeV D on Be*

The radiation response of human tissues to mixed neutron–gamma beams depends on the radiation quality and the dose contribution of both the fast neutron and the gamma-ray components. Microdosimetric investigations of spectral energy deposition in small volumes offer the possibility of determining the relative dose contribution of the gamma component (Menzel *et al.*, 1977; Fidorra and Booz, 1978) and of evaluating parameters which can be used as a physical definition of radiation quality (Booz, 1978). In particular the parameter zeta* (or $y^* \sim$ zeta*) has been proposed as a measure of radiation quality (Kellerer and Rossi, 1972) because according to the Dual Radiation Action hypothesis (DRA), the initial slope of dose response curves is proportional to this parameter. Therefore, y_n^* and $y_{n+\gamma}$ are used to define the saturation corrected dose averaged lineal energy for, respectively, neutrons and neutrons plus gamma-rays.

Studies by several investigators on the dependence of neutron RBE with depth are ambiguous. Some results indicate a constant RBE with depth (Menzel *et al.*, 1978: Bewley *et al.*, 1977, other authors report a small decrease of neutron RBE with depth (Nias *et al.*, 1971; Zeitz *et al.*, 1975). The probable variations of RBE are analysed on the basis of biological data and of measurements of the local distributions of radiation quality of both the fast neutron and the gamma components.

Changes with depth in the central beam: our own results with fast neutrons from the d(15)+Be reaction with a mean energy 6 MeV, show that there is no change of y_n^* with depth until 25 cm (Fig. 1). Also for 14 MeV neutrons, y^* was found to be nearly independent with depth (Menzel *et al.*, 1977; Schmidt and Magiera, 1977). On the basis of these microdosimetric data it may be concluded that up to a depth of 20-25 cm the relative dose contribution of the gamma component is the only parameter which might change the neutron RBE with depth. Therefore, the data of Railton *et al.* (1974), who investigated the RBE for a mixed neutron-gamma field due to two independent sources of ^{60}Co gamma rays and 14 MeV neutrons, can be used for predicting neutron RBE values of $n + \gamma$ fields in the phantom. An excellent agreement was obtained between our calculation using the DRA hypothesis and the data of Railton *et al.* (1974) for the decrease of RBE with increasing gamma dose fraction. Based upon this agreement the DRA hypothesis was used for obtaining the following predictions, which are listed in Table 1.

Table 1. RBE as a function of gamma dose ratio (Booz and Fidorra, 1978)

Cell line	E (MeV)	$RBE_{80\%}$ (0% γ)	$RBE_{80\%}$ (20% γ)	$RBE_{10\%}$ (0% γ)	$RBE_{10\%}$ (20% γ)
CHO	14	5.9	4.8	2.6	2.3
HeLa	14	4.6	3.8	2.0	1.8
HeLa	6	9.2	7.5	3.5	3.0
T	6	5.6	4.7	2.4	2.1

The predicted slight decrease of neutron $RBE_{10\%}$ with depth is qualitatively in agreement with the results of Nias *et al.* (1971) and Zeitz *et*

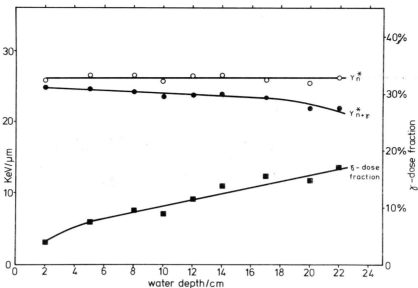

Fig. 1. Radiation quality of 6 MeV neutrons, described by the saturation corrected y*, as a function of water depth. y_n^* and $y_{n+\gamma}^*$ indicate respectively the radiation qualities of only the fast neutrons and the total mixed radiations. In addition, the corresponding gamma dose fraction is shown (Booz and Fidorra, 1978).

al. (1975). Furthermore, if the dose contribution of the gamma component varies between 3 and 15% as shown in Fig. 1 for d(15) + Be neutrons, and also frequently found for other neutron radiations, then the changes in $RBE_{10\%}$ and $RBE_{80\%}$ are small and can perhaps not be detected experimentally.

For water depths larger than 25 cm we found a slight increase in neutron radiation quality (y_n^*).

Lateral displacement in a phantom: for our neutron spectrum and a water depth of 5 cm, with a field size of 8 × 8 cm², the radiation quality of the neutron component (y_n^*) was about 20% larger in the penumbra region than in the beam centre. The mean lineal energy of the mixed radiation ($y_{n+\gamma}^*$) increased up to a maximum of 10% at 7 cm, then decreased to the original value at about 10 cm, and continued to decrease with increasing distance. Hence, for d(15) + Be neutrons, it is to be expected that the biological effectiveness of the scattered neutrons (penumbra region) is slightly larger than that of the primary neutrons. The RBE of the mixed radiation, however, should be approximately constant up to a distance of about 10 cm from the beam axis, and should decrease due to the increasing gamma dose component for larger distances. It must be stressed, however, that this result is valid for the d(15) + Be neutron spectrum and for the described irradiation conditions, and cannot be extrapolated to other irradiation parameters. In a particular case, it will depend on the type of radiation, on the collimator and on the field size whether the increase in neutron effectiveness is compensated by the increasing gamma dose component or whether the latter component is the dominating factor producing a decrease of the total RBE with lateral distance.

References

BEWLEY, D. K., B. CULLEN and B. C. PAGE (1978) Beam profiles measured with physical detectors and a biological dosimeter. In G. BURGER and H. G. EBERT (Eds.), *Proc. Third Symp. on Neutron Dosimetry in Biology and Medicine*, Commission of the European Communities, Luxembourg, pp. 227-234.

BOOZ, J. (1978) Mapping of fast neutron radiation quality. In G. BURGER and H. G. EBERT (Eds.), *Proc. Third Symp. Neutron Dosimetry in Biology and Medicine*, Commission of the European Communities, Luxembourg, pp. 499-514.

FIDORRA, J. and J. BOOZ (1978) The local distribution of radiation quality of a collimated fast neutron beam from 15 MeV deuterons on beryllium. In J. BOOZ and H. G. EBERT (Eds.), *Proc. Sixth Symp. on Microdosimetry*, Commission of the European Communities, Luxembourg (in the press).

KELLERER, A. M. and H. H. ROSSI (1972) The theory of dual radiation action. *Current Topics in Radiat. Res.* **8**, 85-185.

MENZEL, H. G., A. J. WAKER, R. GRILLMAIER and L. BIHY (1978) Radiation quality studies in mixed neutron–gamma fields. In G. BURGER and H. G. EBERT (Eds.), *Proc. Third Symp. on Neutron Dosimetry in Biology and Medicine*, Commission of the European Communities, Luxembourg, pp. 481-496.

NIAS, A. H. W., D. GREENE, and D. MAJOR (1971) Constancy of biological parameters in a 14 MeV neutron field. *Int. J. Radiat. Biol.* **20**, 145-151.

RAILTON, R. D. PORTER, R. C. LAWSON and W. J. HANNAN (1974) The oxygen enhancement ratio and relative biological effectiveness for combined irradiations of Chinese hamster cells by neutrons and γ-rays. *Int. J. Radiat. Biol.* **25**, 121-127.

SCHMIDT, R. and E. MAGIERA (1978) Determination of neutron and gamma spectra in a mixed (n, γ)-field. In G. BURGER and H. G. EBERT (Eds.), *Proc. Third Symp. on Neutron Dosimetry in Biology and Medicine*, Commission of the European Communities, Luxembourg, pp. 443-454.

SEITZ, L., T. R. CANADA, B. DJORDJEVIC, G. DYMBORT, R. FREEMAN, J. C. MCDONALD, J. O'NEIL and J. S. LAUGHLIN (1975) A biological determination of the variation of fast neutron field quality with depth, RBE, and OER. *Radiat. Res.* **63**, 211-225.

G. BURGER and A. MORHART (Neuherberg, Federal Republic of Germany):
Radiation quality for collimated 15-MeV neutron beams

Radiation transport calculations were performed in homogeneous water phantoms by means of a discrete ordinate multigroup transport code (DOT-2). The phantoms were cylinders, the main axis being identical with the beam axis. For a series of ideal geometrical irradiation conditions, as well as two real collimator set-ups in Neuherberg and Rijswijk, neutron and gamma spectra and the corresponding doses inside the phantom were calculated for incoming 15 MeV neutrons.

As a measure for the quality of the neutron field the saturation corrected dose averaged mean lineal energy density y_n^* was calculated. This does not imply any microdosimetric considerations. The only assumption is that mammalian cell killing at doses D_n of 100 rad is roughly proportional to y_n^* computed for targets of micron size. As input function $y_n = f(E_n)$ the recently calculated data of Caswell and Coyne (1978) were used which are based on improved neutron cross-section evaluations.

Local radiation quality depends not only on the neutron spectrum but also on the neutron to gamma-dose ratio. It is quite obvious that the degradation of the neutron field with depth or in the penumbra on one hand, and the build up of the $H(n, \gamma)$-component in these regions on the other hand act opposite with respect to radiation quality and tend to compensate each other. For several collimator situations calculated gamma and neutron doses and a y_γ^* of 2.1 keV/μ for the gamma component were used to derive y_{tot}^* by

$$y_{tot}^* = (D_n/D_{tot})y_n^* + (D_\gamma/D_{tot})y_\gamma^*.$$

The equation is equivalent to

$$D_{eff} = R \cdot D_n + D_\gamma$$

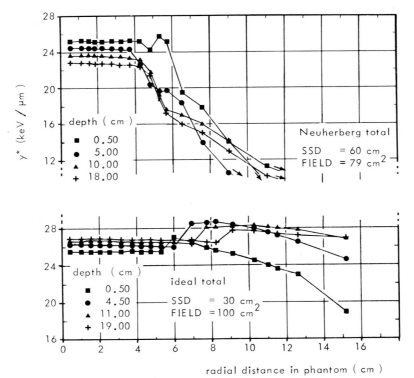

Fig. 2. *The saturation corrected dose averaged mean lineal energy density y_{tot}^* in a water phantom irradiated with 15 MeV neutrons (Burger et al., 1978).*

if one assumes that the y^* values are proportional to corresponding RBEs.

One interesting aspect is, for example, whether in real collimated beams the different tendencies of the two additive terms in the first equation will give rise to a nonmonotonous lateral profile in radiation quality ("edge effect"), as it was recently indicated by proportional counter-measurements (Fidorra and Booz, 1978) and by our own calculations for an ideal aperture. Figure 2 shows the lateral variation of y^*_{tot} in a phantom for an ideal aperture and a collimated beam at Neuherberg (Burger et al., 1978). The occurrence of an "edge effect" depends critically on the relative gamma contribution in the field, and hence on the gamma component from outside and also on the depth in the phantom.

References

BURGER, G., E. MAIER and A. MORHART (1978) Radiation quality and its relevancy in neutron radiotherapy. In J. BOOZ and H. G. EBERT, (Eds.), *Proc. Sixth Symp. on Microdosimetry*, EUR 6064, Harwood Academic Publishers, London, pp. 451-468.

CASWELL, R. S. and J. J. COYNE (1978) Energy deposition spectra for neutrons based on recent cross section evaluations. In J. BOOZ and H. G. EBERT (Eds.), *Proc. Sixth Symp. on Microdosimetry*, EUR 6064, Harwood Academic Publishers, London, pp. 1159-1171.

FIDORRA, J. and J. BOOZ (1978) The local distribution of radiation quality of a collimated fast neutron beam from 15 MeV deuterons on beryllium. In J. BOOZ and H. G. EBERT (Eds.), *Proc. Sixth Sym. on Microdosimetry*, EUR 6064, Harwood Academic Publishers, London, pp. 483-496.

K. GÜNTHER (Berlin, German Democratic Republic): *Theoretical microdosimetry implied in a more general theory of radiation quality*

Microdosimetry deals with single-event size spectra $f_1(z)$ of specific energy z and the corresponding mean values $\bar{z}_F$ and $\bar{z}_D$. As a part of RBE theory, a quite general procedure has been worked out which yields $f_1(z)$ indirectly in terms of the integral

$$A(\eta) = \frac{1}{\bar{z}_F} \int_0^\infty dz\, (1 - e^{-\eta z}) f_1(z)$$

as a function of η.

1. This Laplace transform of $f_1(z)$ is easier to obtain theoretically, allowing compact analytical expressions for any ionizing radiation.
2. It is just this expression, $A(\eta)$, which is needed in the framework of a novel theory of biological radiation effects (Günther et al., 1977)

Utilizing simple numerical methods, one obtains the mean values of specific energy (frequency mean $\bar{z}_F$, dose mean $\bar{z}_D$) as well as the event size spectrum $f_1(z)$ itself from $A(\eta)$:

$$\bar{z}_F = \frac{1}{A(\infty)},\quad \bar{z}_D = \lim_{\eta \to 0} -\frac{d^2 A}{d\eta^2},\quad f_1(z) = L^{-1}[A(\eta)],$$

the inverse Laplace transformation.

Microdosimetric parameters for X-rays, monoenergetic ions and fast neutrons have thus been calculated and tabulated for wide ranges of energies and sensitive site diameters d. The main purpose is to demonstrate the adequacy of this approach by comparison with experimental and other computational data. Figure 3 presents an example.

By specifying only the total energy of absorption events, ordinary microdosimetry is insufficient for the theory of biological effects. One needs a description of the fluctuations of radiation-induced molecular lesions in small volumes, which can only approximately be reduced to the fluctuations of z. The above scheme, however, allows an appropriate generalization which can be expressed formally in terms of "virtual specific energy $\mathbf{z}$". By definition $\bar{\mathbf{z}}_F = \bar{z}_F$, but $\mathbf{z}$ describes "the power of lesion production" of absorption events likewise as the physical energy z describes ionization production power. It is evident that the virtual quantities depend on the particular type of lesion considered, and corresponding modelling will be necessary.

References

BOOZ, J., U. OLDENBURG, and M. COPPOLA (1972) Das Problem der Gewebeäquivalenz für schnelle Neutronen in der Mikrodosimetrie. In G. BURGER, H. SCHRAUBE and H. G. EBERT (Eds.), *Proc. First Symp. on Neutron Dosimetry in Biology and Medicine*, EUR 4896, Commission of the European Communities, Luxembourg, pp. 117-136.

BOOZ, J. and M. COPPOLA (1974) Energy deposition by fast neutrons to small spheres. In J. BOOZ, H. G. EBERT, R. EICKEL and A. WAKER (Eds.), *Proc. Fourth Symp. on Microdosimetry*, EUR 5122, Commission of the European Communities, Luxembourg, pp. 983-1000.

CASWELL, R. S. and J. J. COYNE (1976) Microdosimetric spectra and parameters of fast neutrons. In J. BOOZ, H. G. EBERT and B. G. R. SMITH (Eds.), *Proc. Fifth*

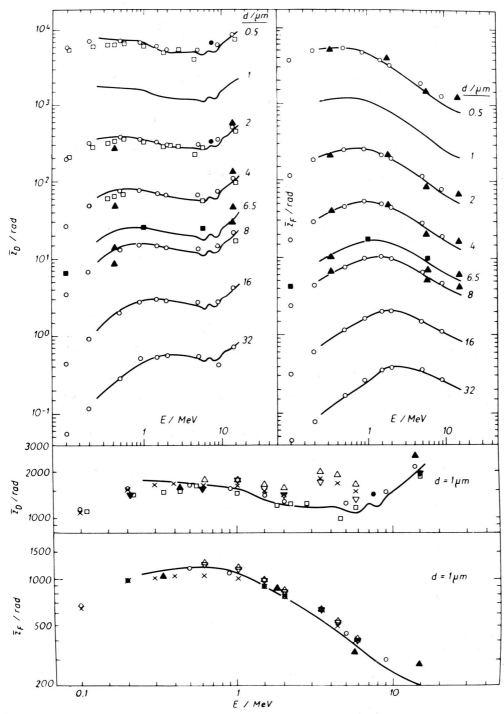

Fig. 3. *Mean values of specific energy per absorption event of fast neutrons versus neutron energy for spherical regions of diameter* d. *Data for comparison: Monte Carlo calculation:* ▽△ *(Booz et al., 1972)*, × *(Booz and Coppola, 1974)*, □ *(Coppola and Booz, 1975)*, ○ *(Caswell and Coyne, 1976) (analytical); measurement:* ▲ *(Rossi, 1966)*, ▼ *(Rodgers et al., 1972)*, □ *(Rodgers and Gross, 1974) (wall-less counter)*, ● *(Menzel et al., 1976)*.

Symp. on Microdosimetry, EUR 5452, Commission of the European Communities, Luxembourg, pp. 97-126.

COPPOLA, M. and J. BOOZ (1975) Neutron scattering and energy deposition spectra. *Radiat. Environm. Biophys.* **12**, 157-168.

GÜNTHER, K., W. SCHULZ, and W. LEISTNER (1977) Microdosimetric approach to cell survival curves in dependence on radiation quality. *Studia Biophysica* **61**, 163-209.

MENZEL, H. G., A. J. WAKER, and G. HARTMANN (1976) Radiation quality studies of a fast neutron therapy beam. In J. BOOZ, H. G. EBERT and B. G. R. SMITH (Eds.), *Proc. Fifth Symp. on Microdosimetry*, EUR 5452, Commission of the European Communities, Luxembourg, pp. 591-608.

RODGERS, R. C., J. F. DICELLO and W. GROSS (1972) Event distributions from monoenergetic neutrons. In *Annual Report on Research Project*, United States Atomic Energy Commission, COO-3243-1, New York, pp. 52-59.

RODGERS, R. C. and W. GROSS (1974) Microdosimetry of monoenergetic neutrons. In J. BOOZ, H. G. EBERT, R. EICKEL and A. WAKER (Eds.), *Proc. Fourth Symp. on Microdosimetry*, EUR 5122, Commission of the European Communities, Luxembourg, pp. 1027-1042.

ROSSI, H. H. (1966) Microdosimetry. In *Biophysical Aspects of Radiation Quality*, Technical Report Series no. 58, International Atomic Energy Agency, Vienna, pp. 81-94.

J. H. HENDRY, D. W. THOMAS, D. GREENE (Manchester, U.K.) and M. BIANCHI (Geneva, Switzerland): *Survival of mouse marrow colony-formers and spermatogonia in the penumbra of the "Hiletron" neutron beam*

Two *in vivo* test systems have been used to quantitate any changes in RBE with distance from the primary 14.7 MeV neutron beam into the penumbra region of the "Hiletron" therapy unit.

A vertical beam was collimated to a nominal 20 × 20 cm² field at 80 cm SSD, where the total dose-rate was 6 rad/min. A long perforated Perspex box, divided into ten equal compartments with area 6 × 6 cm², was placed on the horizontal therapy couch with the centre of the first compartment at the beam axis. Doses were measured at the centre of each compartment using a polythene–ethylene ionization chamber to measure total dose (neutrons plus γ contamination), in conjunction with a carbon-CO_2 chamber to separate the two components of dose. Values of dose-rate and γ-component are shown in Table 2, where limits to the latter values are quoted assuming all neutrons have primary energies (column A), more applicable to the first few compartments, or degraded to fission energies (column B), more appropriate to the last few compartments.

Table 2. Change in dose-rate and γ-component with distance into penumbra

Compartment no.	Distance from beam axis (cm)	Total dose-rate (rad/min)	γ-component (% of dose)	
			A	B
1	0	6.0	4.4	25
2	6	5.7	3.6	24
3	12	3.7	0.64	22
4	18	0.58	5.2	25
5	24	0.32	11	30
6	30	0.20	17	35
7	36	0.14	—	—
8	42	0.11	29	44
9	48	0.093	—	—
10	54	0.083	39	52

A — Calculated assuming all neutrons with primary energy 14.7 MeV.
B — Calculated assuming all neutrons with fission energies.
Values quoted to two significant figures.

The survival of colony-formers (CFU-S) in mouse bone marrow and of spermatogonia was measured in each compartment, after a fixed total dose had been delivered to each. Dose-rate effects are known to be negligible at low doses for both of these systems, and thus any changes in response can be ascribed to differences in beam quality.

Femoral marrow from two mice in each compartment was grafted, soon after each exposure, into lethally irradiated recipient mice for spleen colony development. Survival measurements were transformed into dosage changes, and hence RBE changes, for equivalent survival. At 43 rad neutrons, the RBE did not change significantly from the value of 1.9 in the primary beam, up to 48 cm from the beam axis—only at 54 cm was there a trend towards an increased effectiveness of the radiation.

The testes from other similarly irradiated mice were excised at 96 hr after the exposure, and were fixed, sectioned and stained appropriately. Cross-sections of tubules in stages VI, VII and VIII of development were counted for type B spermatogonia and primary spermatocytes. About forty tubules were scored for each stage at all doses. Measurements of stage frequency in control mice indicate that the data reflect the response of type A2, A3 and A4 spermatogonia, respectively. RBE values at 43 rad neutrons in the primary beam were 2.3 (stage VI) and 2.7 (stages VII and VIII). Changes in RBE with distance were not significant.

These experiments show that up to 48 cm from the central axis of a 20×20 cm^2 field, where the dose-rate is 1 per cent of that in the centre, total doses of 43 rad are about equally effective in these two biological systems. In view of known marked changes in RBE with neutron energy for both tissues, especially for spermatogonia, these results indicate that the near equivalence of "iso-doses" and "iso-effects" is probably fortuitous and could result from a combination of neutron energy degradation (more biologically effective) and increasing γ-component (less effective). Calculations based on the estimates of neutron/γ-ray proportions and on published measurements of sensitivity of both systems as a function of neutron energy support this argument. The above-mentioned equivalence is not expected for biological systems or for dose levels where γ-ray dose-rate effects are significant.

B. Hogeweg, J. Zoetelief and J. J. Broerse (Rijswijk, The Netherlands):
RBE of collimated neutron beams at various positions in a phantom in relation to differences in lineal energy spectra

In the radiotherapy of deep-seated tumours with neutrons, changes in particle spectrum will occur over the irradiated regions due to scattering and absorption interactions of the ionizing radiation with beam-restricting devices and with tissue. Since the biological effectiveness of neutrons is relatively strongly dependent on energy, these changes in spectrum may result in significant variations in the relative biological effectiveness (r.b.e.) with position; this may be relevant for the clinical application of fast neutrons.

To estimate the changes in radiation quality, lineal energy spectra were determined for positions in and outside collimated beams of 15, 6.5 and 0.51 MeV neutrons. The different energy neutrons were produced, respectively, by the $d(.28)+T$, $d(3.5)+D$ and $p(1.34)+T$ reactions and were collimated with the experimental arrangement used for preclinical studies (Broerse *et al.*, 1975; van Peperzeel *et al.*, 1974) at the Radiobiological Institute TNO. The size of the exit field, defined by the tapered steel insert, is 6×8 cm^2. The distributions of the lineal energy (y) were measured with a tissue-equivalent cylindrical proportional counter (Hogeweg, 1978) at three different positions behind the collimator: at the centre of the beam, at the geometrical edge (as defined by the insert) and behind the shielding at a distance of 4 cm from the longest boundary side (equivalent to 7 cm from the centre). The fractional dose distributions derived for centre position and the geometrical edge were almost identical for the three neutron energies used. As a result of the neutron scattering on the inner duct, the distribution for the boundary showed only a slight increase of events with y values around 1000 MeV cm^{-1}. For positions outside the beam behind the shielding, the fractional dose distributions for 15 and 6.5 MeV showed increasing contributions of gamma-rays and attenuated and scattered neutrons. Changes in quality with position were also determined by a biological dosimeter employing cell survival as the biological endpoint. Mammalian cells were irradiated in flasks and tubes at various positions in a cubical water phantom (side lengths 30 cm) with different doses of 6.5 and 15 MeV collimated neutrons. As a typical example, survival data derived for cells irradiated at different depths with 15 MeV neutrons are presented in Fig. 4. Similar survival data have been derived for cells irradiated in positions perpendicular to the beam axis. From these results, it can be concluded that the r.b.e. will not vary significantly in and outside the beam region. For positions in the beam, this is in agreement with the small changes which were observed in the lineal energy spectra. The absence of an r.b.e. change for positions behind the shield is in contrast with the large increase in events having high y values in the fractional dose distribution, but can be explained by the counter-balancing effects of the increasing contribution of gamma rays at these positions.

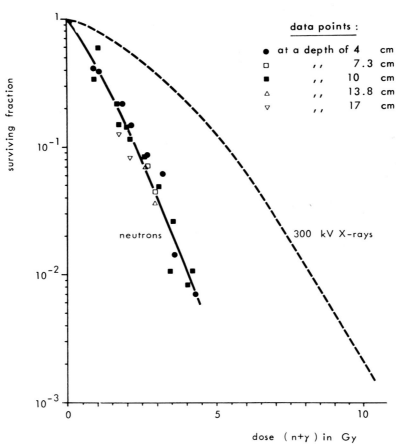

Fig. 4. Survival of cultured T_g cells irradiated in flasks at different depths in a phantom with collimated 15 MeV neutrons. The curve designated as "neutrons" is a mean best-fit survival curve for all data points (Hogeweg et al.).

References

BROERSE, J. J., J. E. BROERS-CHALLISS and B. J. MIJNHEER (1975) Depth–dose measurements of d–T neutrons for radiotherapy applications. *Strahlentherapie* **149**, 585-596.

HOGEWEG, B. (1978) Microdosimetric measurements and some applications in radiobiology and radiation protection. Thesis, University of Amsterdam.

PEPERZEEL, H. A. VAN, K. BREUR, J. J. BROERSE and G. W. BARENDSEN (1974) RBE values of 15 MeV neutrons for responses of pulmonary metastases in patients. *Europ. J. Cancer* **10**, 349-355.

V. E. LEWIS (Teddington, U.K.): *The neutron dosimetry standards facility at the National Physical Laboratory*

The NPL neutron dosimetry facility utilizes three accelerators (see Table 3). A 150-kV SAMES is used to produce $d+T$ neutrons and a 3-MV Van de Graaff for the production of $d+D$, $p+T$ and $d+Be$ neutrons at positions in a low scatter environment. Collimated beams of neutrons from all four reactions, using both Van de Graaff and the purpose-built 150-kV dosimetry accelerator, are available at a position in the Dosimetry Laboratory of the Neutron

Table 3. Fields for neutron dosimetry at NPL (Lewis)

Accelerator	Reaction	Ion energy (MeV)	Neutron energy (MeV)	Environment
SAMES	$^3H(d,n)^4He$	0.15	14.7	Low scatter
Van de Graaff	$^2H(d,n)^3He$	2.3	5.3	Low scatter
	$^3H(p,n)^3He$	2.9	2.1	and
	$^9Be(d,n)^{10}B$	3.0	up to 6	collimated
Dosimetry	$^3H(d,n)^4He$	0.15	14.7	Collimated

Physics building. This is made possible by the use of a bending magnet to direct beams from both accelerators to a common target position. Measurements may be made in air or in a water-filled phantom. Standard cobalt-60 and caesium-137 fields are available for photon calibrations of dosemeters. The sensitivities to thermal neutrons may be measured using the NPL standard thermal flux facility.

JOSEPH C. MCDONALD, I-CHANG MA and JOHN S. LAUGHLIN (New York, U.S.A.): *Calorimetric dosimetry in neutron and charged particle beams*

A portable tissue equivalent (TE) calorimeter, constructed of A-150 plastic, has been employed for the measurement of absorbed dose in several neutron radiotherapy fields and a high-energy proton field. The therapy facilities visited so far include: The University of Washington, Seattle; Fermi National Accelerator Laboratory, Chicago; Texas A and M University, College Station; and the Harvard Cyclotron Laboratory, Cambridge. Comparisons of spherical, cylindrical and thimble-shaped TE ionization chambers have been carried out using either air, or a flow of TE gas in the chambers. The dosimeters were mounted within a 25-cm cubic polystyrene phantom, and measurements were carried out at several depths within A-150 plastic.

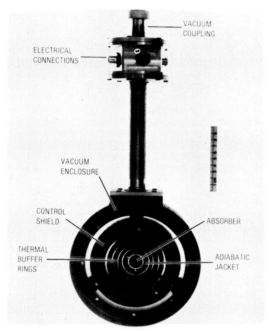

Fig. 5. *Photograph of A-150 plastic calorimeter with cover plates removed (McDonald et al.).*

The design of the instrument is based on the original concept of the local absorbed dose calorimeter, first developed by Laughlin (1952). The calorimeter core is 2 cm in diameter and 2 mm in thickness (see Fig. 5). This element, in which absorbed dose is measured, is thermally isolated from the surrounding Te material by a 0.5-mm vacuum gap around all sides. Therefore, except for the negligible perturbation due to the presence of small vacuum gaps, this calorimetric dosimeter measures absorbed dose in A-150 plastic at a point within a homogeneous TE medium. Specific charge is measured using a cylindrical TE ionization chamber which has dimensions identical to the calorimeter core, or by using either 1 cm^3 or 0.1 cm^3 spherical ionization chambers (E. G. and G., Goleta, California). In this way one can compare the response-sensitivity ratios (ICRU, 1977) for ionization chambers of various types in ^{60}Co gamma-ray and in neutron beams, referenced to the measurement of absorbed dose by the calorimeter. The response-sensitivity ratio for the cylindrical TE ionization chamber filled with methane-based TE gas in neutron fields produced at the Sloan-Kettering cyclotron with mean energies of about 6 and 8 MeV, relative to ^{60}Co, has been determined to be 1.07 ± .02 (McDonald et al., 1977). This quotient is a measure of the product of the mass stopping-power ratio for secondary charged particles between the chamber all and gas ( ) and $\overline{W}_n$, the average energy expended in the gas per ion pair collected. A value of $\overline{W}_n$ for these neutron spectra can be computed by utilizing the Bragg-Gray theory, assuming $\overline{W}_e$ for ^{60}Co to be 29.2 eV/ion pair, and taking ρ to be unity for both radiations. The value we have determined to be applicable for these two broad neutron spectra is 31.2 ± .6 eV/ion pair.

The field experiments carried out to date have been encouraging. The ionization chamber-based dosimetry has been found to be in close agreement to the absorbed dose measured by the calorimeter. Measurements in the 160-MeV proton field at Harvard were carried out at several positions along the spread-out Bragg peak. The dosimeters compared to the calorimeter included ionization chambers, a silicon diode probe and a Faraday cup-based system. The largest differences in dosimetric methods have been generally less than 1-2%. Therefore, it appears that the values employed by the U.S. neutron radiotherapy trial centers for ρ and $\overline{W}_n$ are probably correct to within the limits of experimental uncertainty. In addition, the correction applied to dose obtained using the 1-cm³ spherical chambers in order to account for the material displaced by the collecting volume has been verified to be 0.97 as initially computed by Shapiro *et al.* (1976). This correction can also be made by placing the effective measuring point of the 1-cm³ chamber at approximately $0.8 R$ (where R is the internal radius).

The absolute accuracy of the calorimetric measurement of absorbed dose in A-150 plastic is estimated to be $\pm 2.5\%$. The largest contribution to this quantity is due to the conservative estimate of a $\pm 2\%$ uncertainty in the value of the thermal defect for A-150 plastic, which is taken to be 4% at present (ICRU, 1977). Experiments are now underway (McDonald *et al.*, 1976) to measure the thermal defect in a more complete manner, and this should help reduce the overall uncertainty to better than $\pm 1.5\%$.

References

ICRU (1977) International Commission on Radiation Units and Measurements, *Neutron Dosimetry for Biology and Medicine*, ICRU Report 26.

LAUGHLIN, J. S. (1952) *Biological and Clinical Dosimetry*. Annual Report on Research Project. AEC Contract AT(30-1)-1451.

McDONALD, J. C., J. S. LAUGHLIN and L. J. GOODMAN (1976) Calorimetric dose measurements in fast neutron and cobalt-60 gamma-ray fields. *Proc. of Symp. Measurements for the Safe Use of Radiation*, NBS SP-456, pp. 327-333.

McDONALD, J. C., I. C. MA and J. S. LAUGHLIN (1977) Calorimetric and ionometric dosimetry for cyclotron produced fast neutrons. In G. BURGER and H. G. EBERT, (Eds.), *Proc. Third Symp. on Neutron Dosimetry in Biology and Medicine* Euratom, Brussels, pp. 619-633.

SHAPIRO, P., F. H. ATTIX, L. S. AUGUST, R. B. THEUS and C. C. ROGERS (1976) Displacement correction factor for fast neutron dosimetry in a tissue equivalent phantom. *Med. Phys.* **3**, 87-90.

P. J. MOUNTFORD (Birmingham, U.K.): *Fast neutron spectral measurements at depths in water phantoms*

A fast neutron spectrometer for use in water phantoms has been developed. For good spatial resolution, the detector consists of a small cylindrical 1.5 cm diameter × 1.5 cm length glass cell filled with NE213 organic liquid scintillator. The scintillator cell is remotely coupled to a fast focused photomultiplier tube by a perspex light-guide. Discrimination between fast neutron and gamma-ray events in the scintillator is achieved by pulse shape discrimination based on the zero crossing technique (McBeth *et al.*, 1970). Until an absolute detector response matrix has been derived, accurate measurements pertinent to radiotherapy are limited in energy range.

The system has been used to study, at depths in a water phantom, the effect of field size on the mean energy of a fast neutron spectrum. The fast neutron beam was produced from a ^{238}Pu–^{9}Be isotopic neutron source. The maximum energy from this source was about 11.0 MeV and the minimum unfolded energy was 1.5 MeV. Three field sizes (4.5 × 4.5 cm²; 9.0 × 9.0 cm² and 13.5 × 13.5 cm²) displayed a similar behaviour. The spectrum was a few per cent softer at 2.2 cm depth in water than that at 1.3 cm distance in air from the mouth of the collimator. All three fields continued to soften with distance into the phantom until a depth was reached at which the spectrum started to harden. This depth increased with increase in field size. A fourth field of circular aperture 1.0 cm diameter behaved differently. It appeared to harden continually with depth, but these measurements suffered from large errors due to poor counting statistics. The width of this field was just less than the width of the liquid scintillator cell. Whereas the 1.0-cm diameter field was perhaps more typical of a radiobiology experiment the three larger fields were more representative of a radiotherapy treatment. At the maximum depth studied of 15.0 cm, the mean neutron energy increased with decrease in field size. These results are summarized in Fig. 6. The scatter of the mean neutron energies from a smooth curve for all four fields was small. The reason for this was not immediately clear. It was also not understood why the 4.5 × 4.5 cm² field hardened at a faster rate with depth than the other fields.

Hardening can be explained by the increased cross-section for neutron elastic scattering by hydrogen nuclei preferentially filtering the lower energy component of a fast neutron spectrum. Softening can be explained by the scattering of a high-energy neutron at a point off the axis of the beam, to a lower energy with which it traverses the beam axis at a greater depth. The effect of softening has been shown, as expected, to increase with field size (Mountford *et al.*, 1976). The degree of hardening or softening in each case reported was slight, the maximum difference was only 10%. However, the relative change may well increase with the neutron beams of higher mean energy used in radiotherapy.

From these measurements the mean energy

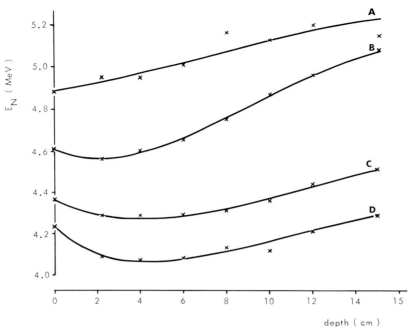

Fig. 6. Variation of mean neutron energy (En) with depth along the central axis of a water phantom for (A) a 1.0-cm diameter circular field aperture: (B) a 4.5 × 4.5 cm² field size; (C) a 9.0 × 9.0 cm² field size and (D) a 13.5 × 13.5-cm field size. The neutron source was a ^{238}Pu–^{9}Be isotopic source (Mountford).

perceived of a fast neutron spectrum depends on the maximum and minimum neutron energy measured, the size of the field, the depth in the phantom, and also the dimensions of the detector in relation to the size of the field. This serves to emphasize the care which must be taken when phantom measurements and radiobiological experiments are extrapolated to radiotherapy.

References

McBeth, G. W., R. A. Winyard and J. E. Luthin (1970) *Pulse Shape Discrimination with Organic Scintillators*, Koch-Light Laboratories Ltd.

Mountford, P. J., B. J. Thomas and J. H. Fremlin (1976) Measurements of neutron spectra at depths in tissue equivalent material for fast neutron beams generated by deuterons on beryllium. *Brit. J. Radiol.* **49**, 630-634.

B. J. Mijnheer, H. Haringa, H. Gorter and B. F. Deys (Amsterdam, The Netherlands): *Measurements of radiation quality of collimated 14 MeV neutrons at different positions in a water phantom*

Variation of beam quality along the central axis or in the penumbra region may occur if a patient or phantom is irradiated with a beam of fast neutrons. These differences in biological effectiveness will result if the neutron spectrum changes or if the relative gamma-ray contribution to the total absorbed dose varies. The changes in radiation quality have been established for the Amsterdam d+T fast neutron therapy facility, which is now routinely used for cancer treatment. In order to study these variations a biological method and physical measurements have been used.

Impairment of the clonogenic capacity of RUC-2 carcinoma cells has been chosen as a biological system. The number of colonies that developed after irradiation on a small area of a culture flask was counted. The measurements have been carried out with cells positioned at the surface and at 10 cm depth in a water phantom. The positions of the cells have been arranged over the central field area and in the penumbra region at neutron dose levels of about 50 per cent and 30 per cent compared to the centre.

The difference in RBE at the surface of the phantom and at 10 cm depth is smaller than about 10 per cent. The ratios of the RBE in the penumbra to that at the central axis are within the experimental uncertainty about equal, which was found at the surface as well as at 10 cm depth.

The changes in relative gamma-ray content of

the beam have been established by means of an energy compensated micro Geiger–Müller tube in combination with a tissue-equivalent ionization chamber.

The changes in neutron energy spectrum have been determined by means of activation and fission counters. A set of twelve reactions was used having threshold energies between about 1 MeV and 12 MeV. The induced radioactivity was determined by means of a calibrated NaI-crystal. The foils were positioned at the same places where the cells have been irradiated.

The neutron energy spectra show a small increase in the number of low-energy neutrons with depth and with distance from the centre. The resulting small increase in biological effectiveness is, however, compensated by the increase in gamma-ray contribution to the total absorbed dose, thus yielding the observed RBE values which remain almost constant with depth and with distance to the centre.

B. J. MIJNHEER, P. A. VISSER (Amsterdam, The Netherlands), V. E. LEWIS (Teddington, U.K.), S. GULDBAKKE, H. LESIECKI (Braunschweig, Federal Republic of Germany), J. ZOETELIEF and J. J. BROERSE (Rijswijk, The Netherlands): *The relative neutron sensitivity of Geiger–Müller counters*

In mixed neutron–gamma fields it is necessary to determine the absorbed dose in tissue of neutrons and photons separately, because of the differences in relative biological effectiveness of these two radiation components. Most instruments that are usually employed to measure photon doses have a relatively large neutron sensitivity which varies also with the neutron energy (ICRU, 1977).

Wagner and Hurst (1961) suggested the use of micro Geiger–Müller counters as photon dosimeters in mixed fields. Their measurements with monoenergetic neutrons showed that the relative neutron sensitivity k_U is less than 0.5 per cent in the energy range from 0.68 to 4.2 MeV. A suitable lead/tin shield was placed around the counter in order to make its response to photons less energy dependent. Thermal neutron response was considerably reduced by an additional shield of ^{6}Li. Recent measurements by Guldbakke et al. (1978), Klein et al. (1978) and Lewis and Hunt (1978) indicate k_U-values of 0.5 per cent to 1.0 per cent from 2.5 to 5.5 MeV. Colvett (1974) determined experimentally that the response for 15 MeV neutrons is predominantly (~90 per cent) due to photons generated locally in the screen resulting in a total relative neutron sensitivity below 0.5 per cent even at this high energy. More recently, however, higher k_U-values have been obtained by Lewis and Young (1977), Hesz et al. (1978), Mijnheer et al. (1978) and Klein et al. (1978). A summary of the new experimental data for k_U (assuming that $h_U = 1$ when calibrated with ^{60}Co gamma-rays) in comparison with other available results is given in Table 4 for the no. 18509 tube (with fixed shield known as ZP 1100) and the no. 18529 tube with different shields as indicated. All counters were irradiated perpendicular to the counter-axis.

Table 4. *Relative neutron sensitivity, k_U, of GM counters with different shield design at different neutron energies (Mijnheer et al.)*

Neutron energy (MeV)	Type of counter	Composition and dimension of shield	k_U (%)	Reference
0.68–4.2	18509	1.35 mm tin + 0.25 mm lead	<0.5	Wagner and Hurst (1961)
2.5	ZP1100	2 mm tin (perforated)	0.49 ± 0.05	Guldbakke et al. (1978)
4.2	ZP1100	,, ,,	0.54 ± 0.06	Lewis and Hunt (1978)
5.0	ZP1100	,, ,,	0.96 ± 0.08	Klein et al. (1978)
5.5	ZP1100	,, ,,	0.57 ± 0.06	Lewis and Hunt (1978)
14.1 (collimated)	ZP1100	,, ,,	2.74 ± 0.40	Mijnheer et al. (1978)
14.7	ZP1100	,, ,,	2.3 ± 0.20	comparison at NPL
15	18509	1.35 mm tin + 0.25 mm lead	<0.5	Colvett (1974)
15.5	ZP1100	2 mm tin (perforated)	2.84	comparison at PTB
15.5	ZP1100	,, ,,	2.86 ± 0.18	Klein et al. (1978)
14.1 (collimated)	18529	1.1 mm tin + 0.5 mm lead	4.9	Hesz et al. (1978)
14.1 (collimated)	18529	1.05 mm tin + 0.55 mm lead (perforated)	1.72 ± 0.25	Mijnheer et al. (1978)
14.7	18529	,, ,, ,,	1.33 ± 0.20	comparison at NPL
14.7	18529	,, ,, ,,	2.15 ± 0.20	Lewis and Young (1977)
15.5	18529	,, ,, ,,	1.71	comparison at PTB
15.5	18529	1 mm tin	1.48	comparison at PTB

The responses of some Geiger–Müller counters of different types and with different shields were compared in the standard $d+T$ neutron fields at NPL (photon component $D_G/D_N = 1.17 \times 10^{-2}$) and at PTB ($D_G/D_N = 0.55 \times 10^{-2}$). At NPL the neutron fluence measurements were performed with an associated particle technique while the photon contribution was determined using Geiger–Müller counters whose k_U-values had been determined by an absolute coincidence technique (Lewis and Young, 1977). At PTB the neutron fluence was determined with a proton recoil telescope and the photon contribution with two different time of flight techniques (Klein et al., 1978). The k_U-values derived from the inter-comparisons are given in Table 4.

From the data in this table it can be concluded that k_U increases with neutron energy and that especially for the higher energies k_U depends on shield design and probably on counter type. For neutron energies up to 5.5. MeV k_U of the ZP 1100 is equal to values between 0.5 and 1.0 per cent. For $d+T$ neutrons in the energy range between 14.1 and 15.5 MeV k_U values of 1.6 and 2.6 per cent appear to be most realistic for the 18529 with Pb/Sn shield and the ZP 1100 Geiger–Müller counters, respectively.

This work was performed within the framework of CENDOS (Cooperative European Research Project on Collection and Evaluation of Neutron Dosimetry Data) and was partially supported by the Commission of the European Communities under contract no. 229-76-10 BIO N.

References

COLVETT, R. D. (1974) Neutron dose response of a Geiger-Müller counter. In *Annual Report Radiol. Res. Lab.*, USAEC Report Coo-3243-3, p. 152.

GULDBAKKE, S., R. JAHR, M. COSACK, H. KLEIN and H. SCHÖLERMANN (1978) A neutron calibration technique for detectors with low neutron and high photon sensitivity. In G. BURGER and H. G. EBERT (Eds.), *Proc. Third Symp. on Neutron Dosimetry in Biology and Medicine*, Commission of the European Communities, Luxembourg. pp. 821-834.

HESZ, A., H. K. KRAUS and H. D. FRANKE (1978) Methods for neutron and gamma dose measurements in a phantom at the d–T neutron therapy facility in Hamburg-Eppendorf. In G. BURGER and H. G. EBERT (Eds.), *Proc. Third Symp. on Neutron Dosimetry in Biology and Medicine*, Commission of the European Communities, Luxembourg, pp. 181-188.

ICRU (1977) Report 26, *Neutron Dosimetry for Biology and Medicine*, ICRU, Washington.

KLEIN, H., S. GULDBAKKE, R. JAHR and H. LESIECKI (1979), the fast neutron sensitivity of a Geiger-Müller counter photon meter by the time-of-flight technique. *Phys. Med. Biol.* **24**, 748–755.

LEWIS, V. E. and D. J. YOUNG (1978) Measurements of the fast neutron sensitivities of Geiger–Müller counter gamma dosemeters. *Phys. Med. Biol.* **22**, 476-480.

LEWIS, V. E. and J. B. HUNT (1978) Fast neutron sensitivities of Geiger–Müller counter gamma dosemeters. *Phys. Med. Biol.* **23**, 888-892.

MIJNHEER, B. J., P. A. VISSER and Tj. WIEBERDINK (1980), to be published.

WAGNER, E. B. and G. S. HURST (1961) A Geiger-Müller gamma-ray dosimeter with low neutron sensitivity. *Health Phys.* **5**, 20-26.

M. OCTAVE-PRIGNOT,* J. VAN DAM,* J. P. MEULDERS† and A. WAMBERSIE* (*Brussels, Belgium, †Louvain-la-Neuve, Belgium): *Dosimetry results for the therapeutic d (50)+Be neutron beam at Cyclone, Louvain-la-Neuve*

Neutron beams used for clinical applications at Cyclone, Louvain-la-Neuve, are produced by bombarding a thick (10 mm) beryllium target with 50 MeV deuterons. Patients are treated at a dose rate of about 0.4 Gy/min (beam current ≈ 4 μA), at a typical target–skin distance (TSD) of 157 cm. The collimation system consists of two parts. The proximal part, in steel, 50 cm thick, determines the maximum available field size (25 × 25 cm²). The distal part consists of interchangeable inserts which determine the actual field sizes. The inserts, 80 cm in length, are made of a mixture of epoxy (25%) and Fe (75%) for the proximal 50 cm, and of a mixture of epoxy (50%) and borax (50%) for the distal 30 cm.

The dosimetric studies included the following parameters for typical treatment conditions: dose rate, depth dose, gamma component, build-up and beam profile. The relative dose rates, as a function of field size, vary from 0.96 to 1.00 and 1.06 for 6 × 8 cm² F, 10 × 10 cm² and 16 × 20 cm² fields, respectively (polystyrene phantom, TSD = 157 cm, depth = 1.25 cm). Variations of dose rate, in air, as a function of TSD, are in accordance with the inverse square law (within about 1%) for TSD between 152 and 167 cm; for smaller distances (and mainly small field sizes;, absorbed dose rates are slightly higher than expected (e.g. 3% at TSD = 147 cm, 6 × 8 cm² field). Depth dose curves in water were measured systematically with an automatic EMI isodose plotter. The data were checked with a TE chamber filled either by TE gas or by air. The depths for 50% absorbed dose ($D_{n+\gamma}$) are given in Table 5.

Table 5. Depths at which dose decreases to 50% of maximum for d (50)+Be neutrons (Octave et al.)

Collimator	Target–skin distance		
	147 cm	157 cm	167 cm
6 × 8 cm²	12.2 cm	12.5 cm	12.8 cm
10 × 10 cm²	13.3 cm	13.6 cm	14.1 cm
16 × 20 cm²	14.9 cm	15.3 cm	15.9 cm

The gamma contribution was measured with a GM counter (Philips ZP1100) calibrated in a ^{60}Co beam. For 10 × 10 cm² collimator, at a TSD of 157 cm, the gamma component was 6.3%, 7.8% and 8.8% at 1.25, 10 and 20 cm depth in water, respectively. Build-up measurements were performed in a polystyrene phantom, with a polystyrene ionization chamber covered with a thin (6 μm) aluminized Mylar wall (see Table 6). Skin sparing improves slightly with increasing TSD.

Table 6. Relative absorbed dose values for the build-up region for d (50)+Be neutrons (Octave et al.)

Depth	Target–skin distance		
	147 cm	157 cm	167 cm
	Collimator–skin distance		
	10 cm	20 cm	30 cm
Surface	62	57	51
1 mm	87	83	80
2 mm	93	91	90
8 mm (maximum)	100	100	100

A typical beam profile is presented in Fig. 7. Variation of the collimator–skin distance, between 10 and 30 cm, does not modify significantly the beam definition: the widths between the 80% and 20% isodoses are 5, 13 and 30 mm at 1.25, 10 and 20 cm in depth, respectively. Flattening filters are foreseen to improve beam homogeneity.

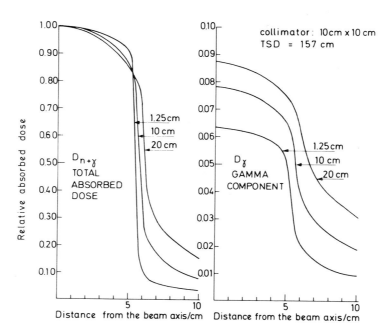

Fig. 7. Beam profiles for the d(50)+Be neutron beam at Louvain-la-Neuve at different depths in a water phantom, 10×10 cm² collimator and target–skin distance of 157 cm (Octave et al.).

J. RASSOW and P. MEISSNER (Essen, Federal Republic of Germany): *Practical dose calculation scheme for neutron irradiations and mathematical description of necessary depth dose and tissue–air ratios*

The practical dose calculation scheme proposed in this contribution is based on a set of five parameters: total absorbed dose (D_t), gamma-ray absorbed dose (D_γ), effective absorbed dose (D_e) and the relative biological effectivenesses for neutrons and gamma-rays (r_n and r_γ) for the specific conditions of the irradiation and geometrical parameters, such as normal treatment distance (s_o = source–isocentre–distance), field size A in distance s_o and depth in the phantom (z for constant source–surface–distance s_o respectively, d for variable source–surface–distance s_o–d in isocentric treatments). All dose calculations are performed with reference to one single absolute total absorbed dose value, respectively, which can easily be calibrated by an external dosemeter. The physical data and radiobiological parameters are strictly separated.

The standardization conditions for the reference total absorbed dose values are the following: field size A_o of 10×10 cm^2: for treatments with constant SSD ($=s_o$) depth in a water phantom z_o of 5 cm $\{D_t(A_o, z_o)\}$, for isocentric treatments with variable SSD $= s_o$–d free-air dose measurement at the isocentre $\{D_t(A_o, 0)\}$.

Quantitative information is provided for the $d(14)$Be–neutron beam produced by the Essen cyclotron facility CIRCE with $s_o=125$ cm for reference depth-dose ratios $p_t(A,z) = D_t(A,z)/D_t(A_o,z_o)$ and $p_\gamma(A,z) = D_\gamma(A,z)/D_t(A_o,z_o)$ for phantom depths from 0.2 cm to 30 cm and field sizes from 5×5 cm^2 to 20×20 cm^2.

For isocentric multiple and moving-field techniques separate calculation and measurements are performed for a constant source–isocentre–distance s_o. Quantitative informations are given for $s_o=125$ cm of the Essen cyclotron facility CIRCE for the reference tissue–air ratios $q_t(A,d) = D_t(A,d)/D_t(A_o,0)$ and $q(A,d) = D_\gamma(A,d)/D_t(A_o,0)$ for phantom depths d between 0.2 cm and 24 cm and field sizes between 5×5 cm^2 and 20×20 cm^2.

The effective absorbed dose can be derived from the measured reference total absorbed-dose values and the reference depth–dose ratios p, respectively, the reference tissue–air ratios q by the relations:

$$\text{SSD} = s_0 : \quad D_e(A,z) = D_t(A_0, z_0)$$
$$\left\{ p_t(A,z) - \frac{r_n - r_\gamma}{r_n} p_\gamma(A,z) \right\}$$

$$\text{SSD} = s_0 - d: \quad D_e(A,d) = D_t(A_0, 0)$$
$$\left\{ q_t(A,d) - \frac{r_n - r_\gamma}{r_n} q_\gamma(A,d) \right\}$$

Equations for the mean total effective absorbed dose $\bar{D}_e$ at the isocentre for n single fixed fields, respectively, a moving field are given with quantitative data from CIRCE as example.

By sets of constants, found by empirical fit of the results of calculations to those of measurements, a description of the dependence of the ratios p and q on field size A and phantom depth z, respectively, d is possible. The ratios p_t and q_t for total absorbed dose are given by the sums of the neutron and gamma-ray contributions ($p_n + p_\gamma$) and ($q_n + q_\gamma$), respectively. The difference between the ratios p and q is mainly represented by a square distance factor.

The ratios p_n and q_n are calculated from a set of six constants using a linear function up to a certain depth and an exponential one behind it. The ratios p_γ and q_γ are calculated from a set of twelve constants using a sum of four exponential functions. A quantitative comparison of measured and calculated values of the ratios p and q is made for $d(14)$Be–neutron beam of CIRCE in Essen. Details about the dose calculation scheme can be found elsewhere (Rassow et al., 1978).

Reference

RASSOW, J., E. MAIER and P. MEISSNER (1978) Proposal for a practical dose calculation scheme for neutron irradiations. *Strahlentherapie* 154, 723–730.

G. SCARPA, M. CAFFARELLI, G. MASSARI and M. MOSCATI (Rome, Italy):
Further investigations on the use of TLD, films and ferrous sulphate in the dosimetry of a mixed field

The present investigation was carried out in a biological facility located within the reflector of the TAPIRO fast reactor at Casaccia Nuclear Centre. The neutron spectrum in the facility is basically a degraded fission spectrum (mean energy: 0.4 MeV). The gamma contamination, as measured by a paired chamber technique, ranges between about 7%, for the "empty-facility" condition, and nearly 12% when three mice are irradiated in the facility.

The thermoluminescent dosimeters used were $\frac{1}{8} \times \frac{1}{8}$-in, ^{7}LiF ribbons, $\frac{1}{4}$-in. BeO discs, $\frac{1}{4} \times \frac{1}{4}$-in. CaF2(Dy) chips and the new sintered $\frac{1}{8} \times \frac{1}{8}$-in. CaSO4(Dy)TLD 900 dosimeters. The last material has been manufactured and made available quite recently in a sintered form by a commercial firm; this material displays some peculiar features such as high sensitivity to gamma radiation (by a factor of nearly 10, as compared to LiF dosimeters of the same size) and a very low sensitivity to fast neutrons.

The gamma energy response of sintered

CaSO$_4$(Dy) was found to be comparable to that of CaF$_2$(Dy), with a peak response around 12 at 30 keV; the energy-dependence curve can be reasonably flattened using a filtration of about 0.4 mm lead. The reproducibility of the readouts of a single dosimeter is approximately ± 6% (at 95% probability level), decreasing to less than 2% when the mean value of 10 dosimeters is used for each experimental point. As photographic detector a Kodak M-54 Industrex film was used.

The sensitivity factors to neutrons (k_U) of the above-listed dosimeters were measured in two different conditions:
(a) with the dosimeter directly exposed to the primary neutron spectrum present in the facility, without any moderating material;
(b) with the dosimeters positioned in the centre of a Mix-D block, simulating a mouse.

The k_U values are shown in a graphical form in Fig. 8. It is easy to observe that all k_U values obtained in condition (a) are significantly lower than those obtained in condition (b), probably due to the presence of a certain amount of thermal neutrons generated within the phantom. In this respect, the extreme case is that of sintered CaSO$_4$(Dy) TLD 900, in which the k_U for primary spectrum is about 1.4, i.e. 40 times higher than that within the phantom (0.035 ± 0.002). This behaviour can suggest the presence of a certain amount of high-cross-section components or impurities in the sintered chips. Beryllium oxide, on the other hand, shows the minimum difference between the two k_U values (0.067 ± 0.003 and 0.042 ± 0.002, respectively) and this feature could be useful when the low-energy tail of the neutron spectrum is largely unknown.

As regards the ferrous sulphate solution (standard Fricke dosimeter) the neutron yield G_N was found 6.6 ± 0.4; independent of the presence or absence of tissue-equivalent material. This value seems in good agreement with the results published by Greene et al. (1973) using a ^{252}Cf source in free air ($G_N = 7.5 ± 15\%$) and in water ($G_N = 6.1 ± 25\%$).

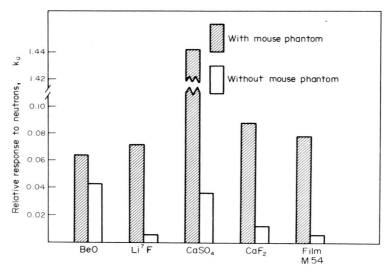

Fig. 8. *Experimental values of the neutron sensitivity coefficient* k_U *for some solid-state dosimeters (Scarpa et al.).*

Reference

GREENE, D., J. LAW and D. MAJOR (1973) The G-value for the ferrous sulphate dosimeter for the radiation from californium-252. *Phys. Med. Biol.* **18**, 800-807.

H. SCHUHMACHER and H. G. MENZEL (Homburg/Saar, Federal Republic of Germany): *Application of experimental microdosimetry to radiation quality with 14 MeV neutrons and cyclotron-produced neutrons**

In radiation therapy with fast neutrons changes of radiation quality within the irradiated patient can be expected. They have to be included in the assessment of both the homogeneity of the biological effective dose to the tumour and the radiation risk for critical

*This work was financially supported by Bundesministerium für Forschung and Technologie (BMFT).

organs in or near the primary beam. Experimental microdosimetry allows a detailed investigation of these local variations. A comparison of two neutron therapy facilities has been made. The influence of beam parameters (e.g. collimation, field size, energy spectrum) and scattering on radiation quality has also been studied.

(simulated diameter 2 μm). Different biological effectivity has to be attributed to each of these components, which represent in a first approximation the dose contribution due to electrons from γ-rays (▲), to protons with energies above 1 MeV (●), to protons with energies below 1 MeV (○) and to α-particles and heavy recoils (■).

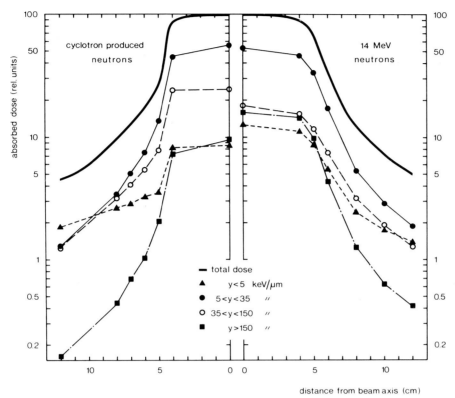

Fig. 9. *Lateral profile of total dose and of four dose components for collimated beams (field size 6 × 8 cm²) of cyclotron-produced neutrons and 14 MeV neutrons at 8 cm depth in a water phantom (Schuhmacher and Menzel).*

Microdosimetric spectra have been measured with a ½-in. Rossi-counter (EG&G) at the two neutron therapy facilities of the Institut für Nuklearmedizin (DKFZ Heidelberg) (Menzel et al., 1978). A broad neutron spectrum ($\bar{E}_n = 8.5$ MeV) is produced at the compact-cyclotron (Schraube et al., 1975) and 14 MeV neutrons are provided by the d+T generator KARIN (Schmidt and Dohrmann, 1976).

At both facilities no marked differences have been observed in the measured energy-deposition spectra with increasing depth in a phantom along the beam axis. Therefore no change of RBE relevant for therapy is expected. Some of the results obtained at lateral distances to the collimated beams are shown in Fig. 9. The upper curves represent the total dose profiles; this total dose has been separated into four components calculated from the microdosimetric spectra

In the beam axis the higher energy of 14 MeV neutrons is reflected by both a higher dose fraction delivered by heavy particles and a lower fraction due to protons below 1 MeV. Also at larger lateral distances the mean neutron energy for the collimated 14-MeV neutron beam is higher than for cyclotron neutrons, which can be seen most clearly in the difference in the heavy ion components in Fig. 9. This is due to a higher energy of scattered neutrons as well as to a non-negligible fraction of 14 MeV neutrons penetrating the shielding as measurements at various depths and in free air have shown.

In the primary beam up to the beam edge at both facilities significant changes of radiation quality are not observed even for the largest field sizes used in radiotherapy. Outside the direct beams the measured spectra show marked changes which are more pronounced with cyclotron-produced neutrons. With increasing lateral distance two effects of opposite consequence for radiation quality are observed:

The components reflecting the fast neutron fraction (● and ■) decrease more rapidly than total dose and the slower neutron component (○). This alone would lead to a higher biological effectivity.

The photon component (▲) decreases more slowly than any other curve. The corresponding increase of the relative γ-dose fraction on its own would lead to a lower biological effectivity.

At present precise quantitative predictions on biological effectivity cannot be given on the basis of microdosimetric spectra. The results obtained at the two therapy facilities permit, however, to exclude an increase of RBE outside the primary beam. To date radiobiological experiments have been carried out only at the compact cyclotron (Grillmaier et al., 1978; Menzel et al., 1978). They have confirmed the statements on biological effectivity based on microdosimetric measurements both inside and outside the primary beam. Further correlated microdosimetric and radiobiological experiments are required to improve the interpretation of energy-deposition spectra with regard to biological effectivity.

References

GRILLMAIER, R. E., L. BIHY, H. G. MENZEL and H. SCHUHMACHER (1978) Chromosome aberration studies and microdosimetry with radiations of varying quality. In J. BOOZ and H. G. EBERT (Eds.), *Proc. Sixth Symp. on Microdosimetry,* EUR-6064, Brussels, 1978, pp. 987-1000.

MENZEL, H. G., A. J. WAKER, R. E. GRILLMAIER, L. BIHY and G. HARTMANN (1978) Radiation quality studies in mixed neutron-gamma fields. In G. BURGER and H. G. EBERT (Eds.), *Proc. Third Symp. on Neutron Dosimetry in Biology and Medicine,* EUR 5848, Munich, 1977, pp. 481-497.

SCHMIDT, K. A. and H. DOHRMANN (1976) Neutronentherapieanlagen im Deutschen Krebsforschungszentrum Heidelberg. Teil II: Die Hochleistungsgeneratorröhre KARIN für 14-MeV-Neutronen. *Atomkernenergie* 27, 159.

SCHRAUBE, H., A. MORHARDT and F. GRÜNAUER (1975) Neutron and gamma radiation field of a deuterium gas target at a compact cyclotron. In G. BURGER and H. G. EBERT (Eds.), *Proc. Second Symp. on Neutron Dosimetry in Biology and Medicine,* EUR 5273, Munich, 1974, pp. 979-1004.

J. B. SMATHERS, R. G. GRAVES (College Station, Texas, U.S.A.), P. R. ALMOND, W. H. GRANT and V. A. OTTE (Houston, Texas, U.S.A.): *Neutron energy spectra for the 41-MeV Be (p,n) and 49-MeV Be (d,n) reactions—A critical comparison*

Due to economic considerations involved in cyclotron design, the Be (*p,n*) reaction has received considerable attention recently in the United States as a fast neutron radiotherapy source. Several medical centers have submitted proposals to the National Cancer Institute for financial support in constructing a neutron treatment facility based on this reaction using protons with a nominal energy of 42 MeV. Others (Cranberg, 1978) seriously question the wisdom of this neutron source selection because of the large low-energy neutron component characteristic of the reaction. To determine the extent to which this concern is warranted, we have measured the neutron spectra for the 49 MeV Be (*d,n*) reaction used clinically for over 5 years and the 41 MeV Be (*p,n*) reaction with particular emphasis placed on characterizing the spectrum in the 2-10 MeV energy range (Graves et al., 1977, 1978).

The 49 MeV Be (*d,n*) measured spectrum agrees well with earlier data (Meulders et al., 1975). The present data have an average energy and integral yield of 19.7 MeV for neutrons with energies greater than 3 MeV and 5.5×10^8 n/sr·nC for neutrons with energies greater than 4 MeV, whereas Meulder's corresponding values are 19.8 MeV and 5.8×10^8 n/sr·nC. The low-energy neutron component of the spectrum is not nearly as significant as has been reported by others for lower-energy deuterons (Cranberg, 1978; Lone et al., 1977). This contribution is further reduced by the polyethylene flattening filters used to compensate for the preferential forward emission of neutrons in this reaction. Based on recent kerma/fluence data, calculations show that the fraction of the total kerma contributed by 2-10 MeV neutrons is reduced from 13% for the open beam to 9% for the filtered beam (Wells, 1978; ICRU, 1977). Measurements of spectra filtered by copper and/or aluminum, of thicknesses comparable to that of cooling and vacuum jacket materials found commonly in radiotherapy target assemblies, indicate little change in spectral shapes or enhancements of lower-energy neutrons relative to the unfiltered spectrum.

The Be (*p,n*) measured spectra differ in the region above 15 MeV from that published by the Davis group (Johnson, 1978, 1977; Heintz et al.,

1977) but agree with that obtained by Waterman (1978). The low-energy region of the neutron spectrum is greatly enhanced over that obtained using the Be (d,n) reaction for similar energy particles and is not effected greatly by filtration through target-holder materials. The spectrum is significantly hardened by filtration through polyethylene. For neutrons with energies greater than 1.9 MeV, average energies are 17.1, 18.4, 19.5 and 20.4 MeV for polyethylene filter thicknesses of 0, 2, 4 and 6 cm, respectively. The beam-hardening process using 6 cm of polyethylene reduces the integral neutron yield to 48% of the unfiltered beam. The effect of 6 cm of polyethylene on the 1.9-10 MeV energy region contribution to neutron kerma has been calculated to be a reduction from 31% to 18%. The latter value is still twice that associated with the 49 MeV Be(d,n) reaction for this energy range.

References

CRANBERG, L. (1978) Neutrons from deuteron bombardment of Be and D. *Phys. Med. Biol.* **23**, 335.
GRAVES, R. G., J. B. SMATHERS, P. R. ALMOND, W. H. GRANT and V. A. OTTE (1978) Letter to Editor. Submitted to *Medical Physics*.
GRAVES, R. G., J. B. SMATHERS, P. R. ALMOND, W. H. GRANT and V. A. OTTE (1977) Thick target neutron yields from the d-Be reaction at E_d = 15.6, 29.9 and 49 MeV. *Proc. of Symp. on Neutron Cross-sections from 10-40 MeV*, Brookhaven National Laboratory, Upton, New York.
HEINTZ, P., S. W. JOHNSEN and N. F. PEEK. (1977) Neutron energy spectra and dose distribution spectra of cyclotron-produced neutron beams. *Med. Phys.* **4**, 250.
ICRU Report No. 26. (1977) *Neutron Dosimetry for Biology and Medicine*, Washington.
JOHNSEN, S. W. (1977) Protron-beryllium neutron production at 25-55 MeV. *Med. Phys.* **4**, 255.
JOHNSEN, S. W. (1978) Polyethylene filtration of 30 and 40 MeV p-Be neutron beams. *Phys. Med. Biol.* **23**, 499-502.
LONE, M. A., C. B. BIGHAM, J. S. FRASER, H. R. SCHNEIDER, T. K. ALEXANDER, A. J. FERGUSON and A. B. MCDONALD. (1977) Thick target neutron yields and spectral distributions from the ^{7}Li(d,n) and ^{9}Be(d,n) reactions. *Nucl. Instrum. Meth.* **143**, 331-344.
MEULDERS, J. P., P. LELEUX, P. E. MACQ and C. PIRART. (1975) Fast neutron yields and spectra from targets of varying atomic number bombarded with deuterons from 16 to 50 MeV. *Phys. Med. Biol.* **20**, 235-243.
WATERMAN, F. (1978) Private communication. Department of Radiology, University of Chicago, Chicago, Illinois, United States of America.
WELLS, A. (1978) Computer evaluation of dosimetry parameters for fast neutron radiotherapy. Doctoral Dissertation, Texas A&M University, College Station, Texas.

J. ZOETELIEF, A. C. ENGELS, J. J. BROERSE (Rijswijk, The Netherlands), B. J. MIJNHEER and P. A. VISSER (Amsterdam, The Netherlands): *Effective measuring point for in-phantom measurements with ion chambers of different sizes**

Tissue equivalent (TE) ionization chambers are the most commonly used instruments for neutron dosimetry in biology and medicine (Broerse *et al.*, 1978). For free-in-air irradiations, it is generally accepted that the geometrical centre of the chamber is the effective measuring point. For the determination of depth-dose distributions, the effective measuring point of the ionization chamber can be displaced because of the replacement of the phantom material by the gas volume of the cavity. It should be realized that, for photon beams (e.g. ^{60}Co gamma-rays), in general relatively small ion chambers can be used and that displacement corrections are only in the order of a few per cent. The penetration of neutron beams with energies below 20 MeV is less than that of ^{60}Co gamma-rays. Due to the relatively large slope of the neutron depth–dose curves and the use of relatively large ion chambers, displacement corrections up to 10 per cent have been used (Broerse *et al.*, 1978). Only limited information is available at present on the determination of the displacement correction for neutrons (Mijnheer *et al.*, 1975; Shapiro *et al.*, 1976.).

Since the displacement correction is shown to be dependent on the geometry of ion chambers (Dutreix and Dutreix, 1966), a set of three spherical TE chambers with internal diameters of 8, 16 and 32 mm, respectively, and a wall thickness of 2.2 mm was constructed.

Measurements with the three ion chambers, flushed with TE gas, were performed for collimated d+T neutrons and ^{60}Co gamma-rays in a water phantom. The chambers were positioned with their geometrical centres at the same depths. The irradiation conditions were as follows: d+T neutrons Rijswijk: SSD 42 cm, field size 6 × 8 cm² (as used during ENDIP (Broerse *et al.*, 1978)); d+T neutrons

---

*This work was partly supported by the Commission of the European Communities, Brussels, Belgium, under contract no. 199-76 BIO N.

Amsterdam: SSD 80 cm and field size 13 × 16 cm² and ^{60}Co gamma-rays Amsterdam: SSD 80 cm and field size 13 × 16 cm². For ^{60}Co gamma-rays, measurements were also carried out with a Baldwin–Farmer thimble-type ionization chamber. For d+T neutrons, measurements with a disc chamber were also performed to allow comparison with previous results (Mijnheer et al., 1975). For each series of measurements, the dosimeters were calibrated at a ^{137}Cs gamma-ray source and corrections were applied for saturation and polarity effects.

For the measurements with ^{60}Co gamma-rays, a displacement correction of $(0.58 \pm 0.06).r$ was derived for the spherical chambers and of $(0.75 \pm 0.09).r$ for the thimble-type Baldwin-Farmer chamber (r being the radius of the gas volume). These values are in agreement with earlier results (Hettinger et al., 1976; Almond et al., 1978).

For d+T neutrons, a preliminary displacement correction of $(0.26 \pm 0.08).r$ was derived for the spherical chambers, which is much smaller than the values reported earlier (Mijnheer et al., 1975; Shapiro et al., 1976). Mijnheer et al. (1975) derived a displacement correction of between r and $\frac{3}{4}r$ for spherical ionization chambers for d+T neutrons from measurements with a disc-type and a spherical-type ion chamber. The results of Shapiro et al. (1976) indicate comparable values of displacement for d (35)+Be neutrons based on measurements with a spherical and a cylindrical ionization chamber. The dose determined with the disc chamber is considerably lower than that derived from the spherical chamber taking the effective measuring point at the position $0.26.r$ to the front. In previous studies, this result was accepted as confirmation of displacement of the effective point of measurement to the inner front wall. However, it now seems more logical to attribute the observed discrepancy to the complex directional response of the disc chamber for free-in-air and in-phantom measurements (Zoetelief et al., 1978). The difference found in this study between displacement corrections for photons and neutrons might be attributed to differences in direction of scattered radiation in the phantom for these two radiation qualities. Extension of the present measurements to lower energy neutrons are in progress. The displacement corrections presently found might account for some of the differences found in ENDIP for in-phantom measurements (Broerse et al., 1978) and could explain the discrepancies between absorbed dose determinations with TE ionization chambers and fission counters (Zoetelief et al., 1978).

References

ALMOND, P. R., M. BEHMARD and A. MENDEZ (1978) Communication, *Med. Phys.* **5**, 63-64.

BROERSE, J. J., G. BURGER and M. COPPOLA (1978) *A European Neutron Dosimetry Intercomparison Project (ENDIP) — Results and Evaluation*, EUR 6004, Commission of the European Communities, Luxembourg.

DUTREIX, J. and A. DUTREIX (1966) Etude comparée d'une série de chambres d'ionisation dans des faisceaux d'electrons de 20 et 10 MeV *Biophysik* **3**, 249-258.

HETTINGER, G., C. PETTERSSON and H. SVENSSON (1967) Displacement effect of thimble chambers exposed to a photon or electron beam from a Betatron, *Acta Radiologica* **6**, 61-64.

MIJNHEER, B. J., J. E. BROERS-CHALLISS and J. J. BROERSE (1975) Measurements of radiation components in a phantom for a collimated d-T neutron beam. In G. BURGER and H. G. EBERT (Eds.), *Proc. Second Symp. on Neutron Dosimetry in Biology and Medicine*, Commission of the European Communities, Luxembourg, pp. 423-439.

SHAPIRO, P., F. H. ATTIX, C. S. AUGUST, R. B. THEUS and C. C. ROGERS, (1976) Displacement factor for fast neutron dosimetry in a tissue equivalent phantom. *Med. Phys.* **3**, 87-90.

ZOETELIEF, J., J. J. BROERSE and B. J. MIJNHEER (1978) Characteristics of ionization chambers and GM counters employed for mixed field dosimetry. In G. BURGER and H. G. EBERT (Eds.), *Proc. Third Symp. on Neutron Dosimetry in Biology and Medicine*, Commission of the European Communities, Luxembourg, pp. 565-578.

Review of RBE data for cells in culture

ERIC J. HALL* AND ALBRECHT KELLERER†

*Radiological Research Laboratory, College of Physicians & Surgeons of Columbia University, New York, N.Y., U.S.A.
†Institut Für Strahlenkunde der Universität, Wurzburg, Germany

Abstract—*The relative biological effectiveness has been compared for ten neutron facilities used for clinical radiotherapy in the United States, Japan, Continental Europe and Great Britain. Mammalian cells in culture were used and in order to exploit the precision of which the in vitro technique is capable, facilities were intercompared in pairs, within a given experiment on the same day. Tables are presented of the relative potency of the various neutron beams. Determinations of the oxygen enhancement ration (OER) have been made for a wide range of neutrons produced by cyclotrons or linear accelerators using the deuteron or proton on beryllium reactions. The OER increases with increasing energy of the charged particle used from 1.5 for 15 MeV $d^+ \to Be$ to 1.9 for 101 MeV $p^+ \to Be$. For a clinically used 14-MeV $d^+ \to T$ generator the OER was found to be 1.6.*

Introduction

The clinical trial of neutrons involves a substantial cost and investment of effort. Consequently, the need for full cooperation between the few centers using these particles has been recognized from the outset. Each of the neutron therapy beams in clinical use is characterized by a different energy spectrum, and as a consequence there are significant variations in their biological effectiveness. To facilitate the pooling of experience and the comparison of clinical results a number of investigators have compared the various neutron beams using a variety of biological systems, and early results have already been published (Hall, 1977; Hall *et al.*, 1978). The present report summarizes the results of a new series of biological intercomparisons performed by three groups of investigators using mammalian cells cultured *in vitro*.

The design of these experiments was based upon two important considerations.

(a) It is a characteristic of cell-culture experiments that variations *within* an experiment are much smaller than those between experiments. Consequently, to exploit the precision of which the *in vitro* technique is capable, neutron facilities were intercompared in pairs, within a given experiment, using cells from a common culture irradiated on the same day.

(b) A standardized treatment fixture was used at all facilities, constructed of lucite, with space for six tissue culture flasks, and provision for an ionization chamber to be inserted into the jig to determine the dose received at the position occupied by the cells. Full build-up was ensured because the cells were overlaid with 2 cm of tissue culture medium. A substantial international effort to achieve compatible dosimetry was mounted by the physicists at the various installations engaged in neutron therapy, as a consequence of which there is agreement to within ± 1.5% for dose measurements in air. The use of the standard treatment fixture was an attempt to extend the compatible dosimetry to a practical set-up for the irradiation of cell cultures.

RBE intercomparison

The neutron facilities visited, together with their principal characteristics, are listed in Table 1. For the most part, this paper will be concerned with our own experiments in the United States, Japan, Britain and Continental Europe, in which Chinese hamster V79 cells were used for RBE intercomparisons. In a typical experiment, appropriate numbers of cells were plated into Falcon tissue culture flasks and allowed to attach by overnight incubation at 37.5°C. The flasks were then filled brimful with medium, sealed, and the temperature lowered to 17°C. For each

[1] Based on work performed under Contract EP-78-S-02-4733 from the United States Department of Energy, and Grant Number CA-18506 awarded by the National Cancer Institute, Department of Health, Education and Welfare.

Table 1. Clinical neutron facilities intercompared

	Facility	Location	Production process	Energy of accelerated particle (MeV)	Mean neutron energy (MeV)
U.S.A.	Fermilab	Batavia Illinois	$p^+ \to Be$	66	25
	Tamvec	College Station, Texas	$d^+ \to Be$	50	19.3
	NRL/Manta	Washington, D.C.	$d^+ \to Be$	35	14.3
	NASA/Glanta	Cleveland, Ohio	$d^+ \to Be$	25	10
	Univ. Wash.	Seattle, Wash.	$d^+ \to Be$	22	8
U.K.	MRC, Hammersmith	London	$d^+ \to Be$	16	7
	MRC, Edinburgh	Edinburgh	$d^+ \to Be$	15	6
Netherlands	Antoni van Leuwenhoek Hospital	Amsterdam	$d^+ \to T$	—	14
Japan	NIRS	Chiba	$d^+ \to Be$	30	12
	IMS	Tokyo	$d^+ \to Be$	15	6

experiment, half of the flasks were transported to the Naval Research Laboratory (NRL) cyclotron in Washington, D.C., and half to one of the other facilities listed in Table 1. The cells were transported in insulated water-jacketed carriers, with the temperature maintained at 17°C. It was found by trial and error that this temperature prevents cell division and progression through the cycle, while maintaining a high plating efficiency (which was characteristically in excess of 80%) for a period of 24 hours. Irradiations were performed simultaneously at the two neutron facilities to be compared, and the cells returned to an incubator for 8 days to assess the proportion able to form colonies.

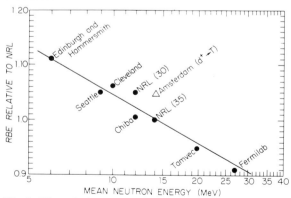

Fig. 1. The relative potency, or RBE relative to the Naval Research Laboratory, as a function of mean neutron energy, for various facilities in clinical use.

The data from each experiment were analysed by a new non-parametric method, which evokes no form for the dose-response relationship. The survival data for the two neutron beams to be compared were fitted by curves of the same shape, the only constraint being that the curve must be convex upwards, and the dose factor necessary to allow a common fit computed. This factor is then the best estimate of the relative potency or RBE difference between the two beams, based on the data from all dose levels studied. The results are shown in Fig. 1; the potency of each beam relative to NRL is plotted as a function of the mean neutron energy. The data points for accelerators using the $d^+ \to Be$ reaction fall close to a common line indicating a gradually increasing RBE with decreasing mean neutron energy.

The data are summarized in a way that is more widely useful in Table 2, which consists of factors relating the RBE's between pairs of machines. This table also includes a summary of data collected by two other investigators. Dr. Paul Todd, Pennsylvania State University, has used T_1 cells of human kidney origin to intercompare a number of neutron beams in clinical use. Dr. Raymond Meyn, M.D. Anderson Hospital, Houston, Texas, used CHO cells for intercomparisons. The details of the experimental methods used by these investigators have been published previously (Gragg *et al.*, 1976; Todd *et al.*, 1978). Both utilized a standardized treatment fixture and most of their recent experiments were designed so that pairs of machines were compared on the same day within the same experiment, though this was not true of earlier studies. The potency factors quoted were calculated at a cell survival

Table 2. *Relative potency of various beams intercompared*
The numbers quoted represent the dose (in Gray) necessary with the facility in the left-hand column to be equivalent to 1 Gray with the facility in the top row. (In each square, the top figure is due to E. J. Hall, the middle to Paul Todd and the lower to Raymond Meyn.)

	IMS	Edin.	Ham.	Amst.	U.W.	Clev.	NIRS	NRL	Tam.	Fermi.
IMS	*	—	—	—	0.96	—	0.82	0.84	—	—
	—	—	—	—	—	—	—	—	—	—
	—	—	—	—	—	—	—	—	—	—
Edin.	—	*	1.00	0.95	0.95	0.95	0.91	0.90	0.86	0.82
	—		—	—	—	—	—	—	—	—
	—		—	—	—	—	—	—	—	—
Ham.	—	1.00	*	0.95	0.95	0.95	0.91	0.90	0.86	0.82
	—	—		—	0.85	0.88	0.76	0.77	0.72	0.84
Amst.	—	1.06	1.06	*	1.00	1.01	0.95	0.95	0.90	0.87
	—	—	—		—	—	—	—	—	—
	—	1.06	1.06	1.00		1.01	0.96	0.95	0.90	0.87
U.W.	1.04	—	—	—	*	—	0.85	0.99	—	—
	—	—	1.18	—		1.04	0.89	0.91	0.85	0.99
	—	1.05	1.05	0.99	0.99		0.95	0.94	0.90	0.86
Clev.	—	—	—	—	—	*	—	0.95	—	—
	—	—	1.13	—	0.96		0.85	0.87	0.81	0.95
	—	1.10	1.10	1.04	1.04	1.05		1.00	0.95	0.91
NIRS	1.22	—	—	—	1.18	—	*	1.00	—	—
	—	—	1.32	—	1.12	1.17		1.02	0.95	1.11
	—	1.11	1.11	1.05	1.05	1.06	1.005		0.95	0.91
NRL	1.19	—	—	—	1.01	1.05	1.00	*	—	—
	—	—	1.30	—	1.10	1.15	0.98		0.93	1.09
	—	1.17	1.17	1.11	1.11	1.12	1.06	1.05		0.96
Tam.	—	—	—	—	—	—	—	—	*	—
	—	—	1.39	—	1.18	1.23	1.05	1.07		1.17
	—	1.22	1.22	1.15	1.15	1.16	1.10	1.10	1.04	
Fermi.	—	—	—	—	—	—	—	—	—	*
	—	—	1.19	—	1.01	1.05	0.90	0.91	0.85	

level of 0.3. In all cases, comparisons have been made on the basis of *TOTAL DOSE*, i.e. neutron and gamma-ray dose. For the European machines, consisting of a $d^+ \rightarrow T$ generator and two lower-energy cyclotrons, the gamma-ray contribution is both larger and more variable than for the higher-energy installations in the United States.

It can be seen from Table 2 that there is, in general, close agreement between the three sets of measurements regarding potency ratios, or RBE differences, between the various neutron beams used clinically. This is particularly true for inter-comparisons between machines within the United States and also for NIRS in Japan. This reflects the greatest effort that has been made to date. By contrast, few direct Transatlantic intercomparisons have so far been completed.

The title of this paper indicates that its scope is restricted to *in vitro* data. However, recent reports (Hall *et al.*, 1978) have summarized RBE intercomparisons performed with *in vivo* systems, such as skin and jejunal crypt cells in mice, as well as for various cultured mammalian cells. The closeness of the agreement observed bears out the suggestion made in the Part II meeting (Hall, 1977) that for intercomparing two neutron beams that differ in energy by only a modest amount, it matters little which biological system is chosen. Systems and endpoints which result in widely different values for the RBE of neutrons relative to X-rays give similar values for the RBE *difference* between two closely related neutron energies. Consequently, the choice of a biological system for intercomparisons should be governed largely by its portability, repeatability and convenience.

OER survey

No review of neutron data for cells in culture would be complete without reference to values obtained for the oxygen enhancement ratio (OER). During the past 2 years a standard biological technique has been used to determine the OER for neutron beams generated by the $d^+ \rightarrow Be$ or $p^+ \rightarrow Be$ processes, where the energy of the accelerated charged particle has ranged from 15 to 101 MeV. For these experiments hamster V79 cells were used and hypoxia produced by crowding a large number of cells into a small volume so that oxygen was consumed by respiration (Hall *et al.*, 1974). The results are shown in Fig. 2. There appears to be a

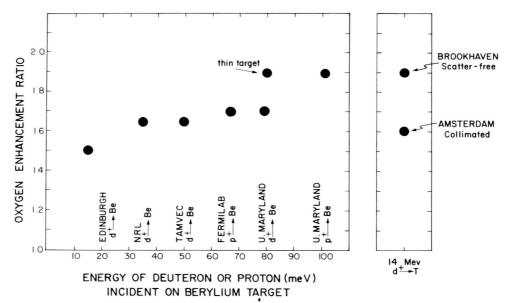

Fig. 2. Values of the oxygen enhancement ratio (OER) for various neutron beams produced by the $p^+ \to Be$, or $d^+ \to T$ reactions. Experiments at the University of Maryland at 80 MeV were repeated with a thick beryllium target and with a thin Be and Al target; the latter resulted in a neutron spectrum with a higher mean energy.

continuous increase in OER with mean neutron energy, and our experience does not confirm the lower OER at high neutron energies previously reported (Harrison et al. 1975, 1976). Using the same biological system, it has been shown also that a *collimated* beam of neutrons from the $d^+ \to T$ generator at Amsterdam is characterized by an OER equal to that of the clinically used cyclotrons in the United States. The value obtained, however, is significantly *lower* than the OER previously reported for 14 MeV $d^+ \to T$ neutrons in a scatter free environment.

Acknowledgements—The implementation of this project was totally dependent upon the cooperation and dedication of the staff at the various neutron facilities where experiments were performed. Special thanks are due to Drs. D. Bewley, J. Parnell and S. B. Field at the Hammersmith Hospital, London; Dr. Peter Bonnett at the Western General Hospital, Edinburgh; Drs. B. J. Mijnheer and K. Breur at the Antoni van Leeuwenhoek Hospital, Amsterdam; Drs. J. Broerse and G. W. Barendsen at the Radiobiological Institute TNO, The Netherlands; Drs. K. Misonon, H. Ohara, T. Inada, K. Kawashima and Dr. Umegaki at NIRS, Chiba; Dr. S. Suzuki at IMS, Tokyo; Drs. Richard Theus, Leon August, and P. Shapiro at the Naval Research Laboratory, Washington, D.C.; Dr. Juri Eenmaa at the University of Washington, Seattle; Dr. James Smathers at the Texas A&M Variable Energy Cyclotron, College Station, Texas; Dr. Miguel Awschalom at the Fermilab, Batavia, Illinois; Drs. Horton and R. Antunez of the Cleveland Clinic.

References

GRAGG, R. L., R. M. HUMPHREY and R. MEYN (1976) The response of Chinese hamster ovary cells to fast neutron radiotherapy beams. 1. Relative biological effectiveness and oxygen enhancement ratio. *Radiat. Res.* **65**, 71-82.

HALL, E. J. (1977) Radiobiological intercomparison *in vivo* and *in vitro*. Proc. of Particles and Radiation Therapy Second International Conference. *Int. J. Radiat. Oncol. Biol. Phys.* **3**. 195-201.

HALL, E. J., S. LEHNERT and L. ROIZIN-TOWLE (1974) Split-dose experiments with hypoxic cells. *Radiology* **112**, 425-430.

HALL, E. J., H. R. WITHERS, J. P. GERACI, R. E. MEYN, J. RASEY, P. TODD and G. E. SHELINE (1978) Radiobiological intercomparisons of fast neutron beams used for therapy in Japan and the United States. *Int. J. Radiat. Oncol. Biol. Phys.* (in press).

HARRISON, G. H., E. B. KUBICZEK and J. E. ROBINSON (1975) OER of neutrons from 80 MeV deuterons on beryllium. *Brit. J. Radiol.* **48**, 409-410.

HARRISON, G. H., E. B. KUBICZEK and J. E. ROBINSON (1976) OER reductions with high-energy neutrons. *Brit. J. Radiol.* **49**, 733.

TODD, P., J. P. GERACI, P. S. FURCINITTI, R. M. ROSSI, F. MIKAGE, R. THEUS and C. B. SCHROY (1978) Comparison of the effects of various cyclotron-produced fast neutrons on the reproductive capacity of cultured human kidney (T-1) cells. *Int. J. Radiat. Oncol. Biol. Phys. (in press).*

RBE values of fast neutrons for responses of experimental tumours

G. W. BARENDSEN

Radiobiological Institute of the Organization for Health Research TNO, Lange Kleiweg 151, Rijswijk, The Netherlands

Abstract—*A review is presented of RBE values of fast neutrons for different types of responses of experimental tumours, induced by relatively small doses or doses per fraction. RBE values derived from data on growth delay, local control and cell survival are compared with RBE values derived from cell survival curves in vitro and with RBE values of published data on growth delay of lung metastases in man. RBE values range from about 2 to 5, depending on the neutron energy, the tumour system and the type of response considered.*

Introduction

Radiobiological properties of radiations with a high linear energy transfer (LET) can only provide an advantage in the treatment of cancer if the relative biological effectiveness (RBE) for tumour eradication is larger than the RBE for severe late damage to the normal tissues, the integrity and function of which must be maintained or restored after treatment (Barendsen, 1971). The ratio of the RBE values for tumour responses and normal tissue tolerance has been denoted therapeutic gain factor (TGF).

Many experimental studies of responses of normal tissues to fast neutrons in comparison with photons have shown that in addition to physical characteristics of the beam, e.g. the neutron energy spectrum, and parameters of the application, such as the dose, fractionation schedule, total time, and treatment volume, biological factors such as the type of tissue and the response considered can influence the RBE values. For instance, RBE values of fast neutrons for damage to the central nervous system of rats and intestine of mice are at small doses per fraction significantly larger than RBE values for damage to bone marrow and skin (Field, 1977; Geraci *et al.*, 1975; v.d. Kogel and Barendsen, 1973). Furthermore, RBE values for late effects on pig skin were suggested to be larger than RBE values for acute desquamation (Withers *et al.*, 1977). Significant differences have also been shown among RBE values for responses of tumours. It is not clear whether these differences are related to the types of tissues from which the tumours have originated, since even among tumours of similar tissue of origin, different RBE values have been obtained (Howlett *et al.*, 1975).

In addition to differences in intrinsic cellular sensitivity, RBE values for responses of tumours can be influenced by the presence of hypoxic cells and reoxygenation during time intervals in fractionated treatments, and by cell proliferation kinetics causing variations in cell cycle distributions and fractions of cycling cells.

Reviews of the reported data on RBE values of fast neutrons for tumour responses and a discussion of various problems involved in the evaluation of these data, have been published recently (Field, 1976; Rasey, 1977). The purpose of the present review is to compare RBE values for growth delay, tumour eradication and cell reproductive death, obtained for different tumours with relatively small doses or doses per fraction. In addition these RBE values will be compared with RBE values for cell reproductive death of mammalian cells *in vitro* and with RBE values for volume changes of lung metastases in patients (Barendsen and Broerse, 1977; van Peperzeel *et al.*, 1974). These comparisons might provide insights in the relative importance of various factors which influence RBE values for tumour responses. Such insights are required in order to provide a basis for the selection of types of tumours in patients for which the application of fast neutrons might provide an advantage over conventional treatments with X-rays. Clinical trials of fast neutron therapy could show a possible advantage more rapidly and with fewer patients if criteria were available for the recognition of types of tumours for which an advantage of neutron therapy is unlikely to be observed, so that these are not included in the trials.

Differences among responses of tumours

The aim of cancer treatments by irradiation is to prevent all tumour cells from continuing with unlimited proliferation. Failure to eradicate a tumour shows that one or more tumour cells must have retained their clonogenic capacity. Thus reproductive death of cells in tumours due to irradiation is intimately linked with the probability of cure, although host factors might play a part in some cases. Tumour growth delay is not only dependent on the number of surviving cells at the end of a treatment, but also on the proliferation kinetics of clonogenic cells. These proliferation kinetics may depend on the dose of radiation and the radiation quality, but in addition reoxygenation patterns and supply of nutrients and oxygen through blood vessels of the host may play a part (Barendsen, 1974).

In fractionated treatments with intervals of 24 hours as commonly applied, it is evident that cell proliferation kinetics and reoxygenation during these intervals might influence the response to subsequent doses. Therefore, it is possible that for the three types of responses most frequently measured, RBE values for fractionated treatments might not always be closely similar. With respect to the design of experiments, studies of tumour growth delay and cell reproductive survival have an advantage over experiments on tumour eradication because of the possibility to investigate a wider range of doses.

RBE values derived from measurements of fractions of clonogenic cells assayed *in vitro* after irradiation of tumours in living animals are for some tumours, e.g. R-1 tumours in WAG/Rij rats, in good agreement with RBE values for tumour control. This does not apply to all experimental tumours, however, as was demonstrated for an EMT-6 tumour in BALB/c mice, where immunologic responses of the host were shown to be significant (Barendsen and Broerse, 1969, 1970; Rasey, 1977).

With respect to the doses for which RBE values of fast neutrons for responses of experimental tumours must be obtained in order to analyse and interpret variations of RBE values obtained in clinical applications, doses of less than 200 rad of neutrons are most relevant, since for treatments of tumours in patients larger doses per fraction are not commonly applied.

Differences in intrinsic sensitivity and RBE values for tumour cells cultured *in vitro*

Responses of tumours to radiation treatments are determined by the sensitivity of the tumour cells and by various factors influencing the cellular environment. It is therefore of interest to review some of the differences among RBE values of fast neutrons for reproductive death of cells from different tumours, irradiated and assayed in standard conditions *in vitro*. Differences among beams of different laboratories due to variation in neutron energy spectra are reviewed by Hall (1979).

In experiments which have been described in detail elsewhere, impairment of the clonogenic capacity of cells has been measured as a function of the dose of 300 kV X-ray or 15 MeV neutrons for nine different cell lines in culture, derived from various types of tumours in rats and mice, including carcinomas and sarcomas (Barendsen and Broerse, 1977). The survival curves show a large variation in intrinsic sensitivity of the cells as well as in the RBE values. For instance, doses of X-rays required to obtain 50 per cent cell reproductive death vary by a factor of 4, from 115 rad for L5178Y cells from a mouse lymphocytic leukaemia to 450 rad for RUC-2 cells derived from a rat ureter carcinoma.

With 15 MeV neutrons, doses required to obtain 50 per cent reproductive death range from 55 rad for L5178Y cells to 190 rad for RSC-1 cells from a rat skin carcinoma, i.e. differences by a factor of up to 3.6 are observed. At larger doses, corresponding to lower fractions of surviving cells, the differences are smaller, as shown in Table 1. It is evident, however, that even for 15 MeV neutrons at higher doses considerable differences in sensitivity are observed between different cell lines.

Consideration of the results presented in Table 1 indicates further that the doses required to yield 50 per cent reproductive death are in general largest for cells derived from the carcinomas, while the most sensitive cells are those derived from a lymphosarcoma and from a lymphocytic leukemia. Cells from mesenchymal sarcomas have an intermediate sensitivity. It is likely, however, that if more cell lines are investigated, the distributions of sensitivities will show overlapping regions, i.e. it cannot be concluded that cells from all carcinomas are significantly more resistant than cells from all sarcomas.

The RBE values presented in Table 1 clearly show a wide spread from 3.0 to 1.8 for a fraction of 0.5 of surviving clonogenic cells, but the differences become less significant with increasing dose. Furthermore, it is evident that RBE values are smaller for lower fractions of

Table 1. Doses of 300 kV X-rays and of 15 MeV neutrons required to attain specified fractions of survival for cells from different types of animal tumours

Origin of cells:	at surviving fraction of 0.5 $\frac{\text{dose X-rays}}{\text{dose neutrons}}$ = RBE		at surviving fraction of 0.1 $\frac{\text{dose X-rays}}{\text{dose neutrons}}$ = RBE	
Rat skin baso-squamous cell carcinoma RSC-1	400/190	= 2.1	980/520	= 1.9
Rat ureter squamous cell carcinoma RUC-1	270/120	= 2.3	760/375	= 2.0
Rat ureter squamous cell carcinoma RUC-2	450/150	= 3.0	1000/490	= 2.0
Rat mammary adenoma RMA-1	160/90	= 1.8	490/280	= 1.7
Rat mammary adeno carcinoma RMA-2	300/120	= 2.5	780/380	= 2.0
Rat rhabdomyosarcoma R-1	210/100	= 2.1	560/290	= 1.9
Rat osteosarcoma ROS-1	240/110	= 2.2	660/330	= 2.0
Mouse lymphosarcoma MLS-1	120/65	= 1.9	390/220	= 1.8
Mouse lymphocytic leukaemia L5178Y	115/55	= 2.1	290/175	= 1.7

surviving cells corresponding to larger doses, as has been observed for many other systems.

It is further of interest to note that at a fraction of 0.5 of surviving cells, the largest RBE values were derived for cells from two carcinomas, namely RUC-2 and RMA-2, and that rather low RBE values were obtained for cells from a lymphosar-coma and a leukaemia. However, for cells of the RSC-1 carcinoma, which is quite resistant to X-rays, the RBE is only 2.1 at a surviving fraction of 0.5.

In radiotherapy of human cancer relatively small multiple doses of radiation are commonly applied with intervals of 24 hours or longer and consequently data for different cell lines obtained at fractions of surviving cells between 0.8 and 0.2 are of greater interest for the interpretation of tumour responses to fractionated treatments than those at lower fractions of surviving cells. This implies that the large variations observed among responses of cells from different tumours to X-rays and to 15 MeV neutrons, and among RBE values of 15 MeV neutrons, must be expected to play a part in causing differences in responses of the tumours which develop after inoculation of these cells in animals and that similar differences must be expected to be observed for responses of tumours in patients.

For the evaluation of possible clinical advantages from fast neutrons, RBE values for effects on tumours must be compared with RBE values of the dose limiting normal tissue in the treated area. If a tumour consists of cells which are as resistant as RUC-2 cells derived from a rat ureter carcinoma, this tumour would be expected to be quite resistant to X-rays and, in addition, because of the relatively large RBE, the application of 15 MeV neutrons might provide an advantage. If a tumour consists of cells which are as sensitive as MLS-1 cells derived from a lymphosarcoma, it would be expected to be quite sensitive to X-rays and because of the relatively low RBE an advantage would not be expected from the use of 15 MeV neutrons.

Ranges of RBE values for responses of experimental tumours

In Table 2 a summary is presented of RBE values for different responses of experimental tumours, derived from published data for doses of 100 rad and 200 rad of fast neutrons.

These data have been derived from the various reports by interpolation between the experimental data and in a few cases by extrapolation on the basis of general characteristics in the data reported. The RBE values given are relative to 200-300-kV X-rays, while doses of gamma-rays have been multiplied by 0.85.

It is evident that the largest number of experimental tumours has been studied with respect to growth delay. From the studies that, for the small doses evaluated, have yielded comparable data for two endpoints, the conclusion can be drawn that RBE values for different endpoints differ generally less than 10 per cent, i.e. agreement is obtained within the experimental errors. The differences described in some reports might well be due to the fact that at larger doses or doses per fraction the influence of hypoxic cells, of changes in cell cycle distributions and of cell proliferation parameters can become significant.

It can further be concluded that significant differences exist among the reported RBE values, ranging from 2 to 5 for the tumours investigated. This variation is even somewhat larger than the range of 1.8 to 3.0 shown in Table 1 for reproductive death of cultured tumour cells irradiated with 15 MeV neutrons and assayed *in vitro*. This implies that in addition to differences in RBE which might be related to intrinsic cellular sensitivity, other factors might also influence the RBE of tumours. It is furthermore evident from Table 2 that the data are not adequate to derive conclusions about the types of tumours for which RBE values are consistently large, and for which a TGF significantly in excess of 1 could be deduced.

The range of RBE values obtained from experimental tumours may also be compared with the range of RBE values obtained from measurement of growth delay of lung metastases in patients, irradiated with 15 MeV neutrons, which were shown to vary from 1.2 to 4 (van Peperzeel *et al.*, 1974). It can be concluded that the variation of RBE values for these metastases is at least as large as observed for experimental tumours. Although all data on human metastases have been derived for tumours which grow with volume doubling times that are a factor of about 10 larger than those of experimental tumours, the RBE values show a similar range of values.

Unfortunately the available data shown in Table 2 do not indicate specific differences to be correlated with histological characteristics of the tumours. The few values for carcinomas appear to be slightly larger than the values for sarcomas but this cannot, on the present evidence, be considered as significant.

Table 2. *RBE values of fast neutrons for different tumour responses at doses or doses per fraction of 100 and 200 rad of fast neutrons.*

Type of tumour and animal species		growth delay		cell reproductive death		tumour eradication	
		dose in rad		*dose in rad*		*dose in rad*	
		100	200	100	200	100	200
Mean neutron energy 5-10 MeV							
RIB5	sarcoma, rat (1)	4.3	3.7				
RIB5	sarcoma, rat (2)		3.0				
RIB5 (c)	sarcoma, rat (3)		2.7				
SSB1 (a)	sarcoma, rat (3)		2.6				
SSB1 (b)	sarcoma, rat (3)		4.2				
SAB2	sarcoma, rat (3)		2.7				
F	sarcoma, mouse (4)		2.6				
NT	carcinoma, mouse (4)	3.6	3.2				
C3H ma.	carcinoma, mouse (5)	3.0	2.9				3.1
C3H ma.	carcinoma, mouse (6)					3.3	3.4
C3HBA ma.	adenocarcinoma, mouse (7)	3.6	3.1				
EMT-6	sarcoma, mouse (7)	3.2	2.7				3
Mean neutron energy 10-20 MeV							
R-1	sarcoma, rat (8)	3.3	2.5	3.0	2.6		
C22LR	sarcoma, mouse (9)			2.9	2.7		
EMT-6	sarcoma, mouse (10)			5	3.7		
—	fibrosarcoma, mouse (11)					2.1	1.9

(1) Field *et al.*, 1967
(2) Field, *et al.*, 1968.
(3) Howlett *et al.*, 1975.
(4) Denekamp *et al.*, 1976.
(5) Fowler *et al.*, 1973.
(6) Fowler *et al.*, 1976.
(7) Rasey, 1977.
(8) Barendsen and Broerse, 1970.
(9) van Putten *et al.*, 1971.
(10) Phillips *et al.*, 1974.
(11) Mason and Withers, 1977.

References

BARENDSEN, G. W. (1971) Cellular responses determining the effectiveness of fast neutrons relative to X-rays for effects on experimental tumours. *Europ. J. Cancer* **7**, 181-190.

BARENDSEN, G. W. (1974) Characteristics of tumour responses to different radiations and the relative biological effectiveness of fast neutrons. *Europ. J. Cancer* **10**, 269-274.

BARENDSEN, G. W. and J. J. BROERSE (1969) Experimental radiotherapy of a rat rhabdomyosarcoma with 15 MeV neutrons and 300 kV X-rays. I. Effects of single exposures. *Europ. J. Cancer* **5**, 373-391.

BARENDSEN, G. W. and J. J. BROERSE (1970) Experimental radiotherapy of a rat rhabdomyosarcoma with 15 MeV neutrons and 300 kV X-rays. II Effects of fractionated treatments applied five times a week for several weeks. *Europ. J. Cancer* **6**, 89-109.

BARENDSEN, G. W. and J. J. BROERSE (1977) Differences in radiosensitivity of cells from various types of experimental tumours in relation to the RBE of 15 MeV neutrons. *Int. J. Radiat. Oncol. Biol. Phys.* **3**, 211-214.

DENEKAMP, J., S. R. HARRIS, C. MORRIS and S. B. FIELD (1976) The response of a transplantable tumour to fractionated irradiation. II. Fast neutrons. *Rad. Res.* **68**, 93-103.

FIELD, S. B., T. JONES and R. H. THOMLINSON (1967) The relative effects of fast neutrons and X-rays on tumour and normal tissue in the rat. I. Single doses. *Brit. J. Radiol.* **40**, 834-841.

FIELD, S. B., T. JONES and R. H. THOMLINSON (1968) The relative effects of fast neutrons and X-rays on tumour and normal tissue in the rat. II. Fractionation, recovery and reoxygenation. *Brit. J. Radiol.* **41**, 597-607.

FIELD, S. B. (1976) An historical survey of radiobiology and radiotherapy with fast neutrons. *Curr. Top. Rad. Res. Quart.* **11**, 1.86.

FIELD, S. B. (1977) Early and late normal tissue damage after fast neutrons. *Int. J. Radiat. Oncol. Biol. Phys.* **3**, 203-210.

FOWLER, J. F., J. DENEKAMP and S. B. FIELD (1973) RBE values for regrowth of C3H mouse mammary carcinomas after single doses of cyclotron neutrons or X-rays. *Europ. J. Cancer* **9**, 853-857.

FOWLER, J. F., P. W. SHELDON and J. DENEKAMP (1976) Optimum fractionation of the C3H mouse mammary carcinoma using X-rays, the hypoxic-cell radiosensitizer Ro-07-0582, or fast neutrons. *Int. J. Radiat. Oncol. Biol. Phys.* **1**, 579–592.

GERACY, J. P., K. L. JACKSON, G. M. CHRISTENSEN, P. D. THOWER and M. S. FOX (1975) Cyclotron fast neutron RBE for various normal tissues. *Radiology* **115**, 459-463.

HALL, E. J. and A. KELLERER (1979) Review of RBE data for cells in culture. In G. W. BARENDSEN, K. BREUR and J. J. BROERSE (Eds.), Pergamon Press Ltd., Oxford, *Proc. Third Symp. on Fundamental and Practical Aspects of the Application of Fast Neutrons and other High-LET Particles in Clinical Radiotherapy*, pp. 171–174.

HOWLETT, J. F., R. H. THOMLINSON and T. ALPER (1975) A marked dependence of the comparative effectiveness of neutrons on tumour line, and its implication for clinical trials. *Brit. J. Radiol.* **48**, 40-47.

KOGEL, A. J. VAN DER and G. W. BARENDSEN (1974) Late effects of spinal cord irradiation with 300 kV X-rays and 15 MeV neutrons. *Brit. J. Radiol.* **47**, 393-398.

MASON, K. A. and H. R. WITHERS (1977) RBE of neutrons generated by 50 MeV deuterons on beryllium for control of artificial pulmonary metastases of a mouse fibrosarcoma. *Brit. J. Radiol.* **50**, 652-657.

PEPERZEEL, H. A. VAN, K. BREUR, J. J. BROERSE, and G. W. BARENDSEN (1974) RBE values of 15 MeV neutrons for responses of pulmonary metastases in patients. *Europ. J. Cancer* **10**, 349-355.

PHILLIPS, T. L., H. H. BARSCHALL, E. GOLDBERG, K. FU and J. ROWE (1974) Comparison of RBE values of 15 MeV neutrons for damage to an experimental tumour and some normal tissues. *Europ. J. Cancer* **10**, 287-292.

PUTTEN, L. M. VAN, P. LELIEVELD, and J. J. BROERSE (1971) Response of a poorly re-oxygenating mouse osteosarcoma to X-rays and fast neutrons. *Europ. J. Cancer* **7**, 153-160.

RASEY, J. S. (1977) The response of experimental animal tumours to neutron radiation therapy. *Int. J. Radiat. Oncol. Biol. Phys.* **3**, 235-242.

WITHERS, H. R., B. L. FLOW, J. I. HUCHTON, D. H. HUSSEY, J. H. JARDINE, K. A. MASON, G. L. RAULSTON and J. B. SMATHERS (1977) Effects of dose fractionation on early and late skin responses to gamma-rays and neutrons. *Int. J. Radiat. Oncol. Biol. Phys.* **3**, 227-233.

Neutron RBE for normal tissues

STANLEY B. FIELD AND SHIRLEY HORNSEY

M.R.C. Cyclotron Unit, Hammersmith Hospital, Ducane Road, London W12 0HS, U.K.

Abstract—*RBE for various normal tissues is considered as a function of dose per fraction. Results from a variety of centres are reviewed. It is shown that RBE is dependent on neutron energy and is tissue dependent, but is not specially high for the more critical tissues or for damage occurring late after irradiation.*

Introduction

The aim of this report is to review three related questions concerning RBE for normal tissues:

1. What is the variation in RBE between tissues?
2. What is the variation in RBE with neutron energy?
3. Are RBE values for late damage different from those for early effects?

Some factors affecting RBE are firmly established. Hypoxia will cause it to increase, so it is extremely important in experiments to avoid causing tissue hypoxia either by the restraining methods or by anaesthesia. Due to the greater degree of recoverable damage after photons, RBE will vary with dose level. Providing cellular repopulation or other time-dependent repair processes are unaltered by radiation quality, RBE should depend only on the dose per fraction. Much analysis has been based on this assumption (e.g. review by Field, 1976), but supporting evidence is mostly indirect and sometimes equivocal. Also there have been some recent reports indicating qualitative differences between the response of tissues to neutrons and X-rays (Hamlet *et al.*, 1976) and also differences in the time dependent repair process (Chauser *et al.*, 1977; Field and Hornsey, 1977).

An important question is whether values for experimental animals are relevant to man. Only in early skin damage has this been firmly established to be the case. RBE values for Ed = 16 (MeV) Be neutrons, with doses in the clinical range, are shown for the skin of four species in Fig. 1. This represents the most complete collection of data available, nevertheless below 100 rad/fraction of neutrons there is considerable uncertainty. For other tissues the uncertainties are even greater.

RBE for different tissues

It is now clear that for any neutron beam there is a range of RBE values for different tissues. Unfortunately comparisons are difficult because the tissue end-points chosen for study are mostly different between centres and in general the results are not very accurate. The majority of results have been published from three cyclotrons [Hammersmith Ed = 16(MeV)Be; Seattle Ed = 22(MeV)Be; TAMVEC Ed = 50(MeV)Be] and from 14-15 MeV D-T neutrons. The available data are shown in Fig. 2. From these results a summary table has been derived in which early damage to mouse intestine is taken as the reference value, since this endpoint has been used with all the neutron beams (Table 1). Table 1 is divided into groups, i.e. those tissues having a lower RBE than for early reactions in mouse intestine, those with higher values and those probably not different.

The RBE for damage to haemopoietic tissues is low, being about one-half of that for gut in all cases. This is because there is little accumulation and repair of sublethal damage in the cells of the haemopoietic system with X-rays and virtually none with neutrons. RBE therefore primarily reflects differences in D_o which may be less profound than differences in ability to accumulate and repair sublethal damage. RBE is also low for killing and reducing the transformation index of circulating lymphocytes (Hedges and Hornsey, 1978). Lung also appears to have a low value. Skin reactions gave a lower value than gut at Hammersmith, but higher at Seattle. Growing cartilage was low at Hammersmith.

Both oesophagus and tail necrosis had high values, but this is almost certainly because of hypoxia in these tissues under the experimental

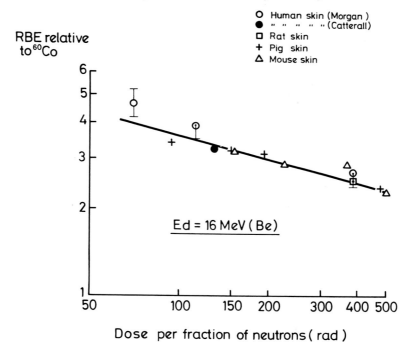

Fig. 1. RBE, relative to ^{60}Co, for skin of four species as a function of dose per fraction of neutrons. 0.9 was taken as the RBE for ^{60}Co.

conditions. Spinal cord gave values not significantly different from those for gut as was the case with the three late endpoints in skin, oral mucosa and kidney from TAMVEC although the tendency was for these to have slightly reduced values. Late gut damage from experiments at Seattle had a higher value but the data are limited and the results not significant. Results from 14 or 15 MeV D-T neutrons suffer from the widest variation, there being differences of about 25% between centres employing different irradiation procedures.

In summary of these results, the RBE for intestinal damage is probably the highest. The only clear exceptions are for tissues which are hypoxic during experiments. The critical tissues and/or those for which damage occurs late after irradiation generally have similar or lower RBE values than for early mouse intestinal injury. There is, however, a marked lack of reliable data, particularly for the same endpoints investigated on more than one machine, from which to draw any firm conclusions.

RBE as a function of neutron energy

It is clearly important to establish RBE values for given endpoints for different neutron beams. Seven beams have been intercompared, D-T neutrons, deuterons from 16 to 50 MeV or protons of 66 MeV on beryllium. In such comparisons accurate dosimetry becomes extremely important as is the need to express neutron dose in the same way. In the present analysis corrections are made to results from Hammersmith to conform with dosimetry from other centres, by reducing Hammersmith quoted RBE values by 5%. A summary of the results is shown in Fig. 3. Most data are for surviving crypts in the small intestine and for skin reactions. It is seen in Fig. 3 that the RBE relative to $E_d = 50$(MeV)Be increases steadily to about 1.35 at $E_d = 16$(MeV)Be, that 14-15 D-T neutrons have RBE values similar to those from $E_d = 25$(MeV)Be. $E_p = 66$(MeV)Be is similar to $E_d = 40$(MeV)Be. There is no indication that such intercomparisons are dependent on fractionation (Withers *et al.*, 1974). Analysis of cell culture experiments yields similar results to those in Fig. 3 (Hall, 1977) but *in vitro* results will be dependent on the shapes of the cell survival curves, which may be different for cells *in vivo*.

RBE for late radiation injury

One of the crucial questions is whether the RBE for late normal tissue injury is greater than that for early damage. Late reactions which have been investigated include damage to lung, skin, spinal cord, oral mucosa, kidney and gut. Data are limited for most of these endpoints but it is seen in Table 1 that all these late endpoints yield similar or lower RBE values than that for early gut damage, with the possible exception of late

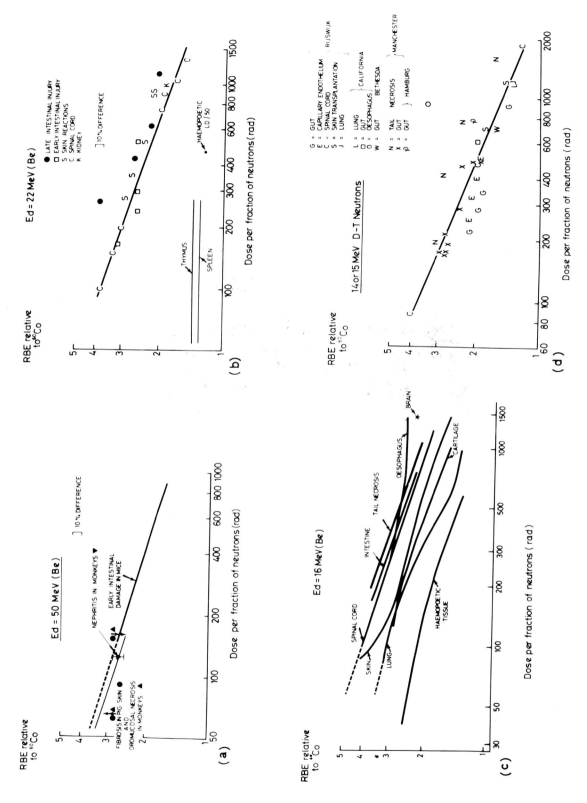

Fig. 2. RBE, relative to ^{60}Co, vs. dose per fraction of neutrons. The RBE for ^{60}Co was taken as 0.9 (a) Data from Withers et al., (1974), Withers et al. (1977), Jardine et al. (1975), Raulston et al. (1978). (b) Data from Geraci et al. (1974, 1976, 1977, 1978), Nelson et al. (1975). (c) Data from White and Hornsey (personal communication), Hornsey et al. (personal communication) and review by Field (1976). (d) Data reviewed by Broerse (1974), from Phillips et al. (1974), Hendry et al. (1975), Hendry et al. (1976) and Zywietz et al. (1978).

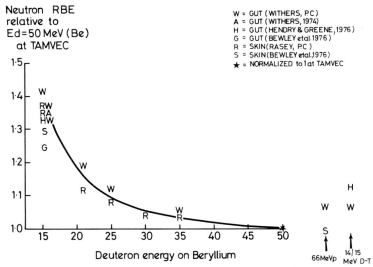

Fig. 3. RBE vs. neutron energy. Data from Withers (personal communication), Withers et al. (1974), Rasey (personal communication), Bewley et al. (1976) and Hendry and Greene (1976).

intestinal injury (Geraci *et al.*, 1977). In some cases early and late injury have been compared in the same tissue. Examples are mice feet (Field, 1969) for which there was no difference in RBE and pig skin for which there was a suggestion of a higher RBE for late skin contraction (Withers *et al.* 1977). The pig skin result may, however, be caused by a higher neutron absorption in subcutaneous fat. Also there was inadequate data on the early pig skin reactions to calculate an RBE with confidence. No increase in RBE for late pig-skin fibrosis was observed at Hammersmith (Bewley *et al.*, 1967). Spinal-cord complications are disastrous and therefore the tissue warrants special attention. Data are limited, but the RBE does not appear to be higher than for early intestinal damage (Table 1). RBE values of 4 or greater have been reported at about 50 rad per fraction of neutrons from Ed = 35(MeV)Be (Bradley *et al.*, 1978; Rogers *et al.*, 1978). These seemingly high values are not inconsistent with RBE for early reactions at the same dose level (see Fig. 2).

The question arises "is early gut injury the

Table 1.

Tissue endpoint	Deuteron energy/beryllium target			D–T 14-15 MeV
	16 MeV	22 MeV	50 MeV	
	Hammersmith	Seattle	TAMVEC	Rijswijk, Manchester and California
Early intestinal injury	1.0	1.0	1.0	1.0
Skin reactions	0.83	(1.10)		
Lung damage	0.70	(0.95)		
Cartilage	0.80			
Haemopoietic tissues	0.53	0.50		0.50
Late skin fibrosis			(0.93)	
Kidney damage		(1.05)	(0.93)	
Oromucosal necrosis			(0.93)	
Late intestinal injury		(1.15)		
Oesophagus	1-1.2	1.3		
Tail necrosis	(1.05)			1.07-1.25
Skin transplantation				(1-1.25)
Spinal cord	(0.95)	(1.0)		(1-1.25)
Capillery endothelium				(0.9-1.1)

correct standard?" The intestinal system has two major advantages: (1) mice are kept for only 4 days after irradiation and (2) it is a cell-survival assay thus giving more information than estimating a gross tissue response. The disadvantages are: (1) there is a large effect of dose rate with photon irradiation so RBE will depend on the photon dose rate, (2) gut is unusual in that the RBE for megavoltage radiation is 1 whereas for most tissues it is about 0.9, and (3) fractionation experiments are complicated by very rapid cellular repopulation. At Hammersmith skin reactions have always warranted most attention and it is known that data from animal skin are comparable with those for man (Fig. 1). However, the RBE for skin is less than for gut at Hammersmith and the situation is reversed at Seattle. At this stage, despite the difficulties mentioned, it is perhaps reasonable to regard mouse intestine as standard but it is important to accumulate more relevant results for intercomparisons.

Slow repair and its influence on late damage.

It has been observed in mouse lung that there is a slow repair process occurring between fractions of X-rays, which is about 100 times slower than repair of sublethal damage, and which does not occur after neutrons (Field and Hornsey, 1977). To what extent this is a general phenomenon is not yet known, but from this result certain effects may be predicted, some of which have been observed, e.g. (1) If there is a slow repair process after X-rays but not after neutrons, then the RBE should increase for protracted fractionation schemes. This has been observed for skin damage with irradiation given weekly for 6 months (Field, 1977). The result implies that the advantage of treating over a long period with photons due to sparing of late damage to normal tissues would not be obtained with neutrons. (2) Slow repair with X-rays may account for residual injury being small, long after photon treatment. After neutrons residual injury would then be expected to be greater and retreatment more hazardous. This has been shown to be the case with tail necrosis (Hendry *et al.*, 1977) and skin damage (Field *et al.*, 1979).

Conclusions

1. The RBE is not the same for all tissues. Early intestinal injury gives amongst the highest values and haemopoietic tissues the lowest.
2. RBE depends on neutron energy in the clinically useful range.
3. RBE for late reactions is not generally higher than for early gut damage, but it may increase for protracted treatments or for retreatment after neutrons.

References

BEWLEY, D. K., B. CULLEN, S. B. FIELD, S. HORNSEY and R. J. BERRY (1976) A comparison for use in radiotherapy of neutron beams generated with 16 and 42 MeV deuterons on beryllium. *Brit. J. Radiol.* **49**, 360-366.

BRADLEY, E. W., P. O. ALDERSON, K. G. MENDENHALL, M. P. FISHER, R. VIERAS and C. C. ROGERS (1978) The effects of fractionated fast neutrons and photons on the canine lung–RBE values from regional pulmonary function studies. *Radiat. Res.* **74**, No. 3, 511-512.

BROERSE, J. J. (1974) Review of RBE values of 15 MeV neutrons for effects on normal tissues. *Europ J. Cancer* **10**, 225-230.

CHAUSER, B., C. MORRIS, S. B. FIELD and P. LEWIS (1977) The effects of fast neutrons and X-rays on the subependymal layer of the rat brain. *Radiology* **122**, (Suppl. 2), 821-823.

FIELD, S. B. (1969) Early and late reactions in skin of rats following irradiation with X-rays or fast neutrons. *Radiology* **92**, 381-384.

FIELD, S. B. (1976) An historical survey of radiobiology and radiotherapy with fast neutrons. *Current Topics in Radiation Research Quarterly*, Vol. 11, 1-86.

FIELD, S. B. (1977) Early and late normal tissue damage after fast neutrons. *Int. J. Radiat. Oncol. Biol. Phys.* **3**, 203-210.

FIELD, S. B. and S. HORNSEY (1977) Slow repair after X-rays and fast neutrons. *Brit. J. Radiol.* **50**, 600-601.

FIELD, S. B., S. MARSTON and R. TOMPKINS (1979) In preparation.

GERACI, J. P., K. L. JACKSON, G. M. CHRISTENSEN, R. G. PARKER, M. S. FOX and P. D. THROWER (1974) The relative biological effectiveness of cyclotron fast neutrons for early and late damage to the small intestine of the mouse. *Europ. J. Cancer* **10**, 99-102.

GERACI, J. P., K. L. JACKSON, G. M. CHRISTENSEN, R. G. PARKER, P. D. THROWER and M. FOX (1976) Single dose fast neutron RBE for pulmonary and oesophageal damage in mice. *Radiol.* **120**, 701-703.

GERACI, J. P., K. L. JACKSON, G. M. CHRISTENSEN, P. D. THROWER and M. MARIANO (1978) RBE for late spinal cord injury following multiple fractions of neutrons. *Radiat. Res.* **74**, 382-386.

GERACI, J. P., K. L. JACKSON, G. M. CHRISTENSEN, P. D. THROWER and B. J. WEYER (1977) Acute and late damage in the mouse small intestine following multiple fractionation of neutrons or X-rays. *Int. J. Radiat. Oncol. Biol. Phys.* **2**, 693-696.

GERACI, J. P., P. D. THROWER and M. MARIANO (1978) Cyclotron fast neutron RBE for late kidney damage. *Radiology*, **126**, 519-520.

HAMLET, R., K. E. CARR, P. G. TONER and A. H. W. NIAS (1976) Scanning electron microscopy of mouse intestine mucosa after cobalt 60 and D–T neutron irradiation. *Brit. J. Radiol.* **49**, 624-629.

HEDGES, M. J. and S. HORNSEY (1978) The effect of X-rays and neutrons on lymphocyte death and transformation. *Int. J. Radiat. Biol.* **33**, 291-300.

HENDRY, J. H. and D. GREENE (1976) Re-evaluation of

published RBE values for mouse intestine. *Brit. J. Radiol.* **49**, 195-196.

HENDRY, J. H., D. MAJOR and D. GREENE (1975) Daily D–T neutron irradiation of mouse intestine. *Radiat. Res.* **63**, 149-156.

HENDRY, J. H., I. ROSENBERG and D. GREENE (1976) Addition of neutron and gamma-ray fractions for intestinal damage. *Radiology* **121**, 483-486.

HENDRY, J. H., I. ROSENBERG, D. GREENE and J. G. STEWART (1976) Tolerance of rodent tails to necrosis after "daily" fractionated X-rays or D–T neutrons. *Brit. J. Radiol.* **49**, 690-699.

HENDRY, J. H., I. ROSENBERG, D. GREENE and J. G. STEWART (1977) Re-irradiation of rat tails to necrosis at six months after treatment with a "tolerance" dose of X-rays or neutrons. *Brit. J. Radiol.* **50**, 567-572.

JARDINE, J. H., D. H. HUSSEY, D. D. BOYD, L. RAULSTON and T. J. DAVIDSON (1975) Acute and late effects of 16– and 50–MeV$_{dBe}$ neutrons on the oral mucosa of Rhesus monkeys. *Radiology* **117**, 185-191.

NELSON, J. S. R., R. E. CARPENTER and R. G. PARKER (1975) Response of mouse skin and the C3H BA mammary carcinoma of the C3H mouse to X-rays and cyclotron neutrons: Effect of mixed neutron–photon fractionation schemes. *Europ. J. Cancer* **11**, 891-901.

PHILLIPS, T. L., H. H. BARSHALL, E. GOLDBERG, K. FU and J. ROWE (1974) Comparison of RBE values of 15 MeV neutrons for damage to an experimental tumour and some normal tissues. *Europ. J. Cancer* **10**, 287-292.

RAULSTON, G. L., K. N. GRAY, C. A. GLEISER, J. H. JARDINE, B. L. FLOW, J. I. HUCHTON, K. R. BENNETT and D. H. HUSSEY (1978) A comparison of the effects of 50 MeV$_{dBe}$ neutron and cobalt-60 irradiation of the kidneys of Rhesus monkeys. *Radiology* **128**, 245-249.

ROGERS, C. C., E. W. BRADLEY, G. W. CASARETT and B. C. ZOOK (1978) Radiation myelopathy in dogs irradiated with fractionated fast neutrons or photons. *Radiat. Res.* **74**, No. 3, 513-514.

WITHERS, H. R., B. L. FLOW, J. I. HUCHTON, D. H. HUSSEY, J. H. JARDINE, K. A. MASON, G. L. RAULSTON and J. B. SMATHERS (1977) Effect of dose fractionation on early and late skin responses to γ-rays and neutrons. *Int. J. Radiat. Oncol. Biol. Phys.* **3**, 227-233.

WITHERS, H. R., K. MASON, B. O. REID, N. DUBRAVSKY, H. T. BARKLEY, B. W. BROWN and J. B. SMATHERS (1974) Response of mouse intestine to neutrons and gamma-rays in relation to dose fractionation and division cycle. *Cancer,* **34**, 39-47.

ZYWIETZ, F., H. JUNG, A. HESS and H. D. FRANKE (1979) Response of mouse intestine to 14 MeV neutrons. *Int. J. Radiat. Biol.* **35**, 63-72.

Fast neutron radiobiology

H. B. KAL AND A. J. VAN DER KOGEL

Radiobiological Institute of the Organization for Health Research TNO, Lange Kleiweg 151, Rijswijk, The Netherlands

It has become increasingly clear during the past several years that possible beneficial results from fast neutron therapy cannot be attributed only to the low oxygen enhancement ratio (OER), and to the presence of hypoxic cells in tumours. Differences with regard to the influence of repair of sublethal and potentially lethal damage among normal tissues and tumours and the reduction of this influence in the case of neutrons might play an equally important role. However, one of the decisive factors for clinical applications of fast neutrons in radiotherapy is whether the relative biological effectiveness (RBE) for late damage in critical normal tissues is lower than the RBE for effects on the tumours, especially at neutron doses of less than 200 rad, that are clinically relevant.

During the past few years, data have been accumulated and reviews of RBE values have been given for cells in culture (Hall, 1979), tumour responses (Barendsen, 1979) and normal tissue tolerances (Field and Hornsey, 1979). Recent results in the area of fast neutron radiobiology are presented in Poster Session C. In this introduction, some aspects of the poster contributions will be considered and some salient discussion items will be summarized.

The effects of fast neutrons on cells cultured in vitro

The first survival curves for mammalian cells obtained with high-LET radiation demonstrated about 20 years ago that the RBE is not constant, but depends on the level of damage or the dose considered (Barendsen *et al.*, 1960). Furthermore, larger RBE values are obtained for fast neutrons of low energies as compared to those obtained with fast neutrons of higher energies (Broerse *et al.*, 1968). In general, the RBE values presented in the poster contributions tend to corroborate these general features and they will be considered in this context.

Chamberlain and Nias determined RBE values for cultured human lymphocytes using the degree of stimulation by PHA as the biological endpoint. In this system the RBE increases with increasing dose. The authors were able to demonstrate that the T lymphocytes consist of two subpopulations with different radio-sensitivity, for which they suggested RBE values of 15 MeV neutrons of 1.7 for the sensitive cells and 7.7 for the resistant subpopulation, based on the ratios of the D_0 values. The increase in RBE with increasing dose, quite in contrast to other cell types, is due to the occurrence of two different subpopulations of T lymphocytes.

Han and Elkind determined RBE values for loss of reproductive capacity of C3H 10T $\frac{1}{2}$ cells, ranging from 2.9 to 2.2, for fission spectrum neutrons, depending on the increasing dose administered. Using the same cell line and neoplastic transformation as the endpoint, RBE values of 2.6 to 10 were obtained. These authors reached the conclusion that the similarity of results of *in vitro* neoplastic transformation and tumour induction in animals indicates that similar mechanisms are involved. This *in vitro* system may therefore be very useful in the study of carcinogenesis. The RBE of 10 for neoplastic transformation at a neutron dose of 20 rad does not necessarily imply a greater production of second tumours after neutron radiotherapy as compared to conventional X-ray therapy. The transformation frequency per surviving cell decreases as a result of fractionation, although that decrease is smaller after neutron irradiation as compared to X-irradiation. However, other factors which may not be observed *in vitro* may play a significant role in living organisms. Withers pointed out that high RBE values observed at low doses are not relevant. It is the absolute effectiveness of the irradiation treatment that counts. At low doses, low-LET radiation is quite ineffective and even high RBE values do not imply a high absolute effectiveness.

Oxygen enhancement ratios were determined by Guichard et al. for neutrons produced by 50 MeV deuterons on Be and 75 MeV protons on Be. The OER values were 1.8-1.9 at the 10 per cent survival level for both neutron energies. Therefore, the high-energy neutrons which have depth-dose characteristics as good as 4 MeV X-rays are candidates for the therapy of deep-seated tumours, without losing the biological advantage of a low OER. The RBE of the neutrons obtained for the 75 MeV protons on Be, however, was 16 per cent lower than the RBE of the (d_{50}-Be) neutrons. This decrease in RBE with increasing neutron energy corresponds with the data of Hall (1979), although the 16 per cent difference quoted is larger than observed in an intercomparison study in the U.S.A. It illustrates that intercomparisons in a single institute with the same cell suspensions and personnel are highly recommended to detect differences of that magnitude.

Rasey studied the repair of potentially lethal damage (p.l.d.) after neutron irradiation of EMT-6 cells in vitro. Repair of p.l.d. occurred in unfed plateau phase cells when subculture was delayed and in exponentially growing cells which were exposed to depleted medium. No p.l.d. repair was seen in fed plateau phase cells. Rasey suggests that post-neutron repair is consistent with the assumption that a repair-promoting substance is released into the medium by plateau phase cells. Raju also observed more p.l.d. repair in plateau phase cells as compared to exponentially growing cells, indicating that G_0 or G_1 cells are involved in the repair of p.l.d. No repair of p.l.d. was observed after 5 MeV alpha irradiation in plateau phase cells or in exponentially growing cells.

Streffer et al. determined changes of nucleic acid synthesis. RBE values were determined for decrease of RNA synthesis in Novikoff hepatoma ascites cells. DNA synthesis was studied in human melanoma cells after neutron doses of 200 and 400 rad. The DNA content per cell was measured by the use of cytofluorometry. After neutron irradiation, a larger number of polyploid cells was observed than are found after X-irradiation.

The effects of fast neutrons on normal tissues

An important aim of studying neutron-induced normal tissue damage is to obtain data on the tolerance of normal tissues to single and fractionated doses for both early and late effects. The application of fast neutrons in radiotherapy introduced by Stone et al. (1940) was abandoned because of unexpected severe late normal tissue reactions. Data for pig skin as well as for rat and mouse skin indicated RBE values which were not different for early and late reactions (Field, 1976). These observations, combined with a better insight in the reasons for the results of the first trial (Sheline et al., 1971), stimulated a renewed interest in the application of fast neutrons at the Hammersmith Hospital, London. In recent publications on early and late reactions of pig skin and mouse intestine, RBE values for late effects were found to be larger than the RBE values for early reactions (Geraci et al., 1977; Withers et al., 1977). Also experiences gained in clinical practice indicated that early reactions of normal tissues were not predictive for the severity of late reactions (Hussey et al., 1974).

Wiernik and Young reported the results of treatments of twenty-five patients with californium-252 neutrons. They did not observe a significant increase in local tumour control rate at an acceptable level of late damage in normal tissues as compared to conventional treatments with low-LET radiations. Because of other aspects, e.g. health physics problems expected to be associated with low neutron doses at low dose rates with an expected high RBE value for cataract induction, they will not continue their pilot study.

Repopulation of the seminiferous tubules of the mouse testis by spermatogenic cells was studied by de Ruiter-Bootsma et al. After a dose of 300 rad of 1 MeV fission neutrons, spermatogenic clones which repopulate the tubules show inhibition of growth, starting at 8 weeks following irradiation. This indicates that complete repopulation is not likely to occur after this dose, in contrast to a complete repopulation occurring after a dose of 100 rad of neutrons. During repopulation after a dose of 150 rad neutrons the stem cells showed increased radiosensitivity when tested with a second dose; this had not returned to its normal value at 12 weeks after irradiation. It would be of interest to compare the effects of neutrons with those of low-LET radiation data to obtain RBE values for the impairment of complete repopulation.

The lung is one of the critical organs in which severe late damage may develop. Travis et al. determined RBE values of 7.5 MeV neutrons for specific histopathological endpoints, i.e. cellular infiltrate, oedema and pathological organization. These values are in general somewhat larger than the RBE for death from lung damage. This study is of interest for a better understanding of the mechanisms of lung damage. So far, RBE values for single doses have been obtained.

Porschen et al. and Zywietz and Jung quoted RBE values for the intestine. Endpoints were the $LD_{50/5}$ and crypt cell survival. RBE values were 2.6 and 2.1 for mean neutron energies of 6 and 14 MeV, respectively. For the bone marrow syndrome, Porschen et al. determined an RBE of 2.1 for 6 MeV neutrons. Zywietz made a plea to standardize the crypt cell survival experiments in order to make intercomparisons more meaningful. He observed that the minimum number of crypts per circumference was obtained earlier after neutron irradiation than after gamma irradiation. Porschen has developed a sensitive biological dosimeter using the IUdR method of labelling of bone marrow cells. An RBE value of 8 for 6 MeV neutrons could be measured at a dose of 0.5 rad.

Streffer et al. studied DNA synthesis in the small intestine after X-irradiation and fast neutron irradiation. As a biological dosimeter the determination of the DNA synthesis appears more sensitive than cell proliferation and is therefore especially useful at small dose fractions.

The effects of fast neutrons on experimental tumours

Responses of several tumours to single doses as well as fractionated doses of fast neutrons were reported. These values must be correlated with normal tissue RBE values for the same neutron energy and neutron doses in order to draw conclusions about a possible therapeutic gain.

To assess responses of cells in tumours after various treatments, tumour cells can be assayed *in vitro*. A large part of the discussions was devoted to questions as to (1) whether RBE values for survival *in vitro* are indicative for the behaviour of tumour cells *in vivo*; (2) the antigenicity of tumours; and (3) cell dispersion methods for selected tumours.

Denekamp et al. obtained RBE values for oxic cells in the carcinoma NT and the sarcomas WHT and S for both X-rays and neutrons combined with the hypoxic cell sensitizer misonidazole. The RBE values obtained in the combination treatments are not above RBE values for skin at any dose level. This suggests that any advantage of neutrons in these two tumours is due to the hypoxic cells present. When all treatments reported are related to a constant degree of skin damage, the therapeutic potential can be compared: fractionated neutron doses or fractionated X-ray doses plus misonidazole are good, but neutrons plus misonidazole might even be better for some schedules. Hyperthermia was used in combination with X-rays. A single dose of X-rays and hyperthermia was found to give a therapeutic gain. For fractionated X-rays and hyperthermia, however, no therapeutic gain could be observed. This type of treatment may be of lesser clinical importance.

In vivo determinations of RBE values of 6 MeV neutrons for responses of solid experimental tumours, the sarcoma 180 and adenocarcinoma EO 771, were made by Porschen et al. They employed a prelabelling technique with iodine-125 or 131. For the average tumour cells labelled 70 hours before treatment (which means that these cells have completed two or three cell cycles after labelling and can then be considered as mostly hypoxic cells) the RBE values are 3.2 and 6 for the two tumours, respectively. For the well-oxygenated cells, these values are 2.7 and 3.5. Growth-delay studies yielded RBE values of 3.4 and 3.3, respectively. Thus, in these two different tumour systems, RBE values are similar for one endpoint; for other endpoints, RBE values may be different. The discrepancy in RBE values observed for the sarcoma 180 with the IUdR method and the classical methods employing TCD_{50} or growth delay can be partly explained by the differences in fractions of anoxic cells in small tumours (10%) and in large tumours (50-60%). For the adenocarcinoma EO 771, the fractions anoxic and euoxic cells are not yet known.

Rasey observed no repair of potentially lethal damage in the mouse EMT-6 tumour *in vivo* after neutron irradiation, in contrast with an observed repair of p.l.d. after X-rays. Only small EMT-6 tumours with volumes between 50 and 200 mm^3 were used. The absence or reduction of repair of p.l.d. after neutron therapy is probably of less importance in the clinical situation where relatively small doses are applied at which p.l.d. repair is very small.

In the discussion, it was stated that p.l.d. repair depends on the tumour size. In general, small tumours show no p.l.d. repair, in contrast to the larger tumours. This might be related to the fraction of hypoxic cells in the tumour, which is size dependent. The differences observed in RBE values for different endpoints, e.g. tumour growth delay, TCD_{50} and *in vitro* survival, were discussed. Rasey stated that the behaviour of tumours treated *in vivo* and left *in vivo* is not necessarily predicted by the behaviour of cells isolated from the tumour and tested *in vitro* for clonogenicity. The antigenicity of the EMT-6 tumour used might play a role in the differences observed. The relatively high total body dose with neutrons might reduce the immune response significantly. McNally, using non-immunogenic tumours treated with hypoxic cell sensitizers in combination with radiation, reported that the enhancement ratios observed were also dependent on the endpoints used. He is of the opinion that methodological differences in the isolation and culture of tumour cells might give rise to the discrepancies. Barendsen pointed out that the discrepancies between RBEs for tumour regrowth or control and tumour cell survival *in vitro* may also be explained by differences in RBE values for the various components of a tumour *in vivo*, e.g. the micro-vasculature or the connective tissue. Also the differences observed between RBEs *in vivo* and *in vitro* may be influenced by the surrounding normal tissues. Rasey reported that the method of tumour disaggregation could influence the shape of the survival curve. Using an enzyme mixture of collagenase, pronase and DNAse as proposed by Brown, a higher cell yield was obtained as compared to the trypsin method. Especially after drug treatment did the two disaggregation methods result in different dose effect curves. Therefore, results obtained with tumours treated *in vivo* and assayed *in vitro* should be used only as a measure of the actual reactions of the tumour if adequate control data are available.

Microdosimetry

In poster Session C, two papers dealt with the problem of correlating energy-deposition processes with responses of the irradiated tissues.

The importance of estimating the RBE in high-LET therapy requires more definitive theoretical guidelines with quantitative predictive power, states Günther. In his theory, molecular characteristics, i.e. relative effectiveness of DNA lesion production, is added to the microdosimetric description of energy absorption at the level of the cell nucleus. This results in a definite scheme for deriving the RBE of any radiation spectrum from the X-ray dose–response relationship and the size of the cell nucleus only. The results obtained so far are in agreement with data obtained for single cell survival as well as for gross tissue reactions, i.e. mortality of mice due to bone marrow and intestinal syndromes.

Pohlit *et al.* stated that, when mammalian cells are irradiated, repairable and irrepairable damage is produced with reaction rate constants η_{AB} and η_{AC}. From experimental results, i.e. from the final slope D_0 and the initial slope of the cell-survival curve, the values of these parameters can be estimated. Since most types of radiation used deliver energy with a spectrum of LET values, the reaction rate constants derived from the experimental data are mean values for a spectrum of LET. Pilot experiments have been performed with radiations of different LET values and the values of the reaction rate constants are plotted versus a single value of LET. From such curves, the mean values of these parameters for any type of radiation with a known distribution of the dose in LET can be reconstructed. The calculated values of η_{AB} and η_{AC} are in good agreement with the experimental results.

References

BARENDSEN, G. W. (1979) RBE-values of fast neutrons for responses of experimental tumours. In G. W. BARENDSEN, K. BREUR and J. J. BROERSE (Eds.), *Proc. Third Symp. on Fundamental and Practical Aspects of Fast Neutrons and other High-LET Particles in Clinical Radiotherapy*, Pergamon Press Ltd., Oxford, pp. 175-179.

BARENDSEN, G. W., T. L. J. BEUSKER, A. J. VERGOESEN and L. BUDKE (1960) Effects of different ionizing radiations on human cells in tissue culture. *Radiat. Res.* **13**, 841-849.

BROERSE, J. J., G. W. BARENDSEN and G. R. VAN KERSEN (1968) Survival of cultured human cells after irradiation with fast neutrons of different energies in hypoxic and oxygenated conditions. *Int. J. Radiat. Biol.* **13**, 559-572.

FIELD, S. B. (1976) An historical survey of radiobiology and radiotherapy with fast neutrons. *Current Topics in Radiat. Res. Quart.* **11**, 1-86.

FIELD, S. B. and S. HORNSEY (1979) Neutron RBE for normal tissues. In G. W. BARENDSEN, K. BREUR and J. J. BROERSE (Eds.), *Proc. Third Symp. on Fundamental and Practical Aspects of Fast Neutrons and other high-LET Particles in Clinical Radiotherapy*, Pergamon Press Ltd., Oxford, pp. 181-186.

GERACI, J. P., K. L. JACKSON, G. M. CHRISTENSEN, P. D. THROWER and B. J. WEYER (1977) Acute and late damage in the mouse small intestine following multiple fractionations of neutrons or X-rays. *Int. J. Radiat. Biol. Phys.* **2**, 693-696.

HALL, E. J. and A. KELLERER (1979) Review of RBE data for cells in culture. In G. W. BARENDSEN, K. BREUR and J. J. BROERSE, (Eds.), *Proc. Third Symp. on Fundamental and Practical Aspects of Fast Neutrons and other High-LET Particles in Clinical Radiotherapy*, Pergamon Press Ltd., Oxford, pp. 171-174.

HUSSEY, D. H., G. H. FLETCHER and J. B. CADERO (1974) Experience with fast neutron therapy using the Texas A&M variable energy cyclotron. *Cancer* **34**, 65-77.

SHELINE, G. E., T. L. PHILLIPS, S. B. FIELD, J. T. BRENNAN and A. RAVENTOS (1971) Effects of fast neutrons on human skin. *Amer. J. Roentgenol.* **111**, 31-41.

STONE, R. S., J. H. LAWRENCE and P. C. AEBERSOLD (1940) Preliminary report on use of fast neutrons in treatment of malignant disease. *Radiology* **35**, 322-327.

WITHERS, H. R., B. L. FLOW, J. I. HUCHTON, D. H. HUSSEY, J. H. JARDINE, K. A. MASON, G. L. RAULSTON and J. B. SMATHERS (1977). Effect of dose fractionation on early and late skin responses to gamma-rays and neutrons. *Int. J. Radiat. Oncol. Biol. Phys.* **3**, 227-233.

POSTER SESSION C

Contributions on fast neutron radiobiology

SUSAN M. CHAMBERLAIN and A. H. W. NIAS* (Glasgow, Scotland): *Variation in neutron RBE values for human lymphocytes*

Lymphocytes have long been assumed to be much more radiosensitive than most other mammalian cells; perhaps suffering interphase rather than mitotic death, following radiation dosage in the therapeutic range. One problem in the investigation of their radiosensitivity is that circulating lymphocytes do not normally replicate so that the customary test for the survival of the reproductive capacity cannot be applied. Another consideration is the fact that lymphocytes consist of different functional populations including "T-cells" and "B-cells" concerned with cell-mediated and humoral immunity, respectively. B-cells are more radioresponsive than T-cells.

Survival data from organ cultures suggest that rat lymphocytes are not particularly radiosensitive; the same applies to lymphocytes in mouse intestine. However, such data depend upon a morphological assay. Transformation by phytohaemagglutinin (PHA), a mitogenic stimulant, of human peripheral blood lymphocytes does enable some measure of replication in T-cells.

Human lymphocytes taken from the peripheral blood were set up in culture medium at a concentration of 10^6 cells/ml. The culture tubes were irradiated with either 250-kV X-rays or $d+T$ neutrons before the addition of PHA. The tubes were incubated for 6 days at 37°C and then labelled with ^{14}C thymidine and prepared for scintillation counting. The results were calculated as percentage stimulation of the cultures compared with a control culture and a dose–response curve was plotted.

Both the X-ray and the neutron curve were seen to have a shoulder, the X-ray curve having the larger shoulder. Also both curves were of a biphasic nature. This clearly indicates that the lymphocytes consist of two subpopulations. PHA is a stimulant of "thymus derived" or T-lymphocytes, and so it is these T-lymphocytes which consist of the two subpopulations. It is possible that the subgroups differ in their degree of radiosensitivity and so show differing gradients in their dose–response curves. Alternatively, they may differ in their affinity for PHA. It is known that PHA confers a slight radioresistance on lymphocytes to which it binds. Thus if subpopulations of T-lymphocytes vary in this affinity they will also vary in their response to radiation.

The change in slope of the X-ray dose–response curve occurs at approximately 500 rad, and for the neutron curve at approximately 250 rad. The curves point to the subpopulations being about 58% more radiosensitive and 42% being less radiosensitive than the average.

Due to these complex dose–response curves the neutron RBE values vary considerably with dose, the values rising with increasing dose, slightly over the sensitive range but more considerably over the resistant range (see Table 1).

Table 1. *RBE of $d+T$ neutrons for transformation of human lymphocytes*

% Transformation	Neutron RBE
90	1.90
80	1.99
70	2.09
60	2.15
50	2.27
40	2.66
35	3.60

J. DENEKAMP, F. A. STEWART and N. H. A. TERRY (Northwood, U.K.): *Fast neutron therapy: comparison with other modes of overcoming hypoxic radioresistance in three murine tumours*

The response of a murine carcinoma (Ca NT) and two sarcomas (WHT Fibro Sa and Sa S) to single doses, 2 and 5 fractions of X-rays or neutrons has been compared using regrowth delay as the assay of damage. RBE values have been derived and compared with those for skin.

*Present address: Richard Dimbleby Department of Cancer Research, St. Thomas' Hospital Medical School, London SE1, England.

Dose–response curves were obtained for the three types of tumour by irradiating subcutaneous tumours with X-rays or fast neutrons and scoring their response *in situ*, as the delay in tumour growth to a specified size after irradiation. The experimental details have been described previously (Denekamp et al., 1976). The X-rays were generated at 240 kV and the neutrons were generated in the MRC cyclotron, Hammersmith Hospital, from 16 MeV deuterons on beryllium. The animals were anaesthetized with 60 mg/kg sodium pentobarbitone for irradiation, and where appropriate they received 670 or 1000 mg/kg misonidazole either 15 or 30 minutes before irradiation. Hyperthermia was applied by immersion in hot water for 1 hour to achieve an intratumour temperature of 42.5°C as described earlier (Stewart and Denekamp, 1978), either immediately or 3 hours after each X-ray fraction.

Dose–response curves were obtained for each treatment schedule. The relative biological effectiveness was derived by comparing the dose of a particular treatment to produce the same level of tumour delay as the same fractionation scheme for X-rays alone, e.g.

$$RBE = \frac{5F \text{ X-ray dose}}{5F \text{ neutron dose}}$$

or

$$RBE = \frac{1F \text{ X-ray dose}}{1F \text{ neutron dose} + \text{misonidazole}}$$

These RBE values are summarized in the first three columns of Table 2. The range indicated has been derived at different dose levels, with the first value being at the lowest dose and the last value at the highest dose tested. The neutron RBE values ranged from 2.4 to 4.3 for the Carcinoma NT, from 2.0 to 3.0 for the WHT Fibrosarcoma and from 0.9 to 3.1 for the Sarcoma S. The extremely low values, around unity, observed at low dose levels for the Sarcoma S have not been observed in any other tumour. They probably relate to the total lack of repair capacity observed in this tumour when treated with X-rays (Denekamp and Stewart, 1979; Denekamp, Terry and Stewart, in preparation). At high levels of tumour damage the RBE values for all three tumours were above the values for skin, but at lower doses they fell on or below the skin RBE data.

It has been postulated that a higher RBE may

Table 2. *Comparison of the effectiveness of different treatment schedules*

Treatment number of fractions	RBE			Tumour growth delay for stated skin reaction (S.R.)		
	Ca NT	Fibro Sa	Sa S	Ca NT S.R. = 0.8	Fibro Sa S.R. = 1.0	Sa S S.R. = 0.6
X-rays alone				Days	Days	Days
1F	1.0	1.0	1.0	18	24	36
2F	1.0	1.0	1.0	24	30	49
5F	1.0	1.0	1.0	34	36	[a]>65
Neutrons alone						
1F	2.4-3.5	2.0-2.6	0.9-2.7	36	28	58
2F	3.0-3.5	2.1-2.5	1.6-3.1	57	33	78
5F	3.4-4.3	3.0-2.4	1.9-2.9	61	42	94
Neutrons + 0582						
1F	3.5-4.2	3.0	—	60	32	—
2F	3.3-3.1	3.0	—	[a]>57	[a]>37	—
5F	4.7-3.3	3.6	—	[a]>61	[a]>42	—
X-rays + 0582						
1F	1.8-2.1	1.3-1.8	>1.3	43	34	[a]>65
2F	1.4-1.6	1.1-1.3	—	50	36	—
5F	1.2-1.4	1.1-1.2	—	[a]>58	51	—
X-rays + heat consecutive						
1F	—	1.4-1.7	—	—	22	—
2F	—	0.9-1.1	—	—	20	—
5F	—	1.3-1.1	—	—	25	—
X-rays + heat 3-hr interval						
1F	—	1.1-1.5	—	—	34	—
2F	—	0.9-1.0	—	—	—	—
5F	—	1.2-1.0	—	—	36	—

[a]The dose ranges for skin reaction and tumour response do not overlap and so these values for days delay in regrowth are considerable underestimates.

be observed in tumours than in normal tissues, either because of tumour hypoxia, or because of differential changes in the survival curve shapes, especially in the shoulder region. We have tested the effectiveness of neutrons on tumours where the effect of hypoxic cells is minimized by administering misonidazole before each X-ray or neutron fraction.

$$\text{RBE "oxic" cells} = \frac{\text{X-ray dose with 0582}}{\text{neutron dose with 0582}}$$

These "oxic" RBE values are not shown in Table 2. At high X-ray dose levels they ranged from 2.1 to 2.3 for Ca NT and from 1.7 to 2.1 for the Fibro Sa. These values are much lower than the values for treatments without misonidazole and are not significantly above the RBE values for skin at any dose level. This indicates that the sole advantage of neutrons on these two tumours is due to the presence of hypoxic cells.

The effectiveness of all these forms of therapy depends upon achieving a greater effect on the tumour than on the normal tissue at risk. The last three columns in Table 2 indicate the tumour effect that can be achieved for a constant level of normal tissue damage (see Hill and Denekamp, 1979). Skin reactions have been used because parallel skin studies have been performed, usually covering the same dose range. The exact level of skin reaction chosen for this analysis was determined by the radiosensitivity and hence the dose range used for each tumour. The longest regrowth delay represents the most effective treatment for each tumour.

For Ca NT no single treatment was best: fractionated neutrons, used alone or with misonidazole, are all good, as are fractionated X-rays with misonidazole. For the Fibro Sa 5 fractions of X-rays with misonidazole gave the longest delay, although fractionated neutrons + 0582 were also effective treatments. Combined X-rays and heat were clearly less effective, particularly when no time elapsed between their application. For the Sa S the best treatment tested to date was 5 fractions of neutrons. Further studies are needed on this tumour for a more complete intercomparison of schedules.

For all three tumours X-rays used with a high dose of misonidazole were similar in effectiveness to fast neutrons, with fast neutrons plus misonidazole being even better for some schedules. The drug dose used is in excess of the clinically tolerated dose but misonidazole clearly has potential as an alternative to fast neutrons, or perhaps as an adjunct to neutron therapy.

Acknowledgements—We should like to thank Dr. S. B. Field, Mr. D. D. Vonberg and other members of the MRC Cyclotron Unit for making possible the neutron irradiations; Roche Products, Welwyn Garden City, for providing the misonidazole; Mrs. L. Hall, Mr. F. Butcher and their staff for exceptional care of the animals; the Cancer Research Campaign for their financial support of this work.

References

DENEKAMP, J., S. R. HARRIS, C. MORRIS and S. B. FIELD (1976) The response of a transplantable tumour to fractionated irradiation. II. Fast neutrons. *Radiat. Res.* **68**, 93-103.

DENEKAMP, J. and F. A. STEWART (1979) Evidence for reduced repair capacity in mouse tumours relative to normal tissues. Submitted to *Brit. J. Radiol.*

HILL, S. A. and J. DENEKAMP (1979) The response of six mouse tumours to combined heat and X-rays: Implications for therapy. *Brit. J. Radiol.* **52**, 209–218

STEWART, F. A. and J. DENEKAMP (1978) The therapeutic advantage of combined heat and X-rays on a mouse fibrosarcoma. *Brit. J. Radiol.* **51**, 307-316.

K. GÜNTHER (Berlin-Buch, G.D.R.): *General theory of radiation quality applied to mammalian cells: prediction of RBE*

This theory (Günther et al., 1972, 1977) is an attempt to combine two factors accounting for radiation quality:

1. *Molecular level.* Primary molecular lesions (presumably DNA lesions), which are responsible for the radiation effect considered, are assumed to depend linearly on dose regarding their mean number per cell. The corresponding yield per unit dose is a function of radiation quality described by the coefficient

r = relative (to ^{60}Co γ-rays) effectiveness of lesion production.

2. *Cellular level.* Microdosimetric characteristics, namely the single-event spectrum $f_1(z)$ of specific energy z for the cell nucleus, determine the statistics of lesions per cell. This makes σ_o, the cell nucleus cross-section, enter the theory. More precisely: $f_1(z)$ has to be replaced by what we call the virtual spectrum $\mathbf{f_1(z)}$.

The response of the cell to a given number of primary lesions is considered as a complex biological phenomenon not described by the theory. This response function may have any non-linear character; it is derived from the empirical dose–effect relation for the particular cell population and environmental conditions under consideration.

Table 3. Final-slope RBE relative to ^{60}Co gamma-rays

D_o/rad	$\sigma_0 = 40/\mu m^2$				$\sigma_0 = 100/\mu m^2$				
	50	100	200	400	50	100	200	400	∞
.5 MeV	2.09	2.67	3.07	3.32	3.00	3.27	3.43	3.51	3.59
1 MeV	1.87	2.40	2.77	2.99	2.66	2.93	3.08	3.16	3.25
6 MeV	1.45	1.68	1.85	1.96	1.78	1.92	2.00	2.05	2.10
15 MeV	1.20	1.42	1.65	1.85	1.54	1.76	1.94	2.05	2.19

$S_o(D)$, the empirical dose–effect relation for γ-rays, has to be expressed mathematically in the form $S_o(D) = \Sigma B_\mu \exp(-\alpha_\mu^{(0)}D)$ with an arbitrary number of exponentials. $S(D)$ for any radiation reads then:

$$S(D) = \sum_\mu B_\mu \exp(-\alpha_\mu D),$$
$$\alpha_\mu = \frac{1}{\bar{z}_F} \int_0^\infty dz (1 - e^{-r\alpha_\mu^{(0)}z}) f_1(z).$$

RBE values are derived from $S(D)$ and $S_o(D)$ by numerical methods.

Application of the procedure to monoenergetic ion bombardment of cells has yielded empirical values of r in a wide LET range. These analyses revealed that r can be considered a universal function for mammalian cells with a LET dependence like that for DNA double-strand breaks. Generalization for any radiation, especially fast neutrons, has been accomplished by means of a target model of lesion formation in terms of ionizations j in a cylindrical nanometer target: $s = \bar{z}_F^{-1} \Sigma \lambda_j f_1(j)$, where $r = s/s_o$. These results are plotted in Fig. 1.

Neutron survival curve calculation based on the γ-ray curve and the estimated cross-sectional area, σ_o, of the cell nucleus was successful, with few exceptions, for both normal and tumour cells. One may also calculate the RBE as a function of neutron energy for particular cell lines at a specified survival level.

The RBE is always a monotonously decreasing function of dose which is determined, according to the theory, by the complete X-ray survival curve. General results can, however, be obtained only for large doses (final-slope RBE) in terms of nothing but the D_o value for X-rays. Table 3 displays the following rule: the more resistant the object, the larger the high-dose RBE.

Mouse leukaemic cells differently treated *in vitro* provide experimental evidence for this rule (Table 4).

If they refer to large doses, it is reasonable to compare RBE data for gross tissue reactions and mortality with theoretical final-slope RBEs (see Table 3) derived from measured stem-cell D_o values. Table 5 demonstrates the good

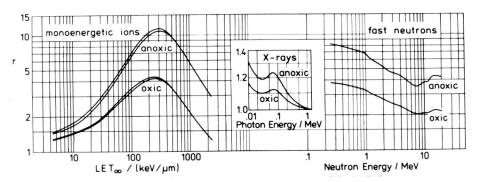

Fig. 1. r, the relative effectiveness of primary lesion production (apparently DNA double-strand breaks) in dependence on radiation quality. Double curves at the left-hand side refer to 3 MeV/nucl (upper curve) and 10 MeV/nucl (lower curve).

Table 4. 15-MeV-neutron RBE for L5178Y cells, $S = 1\%$

D_o/rad	62	85	100	116	123	141	185	238	310
Theory[a]	1.18	1.26	1.34	1.36	1.46	1.39	1.59	1.55	1.61
Exper.	1.09	1.18	1.49	1.30	1.42	1.37	1.60	1.59	1.62
Author	(a)	(a)	(b)	(a)	(c)	(a)	(c)	(a)	(a)

[a] Assumptions: $\sigma_o = 40/\mu m^2$, 5% γ-dose contamination.
(a) Takeshita and Sawada (1974); (b) Broerse and Van Oosterom (1974); (c) Antoku (1975).

Table 5. RBE values relative to 300-kV X-rays for organismic damage

	15 MeV				Cyclotron				Fission	
	A	B	C	D	A	B	C	D	B	D
D_0/rad	?	79[c]	135	167	?	79[c]	135	167	79[c]	167
Theory[a]	—	1.23	1.39	1.45	—	1.50	1.67	1.72	2.02	2.42
Exper.[b]	1.2	1.29[c]	1.4	1.4	1.5	1.63[c]	1.7	2.3	2.07[c]	3.0

A, lung damage, late pneumonitis, mice; B, bone-marrow syndrome, mice; C, skin reactions, mice and rats; D, intestinal syndrome, mice.
[a] Assumptions: $\delta_0 = 40/\mu m^2$, 5% γ-dose contamination.
[b] All data from Broerse and Barendsen (1973) except for A, cyclotron (Field and Hornsey, 1974).
[c] Corrected for dosimetry.

correspondence thus obtained, confirming the well-known fact that the RBE for cell survival *in situ* matches that for gross reactions. These results emphasize the theory to be promising in practical respects, too, since modifications such as neutron energy spectrum, dose, γ-ray contamination, and hypoxia can be simply estimated by calculation, once the response to X-rays of the particular tissue considered has been sufficiently characterized.

References

ANTOKU, S. (1975) Chemical protection of cultured mammalian cells against fast neutrons. *Int. J. Radiat. Biol.* **27**, 287-292.

BROERSE, J. J. and G. W. BARENDSEN (1973) Relative biological effectiveness of fast neutrons for effects on normal tissues. *Curr. Top. in Rad. Res. Quart.* **8**, 305-350.

BROERSE, J. J. and P. M. VAN OOSTEROM (1974) Response of mouse leukaemic cells after irradiation with X-rays and fast neutrons. In *Proc. Symp. on Biological Effects of Neutron Irradiation*, International Atomic Energy Agency, Vienna, pp. 275-282.

FIELD, S. B. and S. HORNSEY (1974) Damage to mouse lung with neutrons and X-rays. *Europ. J. Cancer*, **10**, 621-627.

GÜNTHER, K. and W. SCHULZ (1972) Radiation quality: a theory of action of ionizing radiations on DNA, microorganisms, and mammalian cells. *Studia Biophysica*, **34**, 165-188.

GÜNTHER, K., W. SCHULZ and W. LEISTNER (1977) Microdosimetric approach to cell survival curves in dependence on radiation quality. *Studia Biophysica* **61**, 163-209.

TAKESHITA, K. and S. SAWADA (1974) Lethal effects of 14 MeV fast neutrons on frog eggs and cultured mammalian cells. In *Proc. Symp. on Biological Effects of Neutron Irradiation*, International Atomic Energy Agency, Vienna, pp. 245-255.

M GUICHARD,* J. GUEULETTE,† A. WAMBERSIE† and E. P. MALAISE*
(*Villejuif, France, †Louvain-la-Neuve, Belgium): *Response of V 79 cells to d(50)+Be and p(75)+Be neutrons**

In the perspective of using high-energy neutron therapy for deep-seated tumors, it appears desirable to employ neutron beams with energies in excess of 50 MeV. However, the biological properties of these neutrons are not well known and the limited number of results seem contradictory. Whereas certain studies suggest a reduction in the OER between 35 and 101 MeV (Harrison *et al.*, 1976), others indicate that above 50 MeV the OER is barely modified (Guichard *et al.*, 1978; Hall *et al.*, 1977). In view of these deviating results we performed a biological intercomparison between two high energy neutron beams.

We used on the same day two types of neutrons, d(50)+Be and p(75)+Be, produced by the cyclotron "cyclone" at Louvain-la-Neuve with respective dose rates of 0.60 and 0.06 Gy per minute. V-79 cells were irradiated in suspension in glass ampoules using a technique described elsewhere (Guichard *et al.*, 1978). During irradiation, the ampoules were placed in plexiglass boxes at three different depths (3.5, 7.0 and 12.5 cm) for d(50)+Be neutrons and at two different depths (3.5 and 12.5 cm) for p(75)+Be neutrons. Control ampoules were irradiated with ^{60}Co gamma-rays.

For the two different neutron energies, we found no significant difference in response as function of the depth at which the ampoules were situated. Measured at the survival level of 10^{-1}, the OER was respectively equal to 3.1, 1.9 and 1.8 for gamma-rays, d(50)+Be and p(75)+Be neutrons; at the survival level of 10^{-2}, the OER

*This work was supported by grant 76.7.1686 from D.G.R.S.T.

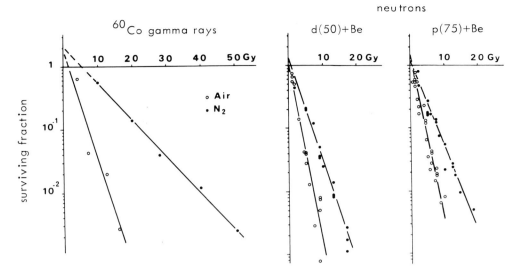

Fig. 2. Survival curves for V 79 cells irradiated with either ^{60}Co gamma rays, $d(50)+Be$ or $p(75)+Be$ neutrons under oxic and anoxic conditions. In oxic conditions, the extrapolation numbers were 1.8, 1.1 and 1.5 and the D_0 values were 2.57, 1.52 and 1.64 Gy, respectively, for gamma-rays, $d(50)+Be$ and $p(75)+Be$ neutrons. In hypoxic conditions, the extrapolation numbers were 2.1, 1.5 and 1.2 and the D_0 values 7.48, 2.64 and 3.18 Gy, respectively.

was 3.1, 1.8 and 1.9 (Fig. 2). We thus found for $d(50)+Be$ and $p(75)+Be$ neutrons an identical mean gain factor of 1.7, comparable to that which we had found previously for $d(50)+Be$ neutrons.

References

GUICHARD, M., J. GUEULETTE, G. LAUBIN, A. WAMBERSIE and E. P. MALAISE (1978) Measurements of the OER and RBE for 70 MeV $p^+ \rightarrow$ Li neutrons. *Brit. J. Radiol.* **51**, 550-551.

HALL, E. J., C. R. GEARD, S. POVLAS and M. ASTOR (1977) The oxygen enhancement ratio for high energy neutrons. *Brit. J. Radiol.* **50**, 679-680.

HARRISON, G. H., E. B. KUBICZEK and J. E. ROBINSON (1976) OER reduction with high energy neutrons. *Brit. J. Radiol.* **49**, 733.

ANTUN HAN and MORTIMER M. ELKIND (Argonne, Ill., U.S.A.): *Neutron RBE of neoplastic transformation* in vitro *and the effects of dose fractionation**

MOUSE embryo-derived cells, C3H 10T1/2, were irradiated with single or fractionated doses of fission spectrum neutrons from the JANUS reactor at the Argonne National Laboratory and 50-kVp X-rays in order to determine the frequency of neoplastic transformation and survival, and to estimate the relative biological effectiveness (RBE) for each end point. Because of the fact that the neutron survival curve of 10T1/2 cells has a relatively large shoulder, as may be judged from the specific shoulder width $(=D_q/D_o)$ of 1.5, there is relatively small dependence of RBE for survival on dose. The RBE is estimated to be 2.9 for small doses (e.g. corresponding to about 0.8 surviving fraction) and progressively decreases with dose, reaching a minimum value of 2.2 at 0.001 survival. Curves for induction neoplastic transformation by both radiations, expressed as the proportion of transformants per surviving cell, are qualitatively similar; an initial straight portion is followed by a plateau. Fission spectrum neutrons produce a maximum of about 6×10^{-3} transformants per survivor. This level is reached at a dose of about 300 rad. The maximum X-ray-induced transformation was about 3.5×10^{-3} transformants per survivor, reached at about 600 rad. The RBE for transformation shows a greater change as a function of dose than that for survival; the RBE decreases from about 10 at doses of 20-50 rad to 2.6 for exposures of about 350 rad.

Fractionation of fission spectrum neutron dose results in only minor changes in survival or transformation. Transformation frequency per surviving cell declines only slightly over a 24-hour period. Compared to X-rays, where an approximately 5-fold reduction in transformation per survivor is observed after dose fractionation, fractionation of the neutron

---

*Work supported by the U.S. Department of Energy.

W. POHLIT, G. ILIAKIS and U. BERTSCHE (Frankfurt (Main), Federal Republic of Germany): *Reparable and irreparable radiation damage in tumour cells after irradiation with fast electrons, fast neutrons and pions*

In irradiated eukaryotic cells reparable (state B) and irreparable radiation lesions (state C) can be observed. The reduction in the number of viable cells N_A (state A) can therefore be described in general by (Kappos and Pohlit, 1972):

$$\frac{dN_A}{dt} = -\eta_{AC} N_A \dot{D} - \eta_{AB} N_A \dot{D} \; \epsilon_{BA} N_B$$

where η_{AC} and η_{AB} are reaction rate constants (unit: Gy^{-1}) for the transition from state A to states C or B and $\dot{D}$ is the absorbed dose rate. ϵ_{BA} is the time constant (unit: hour^{-1}) for the repair transition from B to A.

In shoulder-type dose–effect curves η_{AB} has to be assumed to be a function of absorbed dose D. From experimental values this can be approximated by

$$\eta_{AB} = \eta_{AB}^* \left[1 - \exp(-\eta_r D)\right]$$

where η_{AB}^* is the maximum value approached at high absorbed doses and η_r represents the curvature of the dose–effect curve.

From the measurement of cell survival curves at high absorbed dose rates $\dot{D}$ ($S = N_A/N_{Ao}$) the effective values of $\overline{\eta_{AB}^*}$ and $\overline{\eta_{AC}}$ can be determined averaged over the quality spectrum of the radiation used. The slope of the survival curve at low absorbed doses is given by $\overline{\eta_{AC}^*}$ and at high absorbed doses by $\overline{\eta_{AC}^*} + \overline{\eta_{AB}^*} = 1/D_o$. If the cells are allowed to repair after irradiation, the time constant ϵ_{BA} of this reaction can be estimated, which is independent of the radiation quality.

The determination of $\overline{\eta_{AC}}$ can be done best by plotting $-(\ln N_A/N_{Ao})/D$ as a function of absorbed dose D and extrapolating to $D = 0$. It should be mentioned that $\overline{\eta_{AC}}$ is identical with the quantity α, if a dose effect curve is approximated by

$$S = \exp-(\alpha D + \beta D^2).$$

We have measured values of $\overline{\eta_{AC}}$ for Ehrlich ascites tumour cells exposed to fast electrons, negative pions in the plateau and peak of depth dose curve, 14 MeV neutrons and 3 MeV alpha particles which are shown in Table 6.

In the lower part of Fig. 3 the distribution of absorbed dose in lineal energy y is shown, where $D_{<y}$ is the dose fraction where the energy is imparted in matter with lineal energies smaller than a certain value y. The data are derived from measurements for a sphere of diameter $d = 1\,\mu$m of tissue (Menzel et al., 1978) and from calculations (ICRU, 1970). As can be seen, broad distributions exist for all types of radiations used here with mean values $\bar{y}_D$ ranging from 0.3 to 90 keV/μm.

From these dose distributions in y and the experimental data given in Table 6 the dependence of η_{AC} as a function of y can be determined by mathematical approximation as shown in the upper part of Fig. 3. It should be noted that this quantity η_{AC} would be observed in an irradiation where the absorbed dose is delivered with only one certain value of lineal energy y.

From this curve the effective values $\overline{\eta_{AC}}$ for a certain type of radiation can be recalculated by multiplying the relative dose fraction $D_{<y}/D$ for a certain interval Δy with the value of η_{AC} for this interval and then summing up over all values

Table 6. *Experimental and calculated mean values η_{AC} — mean values over the dose distribution in y (Pohlit et al.)*

	$\overline{\eta_{AC,\mathrm{exp}}}$ /Gy^{-1}	$\overline{\eta_{AC,\mathrm{calc}}}$ /Gy^{-1}	$\bar{y}_D$ /keV · μm^{-1}
30 MeV e$^-$	0.18 ± 0.03	0.21 ± 0.02	0.3
π^--plateau	0.15 ± 0.03	0.15 ± 0.02	0.75
π^--peak	0.46 ± 0.03	0.44 ± 0.02	4.5
14 MeV n	0.58 ± 0.03	0.59 ± 0.02	15
4 MeV alpha	1.30 ± 0.04	1.37 ± 0.02	90

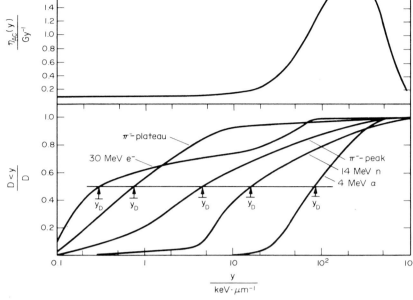

Fig. 3. *Distribution of absorbed dose in lineal energy y, lower part, and dependence of* $\bar{\eta}_{AC}$ *on lineal energy, upper part* (Pohlit et al.).

of Δy. Values of $\bar{\eta}_{AC,calc}$ calculated in this way are also given in the table. As can be seen from the agreement with the experimental values, the same curve for η_{AC} seems to be applicable for all these radiations within the experimental uncertainties. This curve may serve for further quantitative analysis of the biochemical mechanism of the irreparable lesion and its production by ionizing particles of different lineal energy.

In a similar way η^*_{AB} and η_r can be determined as a function of y. These data can be used to predict the influence in biological reactions produced by contamination by sparsely ionizing radiation of high-LET beams, such as gamma-rays in neutron beams or electrons and muons in pion beams.

References

ICRU (1970) ICRU-Report No. 16: Linear energy transfer. Int. Comm. on Radiation Units and Measurements, Washington.

KAPPOS, A. and W. POHLIT (1972) A cybernetic model for radiation reactions in living cells. *Int. J. Radiat. Biol.* **22**, 51-65.

MENSEL, H. G., H. SCHUMACHER and H. BLATTMANN (1978) Experimental microdosimetry at high-LET radiation therapy beams. In J. BOOZ and H. G. EBERT (Eds.), *Proc. Sixth Symp. on Microdosimetry*, Commission of the European Communities, Luxembourg, pp. 563-578.

W. PORSCHEN, J. GARTZEN, H. MÜHLENSIEPEN and L. E. FEINENDEGEN (Jülich, Federal Republic of Germany): In vivo *determination of RBE of cyclotron neutrons for normal tissue and solid experimental tumours by the IUdR prelabelling technique*

THE objective of this study was to determine the relative biological effectiveness of cyclotron neutrons ($\bar{E} = 6$ MeV) compared to 60-Co gamma-rays in normal tissue and the experimental tumours sarcoma-180 and adenocarcinoma EO 771 *in vivo*. A non-invasive labelling technique with 125- or 131-iododeoxyuridine was used. By sequentially labelling the tumour-bearing animals with 125-IUdR and 131-IUdR 50 hours apart, the average tumour cells at the time of the second injection are labelled by 125-IUdR and the euoxic tumour cells are specifically labelled with 131-IUdR. Tumour treatment at this stage of labelling permits the observation of the reaction of euoxic cells and average tumour cells and finally yields data on hypoxic cells. In addition, 125-IUdR was injected after treatment and the relative incorporation rate into the normal tissue and tumour was measured. This information is supplementary to the results from mortality, tumour cure and growth delay.

Using the mortality of 10-week-old NMRI mice as biological end-point, the RBE of cyclotron neutrons was 2.6 ± 0.2 for the intestinal syndrome, 2.1 ± 0.1 for the bone marrow syndrome and 3.0 ± 0.3 for the $TCD_{50/100}$ for sarcoma-180. Based on radiation-induced tumour growth delay, the RBE is equal to 3.4 for sarcoma-180 and 3.3 for the adenocarcinoma EO 771. On the basis of the doses causing a 50% reduction of IUdR incorporation observed 24 hours after irradiation, an RBE of 2.5 ± 0.5 was calculated for the whole-body measurements and 5 ± 1.5 for the sarcoma-180 and the adenocarcinoma EO 771. By sequentially labelling sarcoma-180 cells prior to irradiation, RBE values were calculated to be 3.2 for the activity loss of average tumour cells and 2.7 for euoxic tumour cells. Similarly, RBE values were obtained for adenocarcinoma EO 771 and were found to be 6 for average tumour cells and 3.5 for euoxic tumour cells.

References

BOSILJANOFF, P., W. PORSCHEN, W. PIEPENBRING, H. MÜHLENSIEPEN and L. E. FEINENDEGEN (1977) In-vivo-Untersuchungen über die relative Strahlenempfindlichkeit von hypoxischen Tumorzellen. *Strahlentherapie* 153, 178-189.

FEINENDEGEN, L. E. (1971) Autoradiographische und biologische Untersuchungen der Zellproliferation in vivo. In O. HUG (Ed.), *Präoperative Tumorbestrahlung* (Vorträge Deutscher Röntgenkongreβ, Munich), Urban and Schwarzenberg, Munich, Vienna, p. 12.

PORSCHEN, W. and L. E. FEINENDEGEN (1969). In-vivo-Bestimmung der Zellverlustrate bei Experimentaltumoren mit markiertem Joddesoxyuridin. *Strahlentherapie* 137, 718-723.

PORSCHEN, W. and L. E FEINENDEGEN (1972). In vivo determination of RBE factors of 15 MeV neutrons for different biological effects in normal tissue and sarcoma 180, using cell labelling with ^{125}I-deoxyuridine. In *Radiobiological Applications of Neutron Irradiation*, IAEA, Vienna, pp. 121-134.

PORSCHEN, W. and L. E. FEINENDEGEN (1973) Biologische in-vivo-Dosimetrie von 15 MeV-Neutronen bei normalen und Tumorzellen; Zellmarkierung mit Jod-125-desoxyuridin. *Strahlentherapie* 145, 27-38.

JANET S. RASEY† (Seattle, U.S.A.): *Potentially lethal damage repair in the EMT-6 tumor system treated* in vivo *or* in vitro *with X-rays or cyclotron neutrons**

Several investigators, working with *in vitro* cell lines or tumors growing *in vivo*, have reported that potentially lethal damage (PLD) repair is absent or greatly reduced following irradiation with cyclotron neutrons or alpha particles. This type of repair occurs following exposure to X- or gamma-rays under similar conditions (Gragg *et al.*, 1977; Hall and Kraljevic, 1976; Raju *et al.*, 1977; Shipley *et al.*, 1975). In contrast, Guichard *et al.* (1977) showed that EMT-6 cells growing *in vitro* repaired PLD following exposure to moderately high-LET He ions (10 keV/ μm) to the same extent as after ^{60}Co gamma irradiation (0.25 keV/ μm).

Our investigations with the *in vivo–in vitro* EMT-6 tumor system show that PLD repair occurs following 300 kVp X-ray exposure for all the experimental conditions examined. Survival is increased by delayed sub-culture of 6-day-old unfed or fed (daily medium change) plateau phase cultures; by exposure of 1-day-old low density cultures to depleted growth medium for the first 5-6 hours after irradiation; and by delayed excision of solid tumors exposed *in vivo* and assayed for cell survival by colony-forming ability of cells plated *in vitro*. Under some, but not all, of these growth conditions, EMT-6 cells repair PLD after irradiation with 8 MeV mean energy neutrons produced by the University of Washington cyclotron through the d(21.5)+Be reaction. Furthermore, the extent of repair can be as great following neutron treatment as is the case after X-ray exposure and the time course is also similar (Rasey *et al.*, 1979). However, if X-rays and neutrons are intercompared using the same group of cultures within a single experiment, there is less repair following neutron treatment (Rasey *et al.*, 1979). The results of these repair studies are summarized in Table 7.

The following conclusions may be drawn from these investigations: (1) PLD repair after neutron irradiation occurs in the EMT-6 tumor under some but not all growth conditions. (2)

---

*These investigations were supported by research grant CA-12441 from the National Cancer Institute, Department of Health, Education and Welfare
†Present address: Cancer Research Campaign Gray Laboratory, Mount Vernon Hospital, Northwood, Middlesex HA6 2RN, England.

Discussion

Table 7. *Potentially lethal damage repair in the EMT-6 tumor system* (Rasey)

Growth state: post-irradiation condition to promote PLD repair	PLD repair after X-rays?	PLD repair after neutrons?
Unfed plateau phase cultures: [a] 5 or 6 hr delay in subculture	Yes; SFR [c] = 2.74 at survival = 0.05 for immediate plating sample	Yes; SFR = 1.84 at survival = 0.05
Fed plateau phase cultures [a] 5 or 6 hr delay in subculture	Yes; SFR = 2.06 at survival = 0.05 for immediate plating sample	No
One-day-old exponentially [b] growing cultures: 6 hr exposure to depleted culture medium immediately after irradiation	Yes; SFR = 1.52 at survival = 0.05 in sample not exposed to depleted medium	Yes; SFR = 1.54 at survival = 0.05
Subcutaneous tumors *in vivo*: 6 hr delay in excision for *in vitro* plating of cells	Yes; SFR = 3.86 at survival = 0.058 in immediate plating sample	No

[a] parallel X-ray and neutron experiments were done on the same day with the same batch of cultures handled identically throughout.
[b] X-ray and neutron experiments were done on different days.

[c] SFR = Survival fraction ratio = $\dfrac{\text{survival for cells treated with a given dose and subjected to repair-promoting conditions for 5-6 hr}}{\text{survival for cells treated with the same dose and assayed for survival immediately after treatment}}$

according to Hahn et al., 1973.

The amount of repair can be extensive and variable; the most relevant studies are those which intercompare the same batch of tissue cultures or the same tumor transplant group within a single experiment (Rasey et al., 1979). (3) Post-neutron repair in unfed plateau-phase cultures and in log-phase cultures covered with depleted culture medium for 6 hours is consistent with a repair-promoting substance released into the medium by plateau-phase cells (Little, 1970), or, alternatively, with repair being most facilitated by the least favourable growth conditions. (4) Caution is required in extrapolating data obtained with the *in vitro* variant of an *in vivo–in vitro* tumor system; the behavior of cells removed from the tumor may be not relevant to that of cells left *in situ* (Rasey et al., 1977; McNally, 1975).

References

GRAGG, R. L., R. M. HUMPHREY and R. E. MEYN (1977) The response of Chinese hamster ovary cells to fast neutron radiotherapy beams. II. Sublethal and potentially lethal damage recovery capabilities. *Radiat. Res.* **71**, 461-470.

GUICHARD, M., B. LACHET and E. P. MALAISE (1977) Measurement of RBE, OER and recovery of potentially lethal damage of a 645 MeV helium ion beam using EMT-6 cells. *Radiat. Res.* **71**, 413-429.

HAHN, G. M., M. L. BAGSHAW, R. G. EVANS and L. F. GORDON (1973) Repair of potentially lethal radiation damage in X-irradiated, density inhibited Chinese hamster cells: metabolic effect and hypoxia. *Radiat. Res.* **55**, 280-290.

HALL, E. J. and U. KRALJEVIC (1976) Repair of potentially lethal radiation damage: comparison of neutron and X-ray RBE and implications for radiation therapy. *Radiol.* **121**, 731-735.

LITTLE, J. B. (1970) Repair of potentially lethal damage in mammalian cells: enhancement by conditioned medium from stationary cultures. *Int. J. Radiat. Biol.* **20**, 87-92.

McNALLY, N. J. (1975) A comparison of the effects of radiation on tumour growth delay and cell survival. The effect of radiation quality. *Brit. J. Radiol.* **48**, 141-145.

RAJU, M. R., J. P. FRANK, E. BAIN, T. T. TRUJILLO and R. A. TOBEY (1977) Repair of potentially lethal damage in Chinese hamster cells by X- and alpha-irradiation. *Radiat. Res.* **71**, 614-621.

RASEY, J. S., R. E. CARPENTER and N. J. NELSON (1977) Response of EMT-6 tumors to single fractions of X-rays and cyclotron neutrons: evaluation and comparison of multiple endpoints. *Radiat. Res.* **71**, 430-446.

RASEY, J. S., N. J. NELSON and R. E. CARPENTER (1979) Recovery from potentially lethal damage following irradiation with X-rays or cyclotron neutrons. I. Response of EMT-6 cells *in vitro*. *Int J. Radiat. Oncol. Biol. Phys.* (in press).

SHIPLEY, W. U., J. A. STANLEY, V. D. COURTENAY and S. B. FIELD (1975) Repair of radiation damage in Lewis lung carcinoma cells following *in situ* treatment with fast neutrons and gamma-rays. *Cancer Res.* 35, 932-938.

ANKE L. DE RUITER-BOOTSMA, G. J. M. J. V.D. AARDWEG, M. F. KRAMER and J. A. G. DAVIDS (Utrecht and Petten; The Netherlands): *Proliferation and radiation response of spermatogonial stem cells in the mouse during recovery of the seminiferous epithelium after irradiation with fission neutrons of 1 MeV mean energy*

After a severe cell loss in the testis, inflicted by high doses of fission neutrons, spermatogenetic clones arise from surviving radioresistant stem cells. These clones elongate with time along the walls of the seminiferous tubules. The time relationship of longitudinal clonal growth was studied after two neutron doses: 100 and 300 rad. At different post-irradiation intervals (3, 4, 5, 6, 8, 11, 15 and 20 weeks) the length of the seminiferous tubules covered with spermatogenetic cells (L-rep.) was measured with use of morphometrical methods in histological testis sections. In the sections the percentage of tubules showing spermatogenetic repopulation — the Repopulation Index (RI) — and the spermatogenetic cell types present (spermatogonia, spermatocytes, round and elongated spermatids) were also determined. Between 3 and about 8 weeks after the higher dose of 300 rad, L-rep. increased linearly from 27 ± 3 mm (RI 4%) to 64 ± 6 mm (RI 8%). It can be calculated that in this period the individual clones grew in length at a mean rate of 34 μm per day. At longer post-irradiation times longitudinal clonal growth slowed down; at 20 weeks L-rep. was equal to 90 ± 9 mm (RI 13%). In view of the low RI-values found, the inhibition of longitudinal clonal growth from 8 weeks on cannot result from collision of clones in the tubules. From the data it can be predicted that there will never be complete repopulation after 300 rad neutrons.

Between 3 and 4 weeks after a dose of 100 rad the growth rate of the clones was similar to that after 300 rad and therefore does not seem to depend on the number of surviving stem cells. Thereafter clonal growth became strongly inhibited. Since from 4 weeks on the RI-values exceeded 50%, this early inhibition of longitudinal clonal growth is at least partly caused by collision of clones.

After irradiation surviving stem cells give rise to daughter cells which start the line of differentiation towards spermatozoa. From the cell types present in the repopulated tubular sections at the different post-irradiation intervals it can be concluded that after a dose of 100 rad the time-course of differentiation is not affected. After 300 rad, however, the appearance of young round spermatids was delayed for about 1 week. Whether this delay is due to degeneration of the first waves of spermatocytes reaching meiosis or to a real "meiotic division delay" is not clear at the moment.

During repopulation of the epithelium, the stem cells present in the clones are more radiosensitive than they were prior to irradiation. At 26 days after a dose of 150 rad neutrons the stem cell population is characterized by a D_o value of 49 rad, a value that is about 50% of that prior to irradiation. Although at a post-irradiation interval of 12 weeks the D_o value has increased to 65 rad, the stem cell population has still not returned to its response prior to the first dose. During the first 12 weeks of repopulation there has been an increase in the number of stem cells resulting from stem cell proliferation. The increase is reflected in a rise of the RI values with increasing fractionation interval after the different doses used. Despite the increase in number, the stem cell population has still not returned to its preirradiation size at 12 weeks. Further studies on a return to both qualitatively and quantitatively normal response and the factors influencing this are under way.

C. STREFFER, D. VAN BEUNINGEN, G. SCHMITT, N. ZAMBOGLOU, G. HÜDEPOHL and E. MAIER (Essen, Federal Republic of Germany): *Effects of neutron irradiation on synthesis of nucleic acids in mammalian cells*

SYNTHESIS of nucleic acids is necessary for cell growth and proliferation. It has been demonstrated that DNA synthesis is more sensitive to ionizing radiation with low LET than RNA synthesis. DNA synthesis has been found to be decreased already after low radiation doses of X-rays and this effect shows a good dose relationship. Several authors have therefore proposed to use the radiation induced inhibition of DNA synthesis for biological dosimetry.

It has been the aim of these investigations to study the effects of irradiation with fast neutrons on nucleic acid synthesis and to compare them with those which have been described with low-LET radiation. Especially it is of interest to investigate these processes in such tumor cells which are known to be comparatively radioresistant. Further, we have studied DNA synthesis in the small intestine of mice after irradiation with fast neutrons, as it has been proposed to use cell proliferation in the intestine as a biological dosimeter.

Nucleic acid synthesis was determined by measuring the incorporation of labelled precursors after isolating the nucleic acids. For DNA synthesis ^{3}H-thymidine and for RNA synthesis ^{32}P-phosphate were used. Dose effect curves were determined for the RNA synthesis in Novikoff hepatoma ascites cells after irradiation with fast neutrons. RBE values were determined.

DNA synthesis was studied in human melanoma cells which were irradiated *in vitro* with neutron doses of 200 and 400 rad 24 hours after being brought in culture. The effect of neutrons was compared with the effect of X-rays. A dose of 200 rad neutrons was more efficient to reduce the cell number than 400 R X-rays. In the control cultures as well as after irradiation with 400 R X-rays or with 200 rad neutrons a plateau of the cell number was reached after about 120 hours in culture and the plateau maintained until 168 hours. After 400 rad neutrons the cell number decreased again at the later period.

Furthermore, the DNA content per cell was measured by cytofluorometry with a microscope photometer. These measurements showed that after neutron irradiation a number of polyploid cells occurred which was higher than after X-irradiation. However, the main effect was an increase of hypoploid cells which was dependent on dose and rose with incubation time. These hypoploid cells were apparently formed by a release of chromatin from the cell nuclei which was confirmed by the appearance of numerous micronuclei. Concomitantly the relative number of melanoma cells in G_1-phase decreased with rising neutron doses.

The number of cells in S-phase was only temporarily reduced 48 hours p.r. Furthermore, it was found that the rate of DNA synthesis was constant during the exponential growth of the cells, with a duration of about 13 hours for the S-phase. When the plateau phase was reached, the rate of DNA synthesis decreased to about half the value which was observed at the exponential growth. After neutron irradiation (24-144 hours p.r.) the rate of DNA synthesis in the S-phase cell was not different from the controls. Cell loss can be calculated from these data.

In another set of experiments the DNA synthesis was studied in the small intestine after irradiation with X-rays and with fast neutrons. The results demonstrated that the RBE values are dose dependent as it has been shown for other biological phenomena after irradiation with single doses. In comparison to cell proliferation, measured by the crypt method, the determination of the DNA synthesis may have an advantage in order to use it as a biological dosimeter also for clinical studies, as this biological system already responds after low dose fractions. Therefore this effect can also be measured after fractionated irradiation with dose fractions which are comparable to those used in radiotherapy.

E. L. Travis,* B. Hobson,* S. J. Holmes,* S. B. Field† and R. G. Ahier†
(*Northwood and †London, U.K.): *Lung RBE values from histopathological studies*

The RBE for lung death after single doses or fractionated doses within the clinical range has been shown to be significantly lower than that for other normal tissues (e.g. skin). In the present study, lung damage was quantitated by morphometric methods 20 weeks (140 days) after graded single doses of X-rays or neutrons to the right lung only. Neutrons were produced by the MRC cyclotron at Hammersmith Hospital and had a mean energy of about 7.5 MeV with approximately 4% gamma contamination. Lung damage was assessed quantitatively for histopathological changes and for mast cell numbers. The incidence and severity of three histopathological lesions (cellular infiltrate, oedema and pathological organization) were scored and graded from 1 to 3 depending on the area of the lung section involved.

The cellular infiltrate was characterized by macrophages and desquamated epithelial cells in the air spaces. This lesion exhibited a dose effect between X-ray doses of 5 and 18 Gy and neutron doses of 3 and 13.5 Gy. The severity of this lesion was obscured by oedema after higher doses of neutrons and by pathological organization after higher doses of X-rays. The average RBE for the cellular infiltrate was 1.6-1.8 over the neutron single dose range of 7-13.5 Gy.

Oedema was characterized by a serous exudate in the alveolar spaces and disruption of the alveolar walls. After modest single doses of neutrons the oedema was severe and the severity increased throughout the neutron dose range above a threshold of 11 Gy. In contrast the oedema was not seen after X-ray doses up to 18 Gy. Because there was not a good dose response after X-rays no RBE value could be obtained but it appeared to be in excess of 1.5.

Pathological organization was characterized by fibrin and collagen deposits in the air spaces and in the alveolar walls. This lesion, a prefibrotic phase, exhibited a dose–effect relationship after X-rays and neutrons but the severity of the lesion was obscured by oedema in the high neutron dose groups. The RBE for organization was similar (1.6-1.8) to that for cellular infiltrate, over the same neutron dose range of 7-13.5 Gy.

Mast cells, which are normally present in small numbers and are restricted to the pleura and perivascular areas in unirradiated mouse lungs, infiltrated the alveolar walls after X-rays and neutrons. The number of mast cells/mm^2 increased linearly throughout the X-ray dose range. Mast cells/mm^2 also increased with neutron doses up to 13.5 Gy but were lower after higher doses as the severity of oedema increased. The RBE ranged from 1.7 to 2.0 over the neutron dose range of 7 to 13.5 Gy and was higher than for the other lesions scored.

The findings of the present study indicate that:

1. Histopathological lesions in the lung can be quantitated and can provide dose response curves.

2. The range of RBE values (> 1.5-1.7) for the three histopathological lesions were somewhat higher than the values of 1.2-1.5 reported for death from lung damage (Broerse, 1974; Hornsey and Field, 1974; Phillips and Fu, 1976). There was good agreement between the RBE values for two of the three lesions.

3. RBEs for oedema could not be confidently determined because of the lack of response with X-rays. However, this itself suggests that the RBEs for oedema may be higher than 1.5.

4. Mast cell numbers were the most precise quantitative measurement of lung injury and reflected a combination of injuries in the lung. The RBE values from mast cell counts were higher (1.7-2.0) than those determined from the histopathological lesions.

5. The RBE values for histological evidence of lung damage are closer to those for skin (1.8) than the lung lethality values previously reported.

References

BROERSE, J. J. (1974) Review of RBE values for 15 MeV neutrons for effects on normal tissues. *Europ. J. Cancer* **10**, 225-230.

HORNSEY, S. and S. B. FIELD (1974) The RBE of cyclotron neutrons for effects on normal tissues. *Europ. J. Cancer* **10**, 231-234.

PHILLIPS, T. L. and K. K. FU (1976) Biological effects of 15 MeV neutrons. *Int. J. Radiat. Oncol. Biol. Phys.* **1**, 1139-1142.

G. WIERNIK and C. M. A. YOUNG (Oxford, U.K.): *Late effects of californium-252 therapy*

Twenty-five patients were treated with californium-252 between September 1972 and November 1975, using an afterloading implant technique; the full details concerning these patients have already been published (Wiernik, *et al.*, 1976). All the surviving patients have been followed up to June 1978. Observations were made of tumour recurrence within the treated area and of the late radiation changes that occurred in the normal tissues receiving significant doses of neutron therapy.

Local tumour reccurrence has been assessed relative to the time after implantation. Ten recurrences appeared in the first 13 months and two occurred at 34 and 36 months, respectively. When these recurrences were related to the neutron dose it was found that 75% occurred in those patients receiving a tumour dose of under 5000 effective rad, assuming an RBE of 7.5 for the first three patients and 7.0 for the remainder after comparison of acute normal tissue reactions. These cases were from a group of fourteen patients that had received previous irradiation to the implant site with cobalt teletherapy. Total gamma-ray doses of 3800-4400 rad in 10-13 fractions over 21-52 days had been used. Only two recurrences occurred in a group of eleven patients who had had no previous irradiation. These patients received doses between 5000-6525 effective rad from the californium-252 implant.

Our earlier reports indicated similar acute skin

reactions for the whole group of patients suggesting that a suitable reduction in dose had been made for those patients receiving previous treatment. Patients have now been followed for up to 66 months from the time of their implant. A level of late skin damage approximating to normal tissue tolerance has been observed in both the high and low-dose groups, as judged by the degree of telangiectasis, atrophy, dyspigmentation and dry desquamation.

The californium doses used were low as judged by the maximum degree of acute skin reaction because it was anticipated that late radiation changes would be dose limiting. Too few of the patients have survived long enough to enable us to draw definite conclusions. It should, however, be noted that of the seven patients in whom late normal tissue changes could be assessed, at 3 years or more from the time of implant, two have degrees of late radiation change which could be equated with normal tissue tolerance, whilst the other five showed moderate degrees of late damage. We have, therefore, not observed any evidence to suggest a therapeutic gain from the use of this mode of neutron therapy on the basis of (1) increased tumour control at lower doses than are currently required for photon therapy, and (2) relatively diminished normal tissue damage at the higher tumourcidal doses. No correlation was found between tumour regrowth delay and tumour dose. This was to be expected due to the great differences in human tumour cell kinetics.

Reference

WIERNIK, G., C. H. PAINE, J. B. H. STEDEFORD, R. J. BERRY, H. WEATHERBURN and C. M. A. YOUNG (1976) Californium-252 in radiotherapy. *Da Radiazioni di Alta Energia* **15**, 77-86.

F. ZYWIETZ and H. JUNG (Hamburg, Federal Republic of Germany): *Response of mouse intestine to 14 MeV neutrons after fractionated irradiation*

At the neutron therapy facility of the University Hospital Hamburg–Eppendorf, the survival and the kinetics of regeneration of irradiated jejunal crypts were studied after single and fractionated irradiation with 14 MeV neutrons and Co-γ-rays, respectively. Male (CBA/Rijx-C57BL/Rij)F_1 hybrid mice with weights from 22 to 25 g were whole-body irradiated inside a tissue-equivalent phantom (35.5 cm in diameter, 30.2 cm high) at a depth of 2.1 cm. The field size of the collimated neutron beam was 11×11 cm^2 at the surface of the phantom, placed 80 cm in distance from the target. Details of the experimental arrangement are described elsewhere (Zywietz *et al.*, 1979). The dose-rate amounted to 5.75 rad/min with a gamma-ray contribution to the total dose ($D_n + D_\gamma$) of 9.3%. The number of surviving intestinal crypts per circumference of the jejunum was determined after exposure according to the method of Withers and Elkind (1970). The number of surviving stem cells was calculated on the basis of Poisson statistics.

Figure 4 shows dose survival curves for jejunal crypts and crypt stem cells after exposure to single and fractionated doses of 14 MeV neutrons.

Single-dose irradiations were carried out in the dose range from 450 to 1000 rad. The mice were sacrificed $3\frac{1}{2}$ days after irradiation. The radiation dose necessary to reduce survival to ten crypt stem cells per circumference amounted to 689 ± 19 rad. Since identical effects were observed after a gamma-ray dose of 1449 ± 29 rad (Zywietz *et al.*, 1979) an RBE of 2.1 ± 0.1 is obtained for single-dose irradiations.

Survival of intestinal crypts after fractionated doses of neutrons was measured for two fractions given at a time interval of 8 hours. Here, the total dose per fraction was 250, 300, 350 and 450 rad. The mice were sacrificed 2.0 days after the last exposure. At that time, the number of regenerated crypts remains constant with time. This was shown by measuring the number of crypts per circumference from 12 hours up to 72 hours after irradiation using 300 rad per fraction.

The dose–response curves for the survival of stem cells after neutron irradiation plotted in Fig.

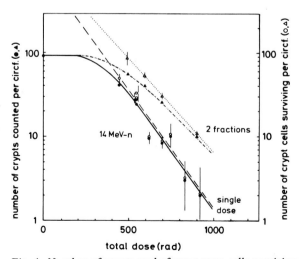

Fig. 4. *Number of crypts and of crypt stem cells surviving per circumference after single and fractionated irradiation with 14 MeV neutrons.*

4 are characterized by a D_o of 152 ± 6 rad and an ordinate intercept of 1100 ± 250 for single doses, whereas a D_o of 191 ± 4 rad and an ordinate intercept of 1191 ± 85 are found after fractionation. Although the slope of the dose–effect curve after fractionated neutron irradiation is less steep than after single doses, the intercepts on the ordinate are identical and equivalent to twelve stem cells per circumference in either case.

After fractionation, the dose necessary to reduce survival to ten crypt stem cells per circumference was determined to be 911 ± 13 rad. Thus, after two fractions of neutrons given at a time interval of 8 hours, the radiation dose has to be 31.5% higher to induce the same radiation damage in the crypts of the jejunum as compared to single-dose exposure.

Acknowledgements—We wish to thank Professor H. D. Franke, Dr. A. Hess and Mr. H. K. Kraus, Department of Radiotherapy, for support and excellent cooperation concerning dosimetry and irradiation.

References

WITHERS, H. R. and M. M. ELKIND (1970) Microcolony survival assay for cells of mouse intestinal mucosa exposed to radiation. *Int. J. Radiat. Biol.* **17**, 261-267.

ZYWIETZ, F., H. JUNG, A. HESS and H. D. FRANKE (1979) Response of mouse intestine to 14 MeV neutrons. *Int. J. Radiat. Biol.* **35**, 63-72.

Dosimetry and radiobiology of negative pions and heavy ions

M. R. RAJU

*Los Alamos Scientific Laboratory,
University of California,
Los Alamos, New Mexico 87545, U.S.A.*

Introduction

Historically, radiation therapy development has been directed toward obtaining more and more penetrating radiations so that the high-dose region can be confined to the treatment volume while minimizing the dose to surrounding normal tissue. While the introduction of fast neutrons in radiotherapy was a step forward in development because of their higher radiation quality, it was a step backward in terms of dose localization. Because of the inherent nature of interactions of fast neutrons in tissue, their dose localization characteristics may never be truly comparable to current megavoltage X-rays used in radiotherapy.

It is clear from presentations of clinical results using fast neutrons during this conference that their poor dose localization may be a limiting factor, especially for deep-seated tumors. In principle, negative pions and heavy ions have radiation qualities similar in certain respects to fast neutrons and, at the same time, have dose localization characteristics superior to even megavoltage radiations because of their Bragg peak ionization characteristics. Unlike fast neutrons and X-rays, negative pions and heavy ions confine dose in the third dimension (at-depth), also because of their Bragg ionization characteristics. The ability to control beam depth necessitates applying inhomogeneity corrections precisely. The availability of CAT scanners provides precise measurement of beam inhomogeneities, at least in principle, and therefore allows taking advantage of the Bragg ionization characteristics of negative pions and heavy ions (Smith *et al.*).

Progress in the application of negative pions and heavy ions since the last neutron conference at The Hague has been phenomenal. Pion radiotherapy programs are currently in progress at three centers: Los Alamos, Vancouver and Zurich. The Stanford University group pioneered the development of a large solid-angle pion-collecting device that permits simultaneous multiport irradiation (Fessenden *et al.*). The Zurich group is incorporating the Stanford design in their therapy facility (Von Essen *et al.*). The Los Alamos and Vancouver groups are using conventional pion-collecting devices. The Los Alamos facility provides a fixed vertical beam and the Vancouver facility a fixed horizontal beam. A clinical program is in progress at Los Alamos and is expected soon at Vancouver and Zurich where pretherapeutic radiobiology programs are being conducted.

Unlike the multiple-pion therapy programs, Berkeley is currently the only location for heavy-ion therapy programs, although there are plans to develop heavy-ion facilities in Dubna (USSR) and Saclay (France). At Berkeley, a helium-ion beam from the 184-in "synchrocyclotron" and heavy-ion beams (carbon, neon, argon) from the BEVALAC are being used in radiotherapy as fixed horizontal beams.

There are eighteen papers in the poster session, approximately half of which cover negative pions and the other half heavy ions. Although the poster contributions represent only a fraction of the recent experimental data on these radiations, they do provide data of interest in therapeutic applications and also show how negative pions and heavy ions compare with fast neutrons. In this brief survey, an attempt will be made to present some aspects of the poster contributions on negative pions and heavy ions and the current status and prospective and retrospective views. Such a view naturally reflects the personal opinion of the rapporteur.

Negative pions

Production of pion beams with intensities adequate for therapy (compared to fast neutrons) is complicated by the requirement that the energy of the primary beam employed to produce pions (proton beams are usually used because of their good production cross-section for pions) must be much higher than 400 Mev, compared to the energy of the primary beam (15 to 50 MeV D^+) used for neutron production by cyclotrons. The primary beam intensity used for pion production must also be very high (~ 20 to 500 μA). The production and collection of pions for radiotherapy applications, although technically achieved, have turned out to be more complicated than originally envisioned.

The depth-dose distribution of pion beams has not been found superior to protons in spite of formation of stars near the end of the range and their Bragg ionization. This is due to enhancement of the dose near the end of the range, compared to the beam entrance from Bragg ionization, being higher for protons than for pions. Star formation near the end of the range for pions approximately compensates for this difference, thereby making the depth distributions of pions and protons comparable (Raju). Early biophysical measurements at Berkeley, CERN (Geneva), and NIMROD (United Kingdom) were very helpful in making the expectations for pions realistic.

Pion radiation quality at the plateau region is comparable to conventional low-LET radiations, and radiobiology results also indicate RBE values close to unity (McEwan et al.; Fessenden et al.; Raju et al.). In the pion-stopping region, the radiation quality increases considerably because of π^--star products. Radiobiology data for negative pions at the Bragg peak position clearly indicate the increase in RBE and the reduction in OER (Baarli et al.; Tremp and Rao; Fritz-Niggli and Blattman). Although the high-LET dose fractions due to charged particles from negative π^--stars are reduced with increasing Bragg peak width, fast neutrons from π^--stars become important and account for approximately 50% of the high-LET dose (Dicello et al.). Raju et al. also reported that, although the RBE values for negative pions at the peak centers decrease considerably with increasing peak width, OER values are nearly identical for different peak widths. The results by Baarli et al. are also consistent with this finding. Even at the Bragg peak position, passing negative pions deposit a large fraction of dose at much lower LET values compared to fast neutrons; hence, the average LET of negative pions is lower (Menzel et al.). As expected, pion radiobiology data have indicated lower RBE values and higher OER values compared to fast neutrons (McEwan et al.; Raju).

Heavy ions

Curtis reported dose-average LET calculations for heavy ions of carbon, neon and argon at various points for 10-cm-wide Bragg peaks. These calculated values, along with values for helium, fast neutrons, and negative pions, are shown in Table 1. The radiation quality of fast neutrons is in between that of carbon and neon ions at the peak region and that of neon ions at the plateau is lower than for fast neutrons. Although dose average LET values are helpful in comparing the radiation quality of particles, it should be stated that this may be an incorrect physical parameter in comparing radiobiological data. For energetic heavy ions, a significant fraction of the dose is deposited by energetic delta rays (low-LET); hence, the microstructure of the track should also be taken into consideration in interpreting the radiobiological data.

The dose average lineal energy of helium ions ranges from 4.4 keV/μm at the plateau to 22.3 keV/μm at the distal end of a 5-cm-wide Bragg peak (Chemtob). Thus, the mean LET value, even at the distal end of the peak, is lower than

Table 1. *Dose average LET values for heavy particles*

	Dose average LET (keV/μm)			
		Peak region (10 cm width)		
Particle	Plateau	Proximal Peak	Central peak	Distal peak
Neutrons	75	75	75	75
Negative pions	6	15	30	60
Helium	5	8	16	30
Carbon	12	30	40	130
Neon	30	70	100	300
Argon	100	200	300	1500

for fast neutrons. Radiobiology results reported by Van Dam et al. and Raju are consistent with LET considerations.

As expected, dose localization of heavy ions has been found to decrease slowly with increasing charge of the heavy ion due to the increasing cross-section for nuclear reactions. One of the pleasant surprises was that the dose localization advantage of heavy ions is still maintained even for neon and argon ions for ranges not exceeding 15 to 20 cm. This is because nuclear secondaries from heavy ion nuclear interactions proceed in the same direction as primary heavy ions with nearly the same velocity and come to rest near the vicinity of the primary beam. Thus, nuclear secondaries were not found to diminish the usefulness of heavy ions considerably.

From studies with 12-day-old spheroids (V79), Luecke-Huhle et al. concluded that the intercellular contact that protects cells after exposure to low-LET radiations is not detected after exposure to heavy ions.

Goldstein and Phillips reported an extensive series of measurements on survival of intestinal crypt cells for carbon and neon ions with single and fractionated doses. They found that single and fractionated doses of heavy ions produce dose-response curves with reduced shoulders but with similar slopes when compared to gamma-rays. As expected, they have found the greatest recovery in the plateau regions. Recovery at the mid-peak region is considerably reduced compared to the plateau but is greater than at the distal end of the peak. These results are consistent with LET values at those respective points. Goldstein and Phillips concluded that fractionated treatments of heavy ions produce an enhanced effect in the peak region compared to the plateau region and could lead to a substantial gain in therapeutic ratio.

Curtis et al. reported a series of tumor (rhabdomyosarcoma) measurements after exposure to carbon, neon, and argon ions. Using some of these results, they calculated a factor of merit defined as the efficiency of killing for hypoxic cells at the peak region compared to oxygenated cells at the plateau region. They found that carbon and neon ions have nearly the same merit for 15-cm penetration. These results indicate that the gain in dose localization for carbon ions compared to neon ions is approximately counterbalanced by the reduction in OER for neon ions compared to carbon ions. For ranges greater than 15-cm, carbon ions are slightly more advantageous than neon ions. Hermens et al. compared the cell proliferation kinetics of rhabdomyosarcoma tumor cells after exposure to neon ions (6 Gy) with 300-kVp X-rays (20 Gy) and concluded that the rate of cell-cycle progression of surviving cells after exposure to neon ions is not significantly different than that for X-rays.

Comparison of heavy particles

Raju reported a series of radiobiological measurements for heavy particles of interest in radiotherapy (p, He, C, Ne, Ar, π^-, n) using the same biological systems. For this comparative study, the depth-dose distributions of all heavy charged particles were modified to 10-cm-wide Bragg peaks. The OER for protons was not significantly different from that for X-rays. The OER values for negative pions, helium ions and carbon ions were larger, for neon ions similar, and for argon ions smaller when compared to fast neutrons. It was disappointing that heavy ions have OER values much higher than the expected values of unity. This could be due to a large delta-ray penumbra associated with energetic heavy ion tracks.

These comparative radiobiological results clearly indicate that there is no unique characteristic in any one particle that is not shared to some degree by other particles. For example, dose localization characteristics of all heavy charged particles are similar. Biological effects such as RBE and OER of fast neutrons appear similar to those for some of the heavy ions. If we are interested only in dose localization without significantly changing the radiation quality from conventional low-LET radiations, the particle of choice is the proton. Ongoing fast neutron therapy trials will answer the question as to whether high-LET is an advantage in treating certain types of resistant tumors. If the results are promising, neon and argon ions may be even more effective. Clinical results with mixed schemes of neutrons and gamma-rays appear promising. If this is confirmed, negative pions, helium ions and carbon ions may be very effective because the radiation quality of these beams is approximately similar to that of the mixed scheme of neutrons and X-rays.

Discussion

Authors of the poster contributions were asked to make specific comments that were not well covered by the rapporteur. Von Essen pointed out that there is a considerable penumbra for negative pions not particularly different from that of fast neutrons because of multiple scattering and nuclear interactions. He hoped that some of this problem could be overcome by using the 60-channel applicator in Zurich. Smith commented that, although the penumbra for large static beams of negative pions is comparable with Co-60 gamma-rays, improvements comparable to the 60-channel applicator can also be made at Los Alamos where they are planning to scan a cylindrical beam through the patient.

Curtis cautioned about using dose-average LET values because they may not be a correct physical parameter for comparing radiobiological results. He also mentioned that the dose-average LET values at the distal end of a broad Bragg peak for heavy ions are point values not relevant for tumors of any significant size.

There was some discussion of LET distribution of negative pions compared to fast neutrons. Fowler concluded that, if high-LET is an advantage, neutrons are better than negative pions. Fessenden stated that, when multiple converging π^--beams are used, the biological effects in the plateau region decrease considerably, which may prove to be an advantage.

Goldstein pointed out the importance of RBE measurements for fractionated treatments and stated that the RBE value using the mouse gut system for carbon ions at the plateau region (30 keV/μm) is not significantly different from the peak region (130 keV/μm) for single doses but that, for fractionated doses, there are large differences due to considerable recovery at the plateau region but lack of recovery at the peak region. Dicello stated that the model he uses would be helpful in understanding Goldstein's data.

Kligerman stated that the biological effectiveness of negative pions for a broad peak they have chosen for therapy remains the same from the proximal to distal side of the peak for single and fractionated doses using multicellular spheroids. Clinical results are consistent with this finding. From these results, it appears to him that a small amount of the high-LET component mixed with the low-LET component interferes with repair of low-LET damage. He felt that these aspects should be taken into consideration.

In comments from the floor, Phillips disagreed with Raju's comment that argon ions may be a progressive form of neutron therapy. He stated that argon ions have disturbing qualities due to poor dose localization and reduction in RBE for aerated cells at the peak compared with the plateau. He also stated that carbon ions and negative pions have similar advantages and that neon ions are radiobiologically similar to fast neutrons but with better dose localization. He suggested that more study is needed before we can choose from these particles for radiotherapy applications. Raju responded that he was not proposing argon ions as the particle of choice for radiotherapy—that he just wanted to point out some of the radiobiological similarities with neutrons. Studies with cultured Chinese hamster cells (V79) using aerobic and hypoxic cells have indicated no significant differences in RBE for aerated cells all through the depth of penetration for the 10-cm-wide Bragg peak, with the exception of the last few millimeters where the RBE is lower due to saturation effects. The RBE for hypoxic cells remains constant with the entire depth of penetration. The dose localization advantages of heavy charged particles also is still maintained for argon ions with ranges up to $\sim$ 15cm.

There was extensive discussion regarding how the OER for fast neutrons compares with the OER for heavy ions. Hall stated that his impression, after seeing the posters, was that the OER for carbon and neon ions for spread-out Bragg peaks is nowhere near as good as for fast neutrons and for argon ions is not much better, while losing the dose localization. After some discussion, Fowler summarized by saying that the OER for neon ions is comparable and that for argon ions is lower compared to fast neutrons. There was some agreement on this summary. Alpen stated that his laboratory has biological data for about five different cellular systems and that there is no way we can make generalizations. For example, there is significant recovery for cultured cells at the neon plateau but no recovery for the gut system. Alpen agreed with Fowler's summary on OER. Broerse suggested that the appropriate way to compare OER data by different investigators is to compare the OER of a given particle with the OER for X-rays by the same investigator for the same system.

POSTER SESSION D

Contributions on dosimetry and radiobiology of negative pions and heavy ions

J. BAARLI, M. BIANCHI, L. FAVALE, C. K. HILL, A. H. SULLIVAN, and J. W. N. TUYN (Geneva 23, Switzerland): *RBE and OER determinations in negative and positive pion beams for growth inhibition of* Vicia faba *roots*

The advantages of negative pi-mesons for radiotherapy are due to their depth-dose distribution combined with a high RBE and favourable OER in the region where the pions come to rest. The RBE and OER values are related to the high-LET component of the dose, which is produced by the secondary radiation from nuclear capture processes occurring at the end of the range of pions. Beam contamination with electrons and muons, together with fast secondaries from in-flight nuclear interactions (neutrons and protons), make the negative pion beam complicated from the point of view of the evaluation of the radiation quality along the beam (Turner *et al.*, 1972).

Some information on the radiation quality may be obtained, however, by comparing the depth-dose distribution of a negative pion beam with that of a positive beam of similar properties (Sullivan and Baarli, 1968). Positive pions are slowed down in much the same way as negative pions, except that they decay into muons at the end of their range instead of forming a nuclear star. The muon has an energy of 4.18 MeV (range of 0.15 g/cm^2 in tissue) and decays into a positron plus neutrinos. The positrons of mean energy of about 40 MeV are essentially minimum ionizing particles and deposit only a small fraction of their energy near the site of interaction. Hence the difference between negative and positive pion depth–dose distributions may be mainly attributed to the high-LET component from negative pion capture.

Figure 1 shows depth–dose distributions of positive and negative pions from the SIN π-E-3 beam. Dose–effect curves, based on 10 days' growth inhibition in *Vicia faba* roots exposed in the peak of the two beams (at points A and B), are also shown in the figure together with the curves for ^{60}Co irradiations.

Table 1 shows that the positive pion beam has RBE and OER values very close to those for ^{60}Co gamma-rays, although the peak RBE appears to be slightly greater than 1. The negative pion beam plateau appears to have a slightly higher RBE than the plateau of the positive beam. Real differences occur in both RBE and OER in the peaks of the negative and positive pion beams.

The data presented here may be of interest for judging whether the suitability of negative pions for radiotherapy can be determined from observation of differences in biological effects between the positive and negative pions. This would be a straightforward test to carry out since the positive pion beam is produced by simply reversing the currents in all beam elements used for the negative pion beam.

References

TURNER, J. E., J. DUTRANNOIS, H. A. WRIGHT, B. N. HANSEN, J. BAARLI, A. H. SULLIVAN, M. J. BERGER and S. M. SELZER, (1972) The computation of pion depth-dose curves in water and comparison with experiment. *Rad. Res.* **52**, 229-246.

SULLIVAN, A. H. and J. BAARLI (1968) Some measurements on the slowing down of π-mesons in tissue-equivalent material. *Phys. Med. Biol.* **13**, 435-441.

Table 1. *RBE and OER in the peak and plateau region of pion beams (Baarli et al.)*

	π^-		π^+		^{60}Co γ-rays	
	RBE	OER	RBE	OER	RBE	OER
Peak	2.6	1.5	1.2	2.7	1.0	2.8
Plateau	1.4	2.6	1.0	2.7	—	—

M. CHEMTOB, V. D. NGUYEN and N. PARMENTIER (Fontenay-aux-Roses, France): *Dosimetry of 650 MeV helium ions*

An important factor for radiobiology and radiotherapy of heavy ions is that the depth–dose curve varies slowly at the beginning of the range and reaches a peak (Bragg peak) near the end. Figure 2 shows a depth–dose curve obtained after modulation of a 650-MeV helium

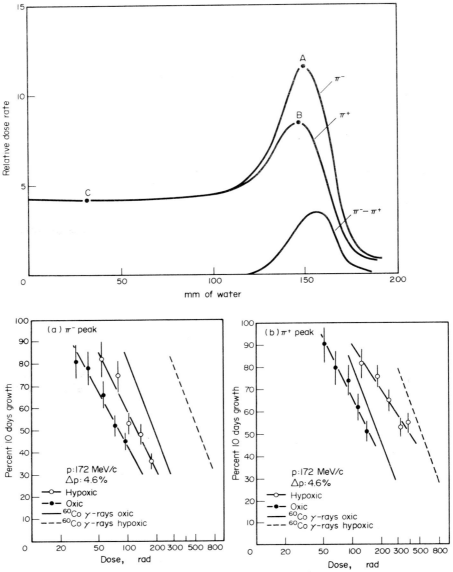

Figure 1. Pi-meson depth dose curve and dose effect curves for growth inhibition of Vicia faba (Baarli et al.).

ion beam by means of a ridge filter (Nguyen et al, 1976). This curve was derived for radiotherapeutic applications (Dutreix and Thevenet, 1976). Moreover, the dose is delivered with a narrow distribution of the lineal energy, y. Figure 3 shows that the spectrum obtained at the beginning of the 650-MeV He ion range is narrower than that obtained near the end of their range. The positions chosen, C_1 to C_5, are shown in Fig. 2 (Nguyen et al., 1976).

In order to carry out radiobiological studies (Parmentier et al., 1976) on the 650-MeV helium ion beam, dosimetric and microdosimetric measurements were undertaken. This paper deals with the problems encountered and the way in which they were overcome. It also points out the importance of the experimental setup for the reproducibility of the dose measurements and the biological irradiations.

The helium ion beam was produced by the SATURNE accelerator. The beam was spread out and collimated by a 8-cm-diameter collimator. In this way, it was possible to obtain a dose homogeneity of ± 5% on a 5-cm diameter field.

Since the dose varies rapidly near the end of the helium ion range, it is necessary to accurately determine the position at which the measurement is made. A parallel-faced water phantom was constructed. It is essentially a cylindrical tank. The lateral walls are built of a stainless-steel bellow. The bases are made from glass of 0.8 mm thickness and a diameter of 8 cm. One base is moveable and remotely controlled. The distance between the two bases is numerically displayed with an accuracy better than 0.01 mm. This accuracy permitted us to measure the dose by steps of 0.5 mm in the Bragg peak region.

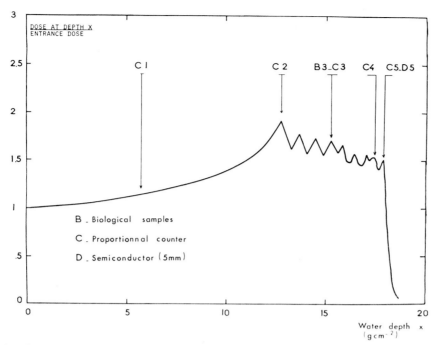

Fig. 2. Depth-dose curve for a 649-Mev helium ion beam with ridge filter (Chemtob et al.).

The dose is measured by using a parallel-plate extrapolation chamber made of TE A150 plastic (Chemtob *et al.*, 1975). The external high-voltage electrode is plane and 1mm in thickness. The collector is plane and parallel to the latter; its diameter is 5mm. The cavity gap is equal to 1mm and is filled with air or TE gas. It is important to point out that the small cavity gap makes it suitable to measure rapid dose variations, as shown in Fig. 2.

Microdosimetric measurements were made with an EGG counter (LET 1/2). The mean dose lineal energy, $\bar{y}_D$, varies from 4.4 to 22.3 keV/μm, corresponding to the extreme positions (C_1, C_5). $\bar{y}_D$ is equal to 9 keV/μm and constant from 126.5 to 152.1 mm and varies rapidly from 174.5 mm to the end of the 650-Mev helium ion spread Bragg peak, reaching 22.3 keV/μm (Nguyen *et al.*, 1976).

These $\bar{y}_D$ values are greater than the $\bar{y}_D$ value

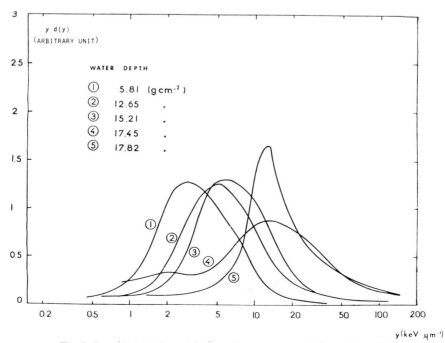

Fig. 3. Dose distribution yd(y) for a 2-μm counter (Chemtob et al.).

obtained in experiments using ^{60}Co gamma-rays. They are smaller than those obtained from 14 MeV neutrons (Lavigne, 1978) or from neutrons produced by d+Be reactions with $E_d =$ 16 to 50 MeV, but the dose distributions for 650 MeV helium ions are narrower than those given by these neutrons (Lavigne, 1978, Nguyen et al., 1977).

An argument in favor of utilizing a heavy ion beam is this narrow y dose distribution. It is sufficient to center this distribution around a value of $\bar{y}_D$ in the range of 100 keV/μm to 120 keV/μm. Since, in the ranges used in radiotherapy, for the same velocity the LET increases proportionally to Z^2 (charge of the incoming particle) nitrogen or Ne ions will give $\bar{y}_D$ values equal to 53.3 and 180 keV/μm, respectively, at the beginning of their range. Then it seems that, for Z higher than 10, the RBE will decrease and the OER will be stabilized around 1. Moreover, the Bragg shape of the depth–dose curve for heavy ions makes it possible to deposit a higher dose value in the tumor region.

References

CHEMTOB, M., C. FICHE, R. MEDIONI, C. NAUDY, V. D. NGUYEN, J. P. NOEL, N. PARMENTIER and A. RICOURT (1975) Chambres d'ionisation opérationnelles pour la dosimétrie des rayonnements à transfert linéique d'énergie élevé, *Proc. Biomedical Dosimetry*, Vienna, IAEA-SM 193/7.

DUTREIX J. and B. THEVENET (1976) Etude des ions He^{++} de 650 MeV en vue de leur application en radiothérapie, *Bull. Inf. Scient. Techn.* no. 220, 79.

LAVIGNE B. (1978) Etudes physique et microdosimétrique des faisceaux de neutrons utilisés en radiobiologie. Thèse, Toulouse, 24 janvier 1978 et Rapport CEA-R-4931.

NGUYEN, V. D., M. CHEMTOB, B. LAVIGNE and N. PARMENTIER (1976), Etude microdosimétrique d'un faisceau d'helions de 649 MeV, in J. BOOZ, H. G. EBERT and B. G. R. SMITH (Eds), *Proc. Fifth Symp. Microdosimetry, Verbania*, Commission of the European Communities, Luxembourg, p. 193.

NGUYEN, V. D., M. CHEMTOB, P. FACHE, N. PARMENTIER and J. LE GRAND (1977), Interprétation en terme de doses des mesures ionométriques pour les neutrons (d, Be - 33 et 50 MeV), in G. BURGER, *Proc. Second Symp. Neutron Dosimetry in Biology and Medicine*, Neuherberg, Commission of the European Communities, Luxembourg, p. 309.

PARMENTIER, N., M. GUICHARD, M. T. DOLOY and J. BRENOT (1976), Modèle biophysique appliqué à la survie des cellules tumorales et à la formation d'aberrations chromosomiques dans les lymphocytes humains, in J. BOOZ, H. G. EBERT and B. G. R. SMITH, *Proc. Fifth Symp. Microdosimetry, Verbania*, Commission of the European Communities, Luxembourg, p. 719.

STANLEY B. CURTIS (Berkeley, Cal., U.S.A.): *Calculations of radiation quality of heavy-ion beams**

A computer program has been developed to calculate physical parameters of interest at depth in a water phantom for high energy and stopping heavy ion beams. Presently, velocity and LET distributions can be calculated both for unmodified beams and for beams modified by a variable thickness absorber (ridge filter). Nuclear interactions are explicitly included in the calculation. Depth–dose curves are calculated for both the degraded primary beam and the secondary particles arising from nuclear interactions of the primaries with the hydrogen and oxygen nuclei of the water absorber. Input to the program are the maximum depth of penetration of the primary beam, the fragmentation cross-sections at high energy for the primaries into each of the secondaries to be considered, the depth of interest where the distributions are to be calculated, and the characteristics of the ridge filter, if desired. Approximations presently in the program include the assumption of no change in velocity or direction between primary ion and secondary fragment in a nuclear interaction and the neglect of tertiary nuclear interactions, neutron production and multiple scattering. The energy dependence of the cross-sections is approximated by appropriately scaling phenomenological expressions of proton–heavy element cross-sections.

Dose-averaged and track-averaged LETs are calculated as a function of depth for the various beams used in the radiobiological experiments performed at the BEVALAC. In the peak regions of the spread-out beams, the values of the dose-averaged and track-averaged LETs differ significantly, the dose-averaged values always being larger. Results of the calculations give dose-averaged LETs for carbon ions at high energy (400 MeV/nucleon) of 12 KeV/μm. As a function of depth, this parameter increases to 30 KeV/μm as the proximal portion of the 10-cm-spread peak region is reached. The value increases slightly to 40 KeV/μm at the midpoint of the peak region and then rises sharply to 130 KeV/μm in the extreme distal portion of the

---

* Research supported by the U.S. Department of Energy.

peak region. For the 400-MeV/nucleon neon ion beam, the high energy (plateau), proximal, midpoint, and extreme distal values for a 10-cm-spread peak are 30, 70, 100 and 300 KeV/μm, respectively. For an argon ion beam with incident energy 570 MeV/nucleon, the corresponding values are 100, 200, 300 and 1500 KeV/μm, respectively. This quantity is only a rough measure of beam quality and must not be considered as equivalent to the LET values used in "track segment" experiments for comparing radiobiological quantities, such as RBE and OER, between ions.

S. B. CURTIS, T. S. TENFORDE and W. A. SCHILLING (Berkeley, Cal., U.S.A.):
Review of the radiobiological response of a rhabdomyosarcoma irradiated with heavy ion beams: In vivo *and* in vitro *results*[*]

A transplantable rhabdomyosarcoma tumor in the WAG/Rij rat is being used to study the radiobiological properties of high-energy carbon, neon and argon beams available at the Berkeley BEVALAC facility. End-points measured are radiation-induced tumor growth delay, $TCD_{50/180}$, cell survival in tumors irradiated *in situ* in both air-breathing and nitrogen-asphyxiated animals, and cell survival for both oxygenated and hypoxic cell populations *in vitro*.

Results to date are the following:

1. In studies of radiation-induced growth delay (using a tumor subline denoted R1/LBL), RBE values for 50-day growth delay were found to be comparable in the distal portions of a 4-cm-spread peak region for the three ion beams. The RBE value for the carbon-ion beam (2.3) is slightly lower than for the neon (2.6) or argon (2.5) beams, but the difference is not statistically significant.

2. The RBE for 50% tumor cure was obtained for the 4-cm neon–ion peak region, and was found to be 3.1 ± 0.6. This value is not significantly different from the RBE for growth delay induced by peak neon ions.

3. Preliminary fractionation studies indicate reduced recovery for R1/LBL tumors exposed to carbon ions in the distal portion of a 4-cm-spread peak. The recovered dose per fraction in a six daily fraction schedule was approximately 5% for the carbon-ion beam, as compared to 15% for X-rays.

4. RBE values at the 10% survival level for tumors irradiated *in situ* and assayed for colony formation *in vitro* were 1.9 and 3.1 for the carbon– and neon–ion beams, respectively. A different tumor subline, designated R2D2, was used in these studies, and yielded RBE values for 50-day growth delay of 2.3 and 2.9 for the carbon– and neon–ion beams, respectively. Here the difference between the carbon– and neon–ion RBE values is statistically significant. From a comparison of the survival curves obtained from nitrogen-asphyxiated and air-breathing animals, the hypoxic fraction for this rhabdomyosarcoma tumor line was found to be approximately 35%.

5. *In vitro* cell survival curves have yielded RBE and OER values for carbon– and neon–ion beams with 15 and 25 cm penetration in water and for spread-out peak widths of 4 and 10 cm. From these data we have calculated a factor of merit for the different heavy-ion beams, defined as the efficiency for hypoxic cell killing in the peak relative to that for oxygenated cell killing in the plateau ionization region. These merit factors have demonstrated that 4-cm-spread peaks have a greater therapeutic efficacy than 10-cm peaks due to a greater peak to plateau physical dose ratio. Carbon beams are slightly more advantageous than neon beams at the deeper penetration (25 cm), but there is no difference between the two ions at 15 cm penetration. In the peak region of the argon beam, the RBE was found to decrease with increasing depth and average LET; thus, the maximum in the RBE vs. LET curve has been reached and surpassed for the argon beam. The OER at 10% survival decreases from about 1.8 in the carbon spread peak to 1.5 in the neon spread peak to 1.2 in the argon spread peak. Clearly, the OER does not drop to unity, even in the argon–ion peak region.

[*] Research supported by the U.S. Department of Energy and by Grant Nos. CA 17411 and Ca 15184 awarded by the National Cancer Institute, DHEW.

J. VAN DAM,* G. BILLIET,* A. WAMBERSIE† and A. DUTREIX‡ (* Leuven and † Brussels, Belgium; ‡ Villejuif, France): *RBE and OER studies in* Vicia faba.

Growth delay in *Vicia faba* at 10 days after irradiation (Hall, 1961) has been used as biological criterion for determination of RBE and OER of several radiation qualities.

Neutron beams were obtained at the cyclotron "CYCLONE", Louvain-la-Neuve, by bombarding a thick (10 mm) beryllium target with 50 MeV deuterons. The doses reported (Table 2) correspond to the total absorbed doses $D_{n+\gamma}$, the ratio $D_\gamma/D_{n+\gamma}$ being 7%. An OER value of 1.5 ± 0.1 was observed; OER for ^{60}Co γ-rays was found equal to 2.6 ± 0.1 (G.F. = 1.8).

Table 2. RBE of d (50)+Be neutrons for growth delay in Vicia faba *bean roots (van Dam et al.) Irradiations at the cyclotron CYCLONE, Louvain-lap-Neuve (reference radiation quality: ^{60}Co γ-rays).*

Absorbed dose/Gy	RBE	(± standard error)
0.25	4.1	(± 0.2)
0.4	3.6	(± 0.1)
0.9	2.8	(± 0.2)

At the same cyclotron, neutrons produced by bombarding a 32-mm beryllium target with 75 MeV protons were studied. Preliminary results show an RBE of 0.9 for these *p* (75)-Be neutrons with respect to *d* (50)-Be neutrons. The OER values observed for both neutron beams were similar (OER = 1.5).

At the Institut Gustave-Roussy in Villejuif, RBE of californium-252 was determined by means of sealed sources used for interstitial therapy by Pr. D. Chassagne. Six sources, active length 52 mm, 1 mm in diameter, each of them containing 1.6 µg of ^{252}Cf, were used. ^{192}Ir wires were used as reference, their activity being adjusted in order to obtain similar irradiation times. Irradiations were performed simultaneously and in similar geometrical conditions; the distance between the sources and the root was 7.5 mm.

Vicia faba were irradiated with ^{252}Cf at a dose rate $(n+\gamma)$ of about 0.15 Gy/h and with ^{192}Ir at a dose rate of about 0.7 Gy/h. Preliminary results indicate an RBE of 6.4 for the $(n+\gamma)$ emission of ^{252}Cf with respect to ^{192}Ir. The contributions of the *n* and γ components of ^{252}Cf to the absorbed dose at the distance of the biological systems were approximately 65% and 35%. RBE derived for the neutron component of ^{252}Cf is then 9.3.

Finally, RBE of a 650 MeV He^{++} beam was determined as a function of depth in the irradiated medium. Irradiations were performed at the synchrotron SATURNE of the CEN-Saclay (Dutreix and Thevenet, 1976). The results presented in Table 3 indicate a progressive increase of RBE as a function of depth, which can be related to an increase of the mean lineal energy y_D (Nguyen *et al.*, 1976).

Table 3. RBE of 650 MeV He^{++} as a function of depth (van Dam et al.) Irradiations at the synchrotron SATURNE, CEN-Saclay (reference radiation quality = 650 MeV He^{++}, in the initial plateau region, at 2 cm in depth).

Absorbed dose/Gy (of the test radiation quality)	RBE	(± standard error)
(A) 650 MeV He^{++} at the beginning of the spread-out Bragg peak (13 cm depth).		
0.4	1.5	(± 0.2)
0.7	1.4	(± 0.1)
1.2	1.4	(± 0.2)
(B) 650 MeV He^{++} at the middle of the spread-out Bragg peak (15 cm depth).		
0.4	1.5	(± 0.2)
0.65	1.5	(± 0.1)
1.1	1.5	(± 0.2)
(C) 650 MeV He^{++} at the end of the spread-out Bragg peak (17 cm depth).		
0.3	2.1	(± 0.3)
0.5	2.0	(± 0.2)
0.85	1.9	(± 0.2)

Acknowledgements—The authors acknowledge with thanks the collaboration of M. Octave-Prignot and J. P. Meulders for the dosimetry at CYCLONE, of A. Mahmoudi and H. Bounik for the dosimetry of ^{252}Cf, and of A. Bridier for the dosimetry at SATURNE.

References

HALL, E. J. (1961) The relative biological efficiency of X-rays generated at 220 kVp and gamma radiation from a cobalt-60 therapy unit. *Brit. J. Radiol.* **34**, 313-317.

DUTREIX, J. and B. THEVENET (1976) Etude des ions He^{++} de 650 MeV en vue de leur application en radiothérapie. *B.I.S.T.* **220**, 79-92.

NGUYEN, V. D., M. CHEMTOB, B. LAVIGNE and N. PARMENTIER, (1976) Etude microdosimétrique d'un faisceau d'hélions de 650 MeV. *Proc. Fifth Symp. on Microdosimetry*, Vol. 1, ed. Commission of the European Communities, Luxembourg, pp. 153-166.

J. F. DICELLO, M. ZAIDER and M. TAKAI* (Los Alamos, U.S.A.): *Some physical characteristics of range-modulated beams of pions*

The physical characteristics of beams of negative pions relevant to radiotherapy cannot be characterized by a single parameter such as the absorbed dose. In a therapeutic situation, such beams contain many types of particles (contaminants and secondaries) distributed over a wide range in energy. One of the problems confronting radiation physicists is to establish that set of physical parameters which is sufficient to uniquely define biological response. An overall program addressing this problem is presently being pursued at Los Alamos.

Beams of negative pions have broad distributions in LET or lineal energy. Microdosimetric data for large treatment volumes of pions have been obtained in order to investigate the differences in the lineal energy spectra. These results are compared with comparable results for neutrons and heavy ions. We propose, on the basis of these data, that microdosimetric spectra for broad beams of pions, including multi-port irradiations, can be generated from the corresponding data for narrow beams from a knowledge of the function used to vary the range of the particles. The present data show that knowledge of the LET distributions along the central axis alone is not sufficient unless the broad and narrow beams have the same lateral dimensions (in the plane perpendicular to the beam axis). This is because significant differences in the distributions are observed because of the differences in secondary particles, particularly, neutrons.

Data have been reported earlier (Amols et al., 1978) on the dose and lineal energy distributions from neutrons produced in the stopping region of a narrow pion beam. This work has been extended to broad beams and a new technique has been developed to directly measure neutron distributions in the stopping region. This method uses activation techniques to measure the $^{39}K(n,2n)^{38}K$ reaction. It is shown that this reaction is a direct result of neutrons from pion stars. A comparison between calculations (Berardo and Zink, 1978) and measurements indicate that there is a correlation between the induced activity and the neutron dose. These results, along with the microdosimetric data, show that as much as 50% of the high-LET dose in the broad pion beams comes from neutron interactions. Furthermore, there is not a direct relationship between the neutron-dose distribution and either the total dose distribution or the pion stopping distribution.

The $^{39}K(n,2n)^{38}K$ activation technique is simple and sufficiently compact so that it can be applied as an *in vivo* dosimetric system in patients.

Another potentially important physical aspect is the distribution of stopping pions. The method adopted by us to obtain this information was originally developed by Laughlin (1965) for electrons and extended to pions by Shortt and Henkelman (1978). This system consists of a small, tissue-equivalent capacitor which measures the net increase in charge from stopping particles. We have obtained large amounts of data for narrow and broad beams of $\pi+$ and $\pi-$. The previously discussed distributions are compared with the corresponding stopping distributions.

Finally, a model (Zaider and Dicello, 1978) has been developed to calculate biological responses for fractionated studies. Single fraction results based on the Theory of Dual Radiation Action (Kellerer and Rossi, 1972) or the Chadwick–Leenhouts model (Chadwick and Leenhouts, 1973) are used as input. In spite of known limitations of the model, it is useful for correlating the radiation quality with biological response. Typical results of this model are presented.

References

AMOLS, H. I., J. N. BRADBURY, J. F. DICELLO, J. A. HELLAND, M. M. KLIGERMAN, T. F. LANE, M. A. PACIOTTI, D. L. ROEDER, and M. E. SCHILLACI, (1978) Dose outside the treatment volume for irradiation with negative pions. *Phys. Med. Biol.* **23**, 385-396.

BERARDO, P. and S. ZINK (1978) Los Alamos Scientific Laboratory, private communication.

CHADWICK, K. H. and H. P. LEENHOUTS (1973) A molecular theory of cell survival. *Phys. Med. Biol.* **18**, 78-87.

KELLERER, A. M. and H. H. ROSSI (1972) The theory of dual radiation action. Current Topics in *Radiat. Res.* **8**, 85-158.

LAUGHLIN, J. S. (1965) Studies of absorption of high energy electron beams. In A. ZUPPINGER and G. PORETTI, *Symposium on High-Energy Electrons*, Springer-Verlag, New York, pp. 11-16.

ZAIDER, M. and J. F. DICELLO, (1978) A FORTRAN program for the computation of RBEs, OERs, survival ratios, and the effects of fractionation using the theory of dual radiation action. Los Alamos Scientific Laboratory, Report LA-7196-MS.

---

*Permanent address: Hamamatsu University, School of Medicine, Hamamatsu, Shizuoka, Japan.

C. F. von Essen, H. Blattmann, Ch. Perret, G. Vécsey and J. P. Blaser (Villigen, Switzerland): *The medical pi-meson project at SIN*

A high-intensity proton beam (600 MeV/20 µA) will produce pi-mesons in a gas-cooled pencil target. The large acceptance optical system (based on a design principle developed at Stanford) will concentrate a considerable pi-meson flux of variable momentum into the volume to be irradiated (Vécsey et al., 1977). Guiding and focusing at the selected momentum is achieved by two large superconducting torus magnets. Each magnet consists of sixty separate pancakes subdividing the total particle flux into sixty beams of identical properties entering the treatment volume radially in a common plane. The intensity of these beams is individually controlled by momentum slits and monitored by ionization chambers. The expected maximum peak dose rate is 50 rad/min into a volume of 1 litre.

To achieve a homogeneous dose distribution with the special geometry of the applicator, two different methods are under investigation. For the ring scan the patient contour will be shaped into a cylinder by means of bolus material. The irradiation procedure involves initially stopping the pions at the central axis of the applicator and then continuously reducing the range of the pions in order to scan the entire treatment volume. When individual beams reach the treatment volume surface they are switched off. For the spot scan the patient will be introduced into a cylindrical water bolus mounted coaxial to the applicator. The momentum of the pions is chosen to stop in a "hot spot" on the axis of the applicator. The patient is then moved within the fixed water bolus so that a homogeneous dose distribution is produced within the treatment volume.

A high degree of precision and reproducibility of patient position and tumor localization is essential to fully exploit the potential geometric advantages of pi-mesons for clinical applications. The patient is positioned in an individually moulded surface upon a semi-cylindrical treatment couch prior to computer tomographic examination. CT scans are performed at narrow slices through the proposed treatment volume following which density data and treatment volume countours are entered into the treatment-planning programme. The patient position is checked and recorded by a precision radiographic simulator on video tape. Finally the patient (on his treatment couch) is positioned in the treatment chamber with a water bolus ring encompassing the treatment sector. Microprocessor directed signals drive the couch in three axes to carry out the planned treatment.

The absorbed dose at specific points in phantoms will be measured with tissue-equivalent ionization chambers. Tissue-equivalent proportional counters will be employed to determine microdosimetric distributions. Treatment plans will be checked by first: measurements of low LET and stopping pions separately with LiF thermoluminescent dosimeters and activation of aluminium, respectively, in phantoms and second: comparison of calculated distribution, biological dose and survival of mammalian cells by irradiating cells in gel in a phantom. Measurement in the patient will be made with LiF thermoluminescent dosimetry and by activation of aluminium.

In order to arrive quickly at a safe yet effective dose schedule for the treatment of human cancers in various parts of the body, four different normal tissues will be investigated in single and fractionated dose experiments. These experiments will be carried out in cooperation with other institutions.

(1) Mouse foot: acute and delayed radiation damage from pions and X-rays (Radiobiological Institute, University of Zurich SBI); (2) mouse gut: (SBI-UniZ); (3) mouse lung: lung damage as determined by mortality following thoracic radiation (MRC Cyclotron Unit, Hammersmith Hospital, London, and SBI-UniZ); (4) spinal cord: determination of the tolerance dose of the central nervous system to peak and plateau pions for different fractionation schemes (Radiobiological Institute TNO, Rijswijk).

For the clinical trials three phases are planned:

1. In order to quickly yet safely proceed from preclinical experiments to definitive therapy of cancer it is proposed to first irradiate patients with multiple small metastases of the skin and subcutaneous tissues with a range of pions and X-ray doses to fields of no larger than 5 cm. An attempt will be made to adjust the dose rate and dose distribution of one modality to the other. Comparable nodules in each patient will be irradiated by one or the other modality. Evaluation of skin reactions will be made independently by trained clinicians using a reaction scale based on those of Bewley et al. and of L. Cohen. Both acute and possibly late effects will be scored and RBEs derived. Evaluations of tumour regression will also be done.

2. Based on phase 1 trial data and on experience from the present clinical trials underway in Los Alamos it is proposed to treat

patients with advanced but generally regionally localized cancers in a variety of locations.

3. Randomized clinical trials comparing the effects of the best available diagnostic information, treatment planning and radiation modality with pi-mesons upon potentially curable cancers will follow.

Reference

VECSEY, G., I. HORVATH and J. ZELLWEGER (1977) Superconducting medical facility at SIN. In *Sixth International Conference on Magnet Technology*, Bratislava, Slovak Academy of Sciences, Bratislava, Czechoslovakia.

P. FESSENDEN,* G. C. LI,* G. LUXTON,* R. TABER,* C. H. YUEN,* H. D. ZEMAN,* M. A. BAGSHAW* and W. HOFFMANN† (*Stanford, CA, U.S.A. †Wuppertal, Federal Republic of Germany): *Dosimetry, microdosimetry and radiobiology with the Stanford medical pion generator (SMPG)*

The Stanford Medical Pion Generator (SMPG) collects and focuses negative pi mesons (pions) in a multiport geometry designed for radiotherapy. The pions are produced as secondaries when 600 MeV electrons are incident on a low-Z production target typically 5 cm long by 1 cm in diameter. The pion transport system contains two sixty-section superconducting toroidal magnets that result in up to sixty pion beams converging simultaneously to the centre of a patient tank (Fig. 4).

Ionization chamber dosimetry shows the dose distribution localized to 3.3 cm and 5.4 cm full width half maxima (FWHM) in the radial (transverse to SMPG axis) and axial directions, respectively, when all 60 beams stop on the axis. The peak dose rate is 3 rad/min per microampere of primary electron beam, and the relative dose at the surface of a 22-cm-diameter phantom is 2%. Further localization is achieved with a shorter production target and/or with a smaller pion momentum acceptance. In order to spread the dose radially, the toroid magnetic field is reduced, literally "sweeping" the pions out from the SMPG axis; axial broadening is achieved by translation of the phantom along the axis. Inhomogeneities are accounted for, and irregular dose distributions are obtained, by selectively

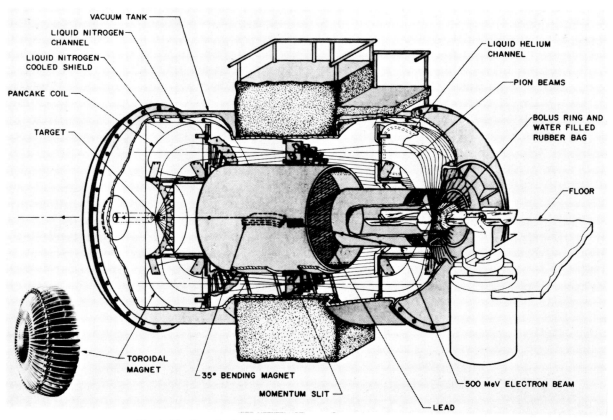

Fig. 4. The Stanford medical pion generator, SMPG (Fessenden et al.).

opening and closing the individual momentum-defining slits associated with each of the sixty beams.

Dose quality has been mapped for various pion beam configurations using a tissue-equivalent proportional counter with 2.0 microns equivalent thickness of gas. The well-localized dose distribution has a dose averaged lineal energy ($\bar{y}_D$) of 55-60 keV/micron throughout a 3-cm-diameter by 5-cm-long region. When the irradiation volume is broadened to give an 8-cm-diameter cylindrical region of uniform dose, the average $\bar{y}_D$ is 48 keV/micron, and varies by $\pm 15\%$ over the volume. The dose quality drops significantly in the plateau region; there, the $\bar{y}_D$ is 4-10 keV/micron, depending on the beam configuration.

Radiobiology studies have been performed with SMPG pions using Chinese hamster cells (HA-1) of ovarian origin suspended in a tissue-equivalent gelatin solution. For the case of sixty beams stopping at the axis (dose rate of 6 rad/min the OER was 1.7 and the RBE was 2.9 and 1.8 at 50% and 20% survival, respectively. The OER and RBE determined for plateau pions (dose rate of 3 rad/min) were nearly the same as those obtained using-250 kVp X-rays, a result consistent with the microdosimetric measurements.

HEDI FRITZ-NIGGLI and H. BLATTMANN (Zurich, Switzerland): *RBE of negative pions in different biological systems for various criteria*

With the biomedical pion channel of the 590-MeV proton-accelerator of the Swiss Institute for Nuclear Research (SIN) we have had, for 4 years, the opportunity to test the theoretical conceptions on several gain factors of negative pions for radiation therapy. Negative pions have the unique particuliarity to act on the tissues in the same treatment with two different types of radiations, in the peak (treatment volume) with high-LET radiation and in the plateau (entrance volume) with low-LET radiation.

The momentum chosen for the experiments was 176 MeV/c with a momentum band width of 6% for the irradiations in the peak region. The beam was contaminated with approx. 11% electrons and 5% muons. The field size was matched for each experiment to the size of the biological object and varied between a circular field of 2.5 cm diameter and a field of 3.5 × 5 cm for the 85% isodose level. The doses were measured with a tissue-equivalent chamber.

As the dose rate in the peak was not more than 10 rad/min only very sensitive biological systems have been chosen to avoid artefacts by a protracted irradiation. The following systems have been used: inactivation of single mammalian cells, induction of chromatid aberrations in Chinese hamster cells; small intestine of mouse (early and late effects); induction of anomalies in mouse embryos; induction of cerebral microvascular damages in neonatal rats; proliferation of Ehrlich-ascites carcinoma cells; induction of different types of mutation in different stages of male germ cells and somatic cells (*Drosophila*). The RBE value in the peak region varies between 0.7-3.3, and are different even in the same system with the same end point but at different cell stages and conditions. For the plateau region the RBE values lie mostly under 1 (compared with 140 kV photons, same dose rate). 3 Facts seem to be highly interesting for clinical purposes. (1) For peak pions the RBE in hypoxic cells (tumor cells) can be much higher than in euoxic systems, to which the healthy cells belong. In radiosensitive euoxic systems the RBE of star pions can be under 1. (2) The RBE-values in the peak for low doses are higher than with high doses. A single fraction of a therapeutic application can be considered as a low dose in comparison with the doses used in our experiments. (3) Not only the RBE of peak pions are interesting, more important seems to be the RBE ratio peak/plateau which varied in the experiments between 1.4-4.0, with the momentum band width of 6%.

L. S. GOLDSTEIN and T. L. PHILLIPS (San Francisco, U.S.A.): *RBE variations with depth for single and fractionated high-LET irradiation*

Charged particle beams have a better physical depth–dose distribution and higher relative biological effectiveness in tumor cells than conventional photon sources. Although the normal tissue may be spared by fractionating the photon dose, it is not known how much sparing will result when irradiation by high-LET particle beams is fractionated. Therefore, we measured the effectiveness of fractionated high-LET radiation relative to fractionated photon radiation in a representative normal tissue.

Male mice were irradiated in either the low-LET (plateau) or high-LET (peak) regions of accelerated helium, carbon or neon beams having clinically relevant Bragg peak widths. Graded doses were given in 5 fractions (carbon and neon) or in 10 fractions (helium) with a 3-hour interval between irradiations to allow for recovery. Similarly treated mice received comparable dose rates with 137-Cs. The number of surviving intestinal crypt cells per circumference was determined $3\frac{1}{2}$ days after irradiation, using procedures modified (Goldstein, Phillips and Ross, subm. 1978) from those of Withers and Elkind (1970).

When compared to irradiation with 137-Cs, both single and fractionated irradiation with accelerated heavy ions produced dose-response curves with reduced shoulders, but with similar slopes.

repairable lesions is found near the end of the track, while the overlying tissue can repair the damage formed by the low-LET in the plateau region at the beginning of the track, then fractionating the dose of accelerated heavy ions enhances the biological depth–dose distribution. Thus, the effective tumoricidal dose to the target volume is increased without added risk to the overlying normal tissue. For example, by fractionating the dose of the carbon beam with a 4-cm spread Bragg peak, the tumoricidal effect is increased by 50% relative to overlying tissue. These data suggest that substantial gains in the therapeutic ratio (relative biological effectiveness in the tumor vs. relative biological effectiveness in normal tissue) may be realized for fractionated doses of high-LET radiation.

Acknowledgements—The authors wish to thank Mrs. G. Y. Ross of the Division of Radiation Oncology and the staff members of the 184-inch synchrocyclotron and Bevalac facilities of the Lawrence Berkeley Laboratory at the University of California for their invaluable assistance. This investigation was supported by National Institutes of Health Grant No. 17227.

Table 4. *Dose to reduce the number of crypt cells to twenty-five after irradiation in the high- and low-LET regions of different charged particle beams*

Beam (energy[a] peak width[b])	Number of fractions	Sample position[c]	Dose, rad	
			Acute	Fractionated
137-Cs	5	—	1300	2160
Carbon, (400,4)	5	PL	1050	1600
	5	DP	1050	1050
Carbon, (400,10)	5	PL	1000	1740
	5	PP	950	1460
	5	MP	950	1320
	5	DP	900	1070
Neon, (425,4)	5	PL	1020	1020
	5	DP	860	760
Neon, (557,10)	5	PL	920	1160
	5	MP	840	930
137-Cs	10	—	1300	2160
Helium, 225.8	10	PL	1250	1920
	10	DP	1100	1600

[a] MeV/amu. [b] in cm. [c] PL = plateau, PP = proximal peak, MP = mid peak, DP = distal peak.

These data in Table 4 indicate that the amount of recovery found after fractionated irradiation depends on the particle beam and the position of the sample in the beam. Recovery was greatest in the plateau region with better recovery found in the mid-peak region than in the distal peak region.

Since the high-LET responsible for non-

References

GOLDSTEIN, L. S., T. L. PHILLIPS and G. Y. ROSS (subm. 1978) Enhancement by fractionation of biological peak-to-plateau RBE ratios for heavy ions. *Int. J. Radiat. Oncol. Biol. Phys.*

WITHERS, H. R. and M. M. ELKIND (1970) Microcolony survival assay for cells of mouse intestinal mucosa exposed to radiation. *Int. J. Radiat. Biol.* **17**, 261-267.

A. F. HERMENS,* S. B. CURTIS,† T. S. TENFORDE,† and G. W. BARENDSEN* (*Rijswijk, The Netherlands; †Berkeley, Cal., U.S.A.): *Analysis of cell proliferation in a rat rhabdomyosarcoma irradiated with neon ions or with 300-kV X-rays*

Ionizing radiations with LET values in excess of 20 keV/μm might provide an advantage in radiotherapy as compared to X-rays, because of a lower OER and an increased contribution of the type of damage that is not dependent on accumulation and is less subject to repair (Barendsen and Broerse, 1969). Responses of tumours after irradiation are also determined, however, by repopulation of clonogenic cells which may show an enhanced progression through their cell cycle (Hermens and Barendsen, 1977). Because the oxygenation status of surviving cells, their distribution within the tumour and their capacity for rapid recruitment and proliferation might be different after exposure to high-LET radiation as compared to low-LET radiation, studies of the relation between tumour growth delay and cell proliferation kinetics may yield insight in the relative influence of these factors.

Studies on cell proliferation kinetics and growth delay after exposure to neon ions have been carried out for a subline of the R-1 rhabdomyosarcoma, designated R-1/LBL, transplantable in WAG/Rij rats (Curtis *et al.*, 1979). Tumours of 500-1100 m³, growing in the flanks of rats, were irradiated with a dose of 6 Gy of neon ions from the Berkeley BEVALAC facility. Effects on tumour volume changes and on cell proliferation kinetics are compared with data obtained earlier for R-1 tumours irradiated with a dose of 20 Gy of 300-kV X-rays at TNO Rijswijk. Comparison of the two series of results was expected to be justified because the ratio of these radiation doses correlates rather well with RBE values of 2.6 to 3.3 derived for growth delay (Barendsen and Broerse, 1969; Curtis *et al.*, 1979).

Effects on tumour growth were analysed from measurements of tumour volumes before and after irradiation, while effects on cell proliferation were studied in tumours irradiated 8 days before analysis and in non-irradiated controls by using the ^{3}HTdR labelling technique (Hermens and Barendsen, 1969). Results on cell-proliferation kinetics obtained so far for the R-1/LBL tumours are limited to data collected from the 2-mm-wide peripheral tumour shell.

The doubling time (T_d) of non-irradiated R-1/LBL tumours measuring 740 mm³ is about 120 h (= 5.0 d). This is about a factor of 1.4 longer as compared to the T_d value reported earlier for R-1 tumours measuring 1500 mm³ (Hermens and Barendsen, 1969). The fractions of proliferating cells (F) derived for these R-1/LBL and R-1 tumours are $F = 19\%$ and $F = 33\%$, respectively. In contrast to these differences in T_d and F between the two sublines, there are no significant differences between corresponding phase durations and cell cycle times derived for cycling cells of R-1/LBL and R-1 tumours.

Irradiation of R-1/LBL tumours with 6 Gy of neon ions induced a growth delay (TGD) of 204 h (= 8.5 d) and resulted in a regrowth with an average value of $T_d = 228$ h (= 9.5 d) between day 8 and 14 after irradiation. These values can be compared with values of TGD = 336 h (= 14.0 d) and an average $T_d = 156$ h (= 6.5 d) between days 8 and 14 measured for R-1 tumours exposed to 20 Gy of 300-kV X-rays. Values derived for the growth fractions of R-1/LBL and R-1 tumours at day 8 after irradiation are equal to $F = 29\%$ and $F = 24\%$, respectively. The phase durations of the cell cycle and the cell cycle times measured in irradiated R/1/LBL tumours are not significantly different from those measured in irradiated R-1 tumours. Interestingly, in both sublines of irradiated tumours the cell cycle time is about 4 h shorter as compared to that measured in unirradiated controls.

The growth delay of 8.5 day observed in the R-1/LBL tumours after 6 Gy of neon ions is considerably shorter than was expected on the basis of an RBE = 2.6 to 3.3 for growth delay measured in other experiments (Curtis *et al.*, 1979) and a TGD = 14 days induced by 20 Gy of X-rays. This shorter growth delay might be correlated with a 10% increase in the fraction of cycling cells in the R-1/LBL tumours exposed to neon ions as compared to that before irradiation, in contrast to the 10% decrease in the fraction of cycling cells observed in X-irradiated R-1 tumours.

Finally, it may be concluded that the rate of cell cycle progression of surviving cells is not influenced differently by high-LET neon ions as compared to X-rays.

The interpretation of the differences observed requires analysis of cell-proliferation kinetics in X-ray-treated R-1/LBL tumours and in other regions of these tumours.

References

BARENDSEN, G. W. and J. J. BROERSE (1969) Experimental radiotherapy of a rat rhabdomyosarcoma with 15 MeV neutrons and 300 kV X-rays. I. Effects of single exposure. *Europ. J. Cancer* 5, 373-391.

CURTIS, S. B., T. S. TENFORDE and SCHILLING (1979) Review of the radiobiological response of a rhabdomyosarcoma irradiated with heavy ion beams: *in vivo* and *in vitro* results. See: abstract of poster no. 4, Poster Session D.

HERMENS, A. F. and G. S. BARENDSEN (1969) Changes of cell proliferation characteristics in a rat rhabdomyosarcoma before and after X-irradiation. *Europ. J. Cancer* 5, 173-189.

HERMENS, A. F. and G. W. BARENDSEN (1977) Effects of ionizing radiation on the growth kinetics of tumours. In B. DREWINKO and R. M. HUMPHREY (Eds.) *Growth Kinetics and Biochemical Regulation of Normal and Malignant Cells*, The University of Texas System Cancer Center, M. C. Anderson Hospital and Tumor Institute, 29th Annual Symp. on Fundamental Cancer Research. Williams and Wilkins Co., Baltimore, pp. 531-546.

C. LÜCKE-HUHLE (Karlsruhe, Federal Republic of Germany), E. A. BLAKELY and C. A. TOBIAS (Berkeley, Cal., U.S.A.): *The influence of intercellular contact on mammalian cell survival after heavy-ion irradiation*

Heavy-ion beams of carbon, neon and argon produced by the Bevalac at the Lawrence Berkeley Laboratory were used to study the killing efficiency of high-LET radiations on 12-days-old V79 spheroids. Spheroids are a multicellular tumor-like system whose cells are more resistant to low-LET radiations such as γ- and X-rays than mono-layer cultures of the same cell line (Dertinger and Lücke-Huhle, 1975; Sutherland and Durand, 1976). The range of the heavy-ion particles was modulated by a leaden ridge filter in order to spread out the Bragg peak. Spheroids were irradiated either at a plateau position or in the middle of the 4-cm spread peak under conditions which guaranteed sufficient oxygen and nutrient supply and constant temperature of 37°C. After irradiation, spheroids were immediately trypsinized and plated for the survival assay.

Figure 5 compares survival after X-ray with survival after exposure to 400 MeV/N carbon ions, 425 MeV/N neon ions and 570 MeV/N argon ions at a plateau and midpeak position, respectively. The greater resistance to X-ray of spheroid cells due to intercellular contact is indicated by the pronounced initial shoulder of the X-ray survival curves. Exposure of spheroids at the carbon and neon plateau also yielded survival curves with shoulders. The RBE value at 10% survival for the carbon plateau was 1.0, for the neon plateau 1.5, comparable to the RBE described for peak pions (Lücke-Huhle et al., 1977). Bragg peak radiation, however, produced RBE values of 4.1-4.2 and the curves showed no shoulder. The RBE values for spheroids irradiated at the peak positions were high in comparison to those obtained with monolayer cultures of the same cell line and under identical

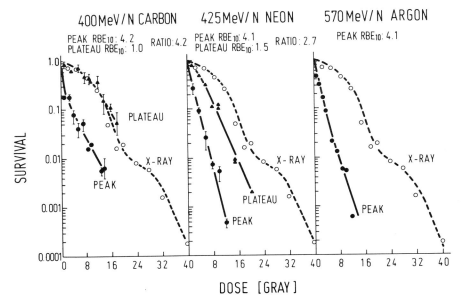

Fig. 5. Comparison of survival of V79 spheroid cells after X-ray (open circles) with that after heavy ions at the plateau (closed triangles) or at the middle of a 4-cm spread peak (closed circles).

experimental conditions. The RBE values for V79 monolayers were 1.8, 2.4 and 2.1 for carbon, neon and argon, respectively (data were obtained together with F. Ngo, LBL). Since RBE values are based on X-ray data for reference, the fact that the increased resistance of spheroid compared to mono-layer cells is absent, must lead to a higher RBE value for spheroids.

In conclusion, while three-dimensional cell contact benefits survival after low-LET radiation, this increased resistance of the multicellular system was not detected after irradiation with heavy ions at the spread peak. With respect to radiotherapy, the peak/plateau RBE ratio was most advantageous for the carbon beam. Healthy tissue in front of a tumor would be protected by a factor of 4.2 as compared to cell killing of the tumor cells centered within the spread carbon peak.

References

DERTINGER, H. and C. LÜCKE-HUHLE (1975) A comparative study of post-irradiation growth kinetics of spheroids and monolayers. *Int. J. Radiat. Biol.* **28**, 255-265.

LÜCKE-HUHLE, C., H. DERTINGER, H. SCHLAG and K. F. WEIBEZAHN (1977) Comparative studies on the effect of negative pions, ^{60}Co-Gamma irradiation and hyperthermia on an *in vitro* tumor model. In *Proc. International Symp. on Radiobiological Research needed for the Improvement of Radiotherapy*, vol. 1, International Atomic Energy Agency, Vienna, pp. 279-287.

SUTHERLAND, R. M. and R. E. DURAND (1976) Radiation response of multicell spheroids—An *in vitro* tumour model. *Curr. Topics Radiat. Res. Quart.* **11**, 87-139.

SHEILA MCEWAN and A. H. W. NIAS* (Glasgow, U.K.) M. A. HYNES, D. H. READING and W. S. SPINKS (Didcot, U.K.): *Negative pion depth–dose profile examined by means of HeLa cell survival curves*

HeLa cells were irradiated at liquid nitrogen temperature; a condition where there is no recovery from sublethal damage and the effects of radiation delivered at low dose rates can be compared without the complication of proliferation during the actual irradiation. Complete dose–response curves were determined for frozen cells given single doses of pions at fifteeen positions along the depth–dose profile of the beam from the 8-GeV/c proton synchrotron NIMROD at the Rutherford laboratory. The pion beam had a 15% full width at half height momentum acceptance and was tuned to 160 MeV/c. This gave a dose peak to plateau ratio of 1.4 in the phantom used.

After irradiation the frozen cells were returned to Glasgow where they were thawed and assayed for survival of colony-forming ability. Dose–response curves were fitted to the survival data pooled from repeated determinations, using a computer programme for Puck curves. All the curves could be fitted to the same extrapolation number of 1.91 ± 0.09 so that the computed D_o values permit a simple comparison of the radiosensitivity of the frozen HeLa cells at the different positions along the depth dose profile. These D_o values are plotted in Fig. 6 with their standard errors. A recent D_o value for 250-kV X-rays is also shown, for comparison.

These data are in agreement with our previous observations of frozen HeLa cells irradiated with pions at the ionization peak ($D_o = 3.64 \pm 0.24$ Gy) and with 300-kV X-rays ($D_o = 6.77 \pm 0.31$ Gy) which indicated an RBE value of 1.86 (Nias *et al.*, 1974). We had also observed an RBE

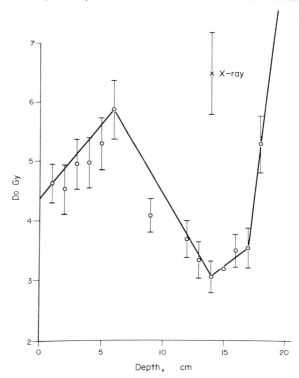

Fig. 6. D_o values of survival curves for frozen HeLa cells irradiated at different positions in a pion beam (McEwan et al.).

---

*Present address: Richard Dimbleby Department of Cancer Research, St. Thomas's Hospital Medical School, London, SE1, England.

value of 1 in the "plateau" region which is also in agreement with the current data at the 6-cm position.

The other data in Fig. 6 shows a significant rise in the D_o value of cell samples irradiated at positions in the "tail" beyond the peak positions of the pion beam (12-17 cm), as might be expected. (Cell survival at positions beyond 18 cm was very high.) In addition, however, the data indicate an increase in radiosensitivity at the "surface", over at least the first 2 cm of depth.

Reference

NIAS, A. H. W., D. GREENE, D. MAJOR, D. R. PERRY and D. H. READING (1974) Determination of RBE values for fast neutrons and negative pi-mesons using frozen HeLa cells. *Brit. J. Radiol.* **47**, 800-804.

H. G. MENZEL, H. SCHUHMACHER (Hamburg Saar, Federal Republic of Germany) and H. BLATTMANN (Zurich, Switzerland): *Microdosimetric characteristics of a biomedical pion beam**

One important aspect of the application of negative pions in radiotherapy and radiobiology is the considerable and complex variation of biological effectivity within pion irradiated objects. Beam parameters such as momentum, momentum spread, collimation and focusing have distinct influences on the local variations of radiation quality and have to be varied in practical radiotherapy. The required knowledge on these changes is obtained by experimental microdosimetry.

At the biomedical pion beam of the Swiss Institute for Nuclear Research (SIN) (Haefeli, 1974) microdosimetric spectra have been measured at different positions inside a water phantom (Menzel *et al.*, 1977). For the experiments a ½-inch spherical tissue-equivalent proportional counter (EG&G) has been used. The tissue diameter simulated was 2 μm.

Microdosimetric spectra have been measured at eleven different depths in phantom for a pion beam with a momentum 180 MeV/c and a momentum spread of $\Delta p/p = 4\%$ FWHM. This and similar beam settings have been frequently used for radiobiological experiments. The resulting depth–dose curve has its maximum at 18.3 g/cm^2 and a peak-to-plateau dose ratio of 1:2.5.

In Fig. 7 results obtained at seven different depths are presented in the form of integral (or cumulative) dose distributions $(1-D(y))$ in lineal energy y. The ordinate values represent the fraction of dose due to events with lineal energies greater than a given value of y.

Near, and particularly beyond, the dose peak (18.3 g/cm^2) star-produced particles lead to a significant dose contribution at high lineal energies. The small fraction of dose delivered above 10 keV/μm in the plateau (curve (a): 7%; curve (b): 8.5%) is mainly due to pions captured in flight and partly to star-produced neutrons. At 24.3 g/cm^2 where all pions have been stopped the measured dose distribution reveals the presence of contaminating e^- and μ^-.

The narrow momentum spread used in the experiment results in a narrow region of peak dose. In practical radiotherapy this region has to be broadened to account for the actual size of the tumors. Within limits the width of the dose peak can be increased by using larger momentum spreads. Microdosimetric measurements in beams with increased momentum band width have shown that in the peak region the dose fraction of star produced high-LET particles is lower than in the case of narrow beams. Moreover, the maximum contribution of star products to dose is observed at a greater distance to the maximum of total dose (Menzel *et al.*, 1978). It must therefore be expected that the biological effectivity in the tumor region for broad beams to be applied in radiotherapy is lower than that found in the dose peak region in radiobiological experiments with narrow beams.

Microdosimetric spectra provide a suitable basis for the comparison of different types of high-LET radiations applied or investigated for radiotherapy. A comparison between fast neutrons (cyclotron-produced neutrons and 14 MeV neutrons) and negative pions in the dose peak region has revealed distinct differences (Menzel *et al.*, 1978). The main difference is that pions deposit a clearly larger dose fraction at lineal energies below 5 keV/μm. This is due to the presence of non-stopped pions, electrons and muons and a high mean energy of star-produced protons. The correspondingly expected smaller RBE of negative pions is in agreement with a number of radiobiological results.

---

*This work has been financially supported by the German Bundesministerium für Forschung und Technologie (BMFT).

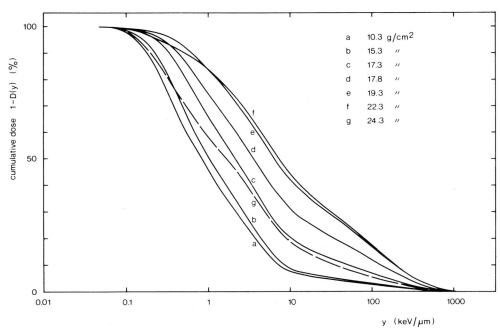

Fig. 7. Integral dose distribution in lineal energy y (1-D(y) vs. log y) for negative pions measured at various depths in a water/lucite phantom along the beam axis (Menzel et al.).

Further microdosimetric measurements are required for therapy conditions. Moreover, an already initiated investigation of radiation quality outside the primary beam will be of importance.

References

HAEFELI, P. (1974) Der Pionenstrahl für biologische und medizinische Anwendungen am Schweizerischen Institut für Nuklearforschung (SIN). *Fortschr. Röntgenstr.* **121**, 101-105.

MENZEL, H. G., H. SCHUHMACHER and H. BLATTMANN (1978) Experimental microdosimetry at high-LET radiation therapy beams. In J. BOOZ and H. G. EBERT (Eds.) *Proc. Sixth Symp. on Microdosimetry*, Commission of the European Communities, Luxembourg. pp. 563-578.

MENZEL, H. G., H. SCHUHMACHER, A. J. WAKER, H. BLATTMANN and W. HETT (1977) Experimental microdosimetry at the biomedical pion beam. *SIN Newsletter* **9**, 54-55.

M. R. RAJU (Los Alamos, U.S.A.): *A heavy-particle comparative study: p, He, C, Ne, Ar, π^-, n*

Progress in applying heavy particles in radiotherapy has been phenomenal during the past 5 years. If comparative radiobiological data for different heavy particles of interest in radiotherapy using the same biological systems were available, the complementary therapeutic potential of these particles could be evaluated. Such a study using protons, helium, carbon, neon and argon ions, negative pions, and fast neutrons is in progress. Since the biological effects of heavy charged particles at the Bragg peak position depend upon the width of the peak, all heavy charged particle beams used in this study were modified to give the same Bragg peak width of 10 cm in water. Negative pions from the Los Alamos Meson Physics Facility (LAMPF), protons from the Harvard cyclotron (HC), helium ions from the Berkeley 184-in. cyclotron, heavy ions (C, Ne, and Ar) from the Berkeley BEVALAC, and fast neutrons (50-MeV deuterons on beryllium) from the Texas A & M variable energy cyclotron (TAMVEC) were used. The following measurements were made for all particle beams: (1) depth–dose distribution; (2) cell survival (cultured human kidney cells, T_1) as a function of depth; (3) cell survival (cultured Chinese hamster cells, V79) curves under aerobic and hypoxic conditions at the beam entrance (plateau) and in the peak region; and (4) early and late reactions using the mouse foot system RF/J mice) and residual injury studies in mice exposed at the beam entrance and peak center.

The depth–dose distributions of all heavy charged particle beams are quite similar. If one is interested in dose localization alone, protons would be the particle of choice, because proton beams can be produced relatively inexpensively compared to negative pions and heavy ions. Dose localization of heavy ions is reduced somewhat with increasing charge of the heavy ions, and dose localization of heavy ions such as neon and argon is still favorable if the treatment volumes are located at depths not exceeding about 15 to 20 cm.

The data on cell killing with depth of penetration are remarkably similar for all heavy charged particle beams (p, He, C, Ne, Ar, π^-). No significant enhancement in cell killing at the peak region, compared to the entrance, has been observed for single-port exposures of heavy charged particles when the Bragg peaks are broadened to cover a width of 10 cm. However, when two opposed and overlapping fields are used, a large enhancement in cell killing at the peak region is obtained, compared to outside the peak region.

The differences in relative biological effectiveness (RBE) between the peak center and plateau (beam entrance) are not very large when the Bragg peaks are broadened to 10 cm in width. The RBE for argon ions at the peak center is nearly identical to that at the beam entrance but is slightly lower at the distal end of the peak because of saturation effects of high-LET. The oxygen-enhancement ratio (OER) for protons is not significantly different from that for X-rays. The OER for helium ions, carbon ions and negative pions is larger, for neon ions it is similar, and for argon ions it is smaller when compared to fast neutrons. The OER values for heavy ions are higher than expected and could be due to a large delta-ray penumbra associated with the energy deposited by energetic heavy ions.

The time-course of skin-reaction development and subsequent healing after exposure to neutrons or heavy ions is remarkably similar to that after exposure to ^{60}Co gamma-rays, suggesting that skin damage and subsequent epithelial repopulation after exposure to high-LET radiations are no different from ^{60}Co gamma-rays. The RBE at the peak center, compared to the plateau, is significantly higher for carbon ions and nearly identical or even lower for neon and argon ions because of saturation effects at high-LET. The RBE for fast neutrons is comparable to carbon ions at the peak. The relationship between early skin reaction and foot deformity remains the same for all particles.

The mice that did not develop significant deformity after exposure to various particles were re-exposed to a fixed dose of X-rays (1750 or 2000 rad) 8 months after the first exposure to determine the residual injury from the first exposure. Acute skin reaction and foot deformity (8 months after the second exposure) were scored. The results indicate that there is no significant memory of the first exposure for acute skin reaction for all particles but that, for foot deformity, there is a memory of the biologically effective dose.

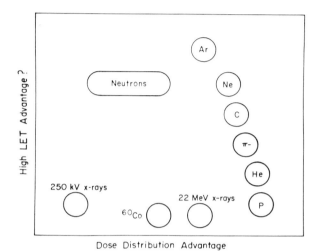

Fig. 8. Schematic comparison of heavy particles. This elegant manner of presentation was first introduced by Koehler (Raju).

The question arises that, from the results of a comparative type of study, can one identify the best particle for radiotherapy. The results clearly indicate that there is no unique characteristic in any one particle that is not shared to some degree by the other particles. For example, dose localization characteristics of all heavy charged particles are similar. Biological effects such as RBE and OER produced by fast neutrons appear similar to some of the heavy ions. Figure 8 shows a schematic comparison of different types of radiations using data obtained mostly from this comparative study. It should be emphasized that this comparison, although based on experimental results, is oversimplified for clarity. If we consider only physical factors of dose localization without significantly changing the radiation quality from conventional low-LET radiations, the particle of choice is the proton. Fast neutrons have no advantage in dose distribution compared to megavoltage low-LET radiations but do have a higher LET. Ongoing fast neutron therapy trials will answer the question of whether high-LET is an advantage in treating certain types of resistant tumors. If the results turn out to be promising, neon and argon ions may be even more effective. Clinical results with the mixed schemes of neutrons and gamma-rays appear promising. If this is confirmed, negative pions, helium and carbon ions may be very effective because the radiation quality of these beams is approximately similar to that of a mixed scheme of neutrons and X-rays.

This work was performed in collaboration with many colleagues at the Los Alamos Scientific Laboratory and at other particle facilities under grant CA17290 from the National Cancer Institute.

M. R. Raju, H. I. Amols, E. Bain, S. G. Carpenter, R. A. Cox, J. Dicello, J. B. Robertson, N. Tokita and C. Von Essen (Los Alamos, U.S.A.): *Pion Radiobiology Studies**

The effects of negative pions on cultured cells and mouse skin (foot) were investigated to answer some of the questions that are of interest in using negative pions in radiotherapy.

Cell survival as a function of depth for modulated negative pion beams.—When negative pion beams are used in radiotherapy, it is necessary to modify the dose distribution to produce a uniform effect in the region of interest because of changes in radiation quality. Figure 9 shows the depth–dose distribution (top panel) for three pion beams modulated in different ways and the corresponding cell survival data (cultured human kidney cells, T_1). A uniform dose distribution in the peak region produces more cell killing at the distal side of the peak, and a uniform negative pion stopping distribution in the peak region produces more cell killing at the proximal side of the peak. Cell killing in the peak region is nearly uniform for depth–dose distribution "C" in the top panel.

OER and RBE for pion beams of different peak widths.—Cell survival curves (V79 cells) under aerobic and hypoxic conditions were obtained for pion beams at the peak centers of 1.3, 7.8 and 10.5 cm in water and at the plateau. The RBE (reference radiation 250-kVp X-rays) at the beam entrance is approximately 1. Results at the peak positions indicate that, although the RBE (at the 50% survival level) at the peak center of the 10.5-cm-wide peak (1.2) is significantly lower when compared to the peak centers of the 1.3-cm (1.7) and 7.8-cm-wide peaks (1.6), the OER values (~ 2.2) are similar for all peak widths used in this study.

Effect of negative pions on cells plated on glass and plastic surfaces.—It is of interest to know whether significant differences in cell killing are expected at interfaces of materials of different atomic composition such as bone and soft tissue.

The effective atomic number of bone is close to that of glass and that of soft tissue is similar to

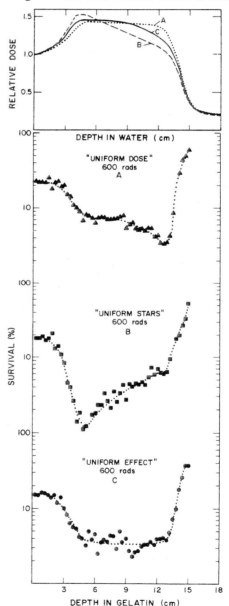

Fig.9. *Depth-dose distribution and cell killing as a function of depth for modulated pion beams (Raju et al.).*

*This work was supported by grant CA17290 from the National Cancer Institute.

plastic. No significant differences in cell killing have been observed when cells plated on plastic and glass surfaces are exposed to pions at the Bragg peak position. These results do not rule out the differences in cell killing as a consequence of inherent differences in radiosensitivity of cells at bone and normal tissue interfaces.

Age response for CHO cells exposed to negative pions. The cells were synchronized first by mitotic selection and resynchronized in the region of late G_1/early S by treatment with hydroxyurea (10^{-3}M). The cells were exposed to a fixed dose of X-rays or negative pions at different stages in the cell cycle. The results indicate that the variation (ratio of maximum survival to minimum survival) after exposure to negative pions at the peak position (the width of the peak at the 80% dose level 1.2 cm in lucite) is 5-fold compared to 10- to 15-fold for X-rays (300 kVp) at doses that produce comparable survival levels.

Acute skin response (mouse foot) after exposure to single and fractionated doses of negative pions compared to 300-kVp X-rays. Several groups of mice (6 mice per group) were exposed to one, two and five daily fractions of pions at the peak position (the width of the peak at the 80% dose level 1.2 cm in lucite) and to X-rays (300-kVp). The preliminary results indicate that the RBE values for pions are 1.1, 1.3 and 1.4 for one, two and five fractions, respectively.

A. SMITH,* K. HOGSTROM,* S. SIMON,* P. BERARDO,† S. ZINK†, J. SOMERS,* M. KLIGERMAN* and H. TSUIJII* (*Albuquerque and †Los Alamos, New Mexico, U.S.A.): *Dosimetry and treatment planning for pion radiotherapy at LAMPF*

Cooperative studies to test the potential of negative pi-mesons (pions) in radiation therapy for advanced solid tumors are being conducted by the University of New Mexico Cancer Research and Treatment Center, Albuquerque, and the Los Alamos Scientific Laboratory, at Los Alamos, New Mexico.

Pion beams of three different energies (penetrations of approximately 13, 18 and 25 cm in water) have been developed for patient treatment. The tuning of the channel has been described by Paciotti *et al.* (1975). For each energy there are two different beam tunes with treatment areas of approximately 80 and 200 cm². Target volumes requiring larger portals than 200 cm² are treated using abutting fields; fields abutted at the 50% isodose level give very uniform dose profiles. The pion Bragg peak is spread in depth by use of a dynamic range shifter (Amols, *et al.*, 1977) utilizing thirteen different range-shifter functions which produce spread peaks equally spaced from 3 to 14 cm. The slope of the spread peak is determined by the high-LET component of the beam which is maintained, for radiobiological reasons, at a minimum of 10% of the total peak dose throughout the stopping pion region. This criteria results in spread peaks that are flat in total dose for peaks spread up to 8 cm whereas for those larger than 8 cm the total dose is sloped across the stopping region in order to maintain the high-LET criteria. This slope is maximum for the 14-cm range-shifter peak, ranging from 100% at the proximal peak to 65% at the distal peak. All but a very few treatments are given using opposing ports with the range-shifted peaks completely overlapped. This results in a composite depth-dose profile in which the total dose and high-LET dose are uniform throughout the target volume. The skin (plateau) dose for these treatments is approximately 50% of the peak dose.

Treatment planning for pion patients requires data obtained from CT scans to correct for tissue inhomogeneities in the beam path and irregularly shaped target volumes. To utilize the information obtained from CT scans the patient must be scanned in the same position as that used for treatment. This is accomplished by immobilizing the patient in a rigid fiber-glass form (Multiclinic Evaluation Group, 1974) during the CT scan and in the larger immobilization module during treatment.

CT scans are usually taken at 1.0-cm intervals throughout the volume of interest. The patient is tattooed at the time of the CT scan with tattoos marking a specific reference scan and also defining the axis along which sequential scans are taken. Localization X-ray films are also taken with the patient in the scanning-treatment position with opaque markers on the tattoos so that the tattoos can be referenced to internal anatomy.

A photographic transparency of the CT scan image is used to define the target volume for pion therapy. The tape containing CT scan data is read into the bolus design file. A particular file (scan) is displayed on a CRT screen and the target volume is entered from a magnified CT image using a digitizing pen. Additional input data are the appropriate pion range and range-shifter function. The bolus (paraffin) required to stop the pions at the prescribed target-volume boundary is then calculated automatically by the computer program by examining the CT data. This calculation utilizes correlations of relative (to water) pion linear stopping powers vs. relative photon linear attenuation coefficients. These data are from measurements of various tissue-equivalent materials and bone substitutes taken on CT scanners and in the pion beam (Hills, *et al.*, 1978).

The bolus design for each CT scan is then transferred to a styrofoam template. All the templates for a treatment port are then combined using alignment reference axes to produce a styrofoam bolus model. The model is encased in aliginate (a dental impression material) to form a mold into which paraffin is poured to produce a final bolus. This bolus is fitted to the patient using low-density polyurethane foam. In some cases the bolus is designed to fit into and be supported by the collimator. The collimator is designed from the target volumes marked on each CT scan and is fabricated using a low melting-point alloy (50% bismuth, 26.7% lead, 13.3% tin, 10.0% cadmium).

Dose measurements, using an ionization chamber, are made in a water phantom in a mock-up of the patient treatment, i.e. using the beam tune, range-shifter function, collimator and geometry of the actual patient treatment. These measurements are along the central axis and in a plane perpendicular to the central axis and located at the center of the spread peak. Smith *et al.* (1977) have described the dosimetry of pion beams at LAMPF. From these measurements, isodose distributions in water are calculated for planes parallel to the central axis, i.e. planes corresponding to the CT scans. These isodose distributions are then corrected for the effect of the bolus and tissue inhomogeneities using the relative linear stopping powers for paraffin and for the various tissues, incorporating the computer files generated from the patient's CT data when the bolus was designed. Thereby, patient isodose distributions for each CT scan can be calculated. These distributions can then be combined to produce isodose distributions for saggital, coronal or arbitrary planes, if desired. The procedures described here have been reported by Hogstrom *et al.* (1978).

In vivo measurements using an intra-cavitary ionization chamber (total dose), silicon detectors and Al activation pellets (high-LET dose) (Hogstrom *et al.*, 1977) are made whenever possible to verify the calculated doses. When *in vivo* dosimeters can be placed in the target volume at known positions the measured dose has agreed very well with that predicted by calculated isodose distributions.

References

AMOLS, H. I., D. J. LISKA and J. HALBIG (1977) Use of a dynamic range-shifter for modifying the depth–dose distributions of negative pions. *Med. Phys.* **4**, 404-407

HILLS, J., W. R. HENDEE and A. SMITH (1978) Converting CT numbers to stopping powers for pion therapy. *Med. Phys.* **5**, 325.

HOGSTROM, K. R., A. R. SMITH and J. W. SOMERS (1977) *In vivo* dosimetry for negative pion therapy. *Int. J. Radiat. Oncol. Biol. Phys.* **2**, 134.

HOGSTROM, K. R., A. R. SMITH, C. A. KELSEY, S. L. SIMON, J. W. SOMERS, R. G. LANE, I. I. ROSEN, P. A. BERARDO and S. M. ZINK (1978) Static pion beam treatment planning of deep seated tumors using CT scans at LAMPF: physical principles. Submitted to *Int. J. Radiat. Oncol. Biol. Phys.*

MULTICLINIC EVALUATION GROUP (1974) New fiber-glass casting system. *Clin. Orthop. Rel. Res.* **103**, 109-117.

PACIOTTI, M. A., J. N. BRADBURY, J. A. HELLAND, R. L. HUTSIN, I. A. KNAPP, O. M. RIVERA, H. B. KNOWLES and G. W. PFEUFFER (1975) Tuning of the first section of the biomedical channel at LAMPF. *IEEE Trans. Nucl. Sci.* N.S. 22.

SMITH, A. R., I. I. ROSEN, K. E. HOGSTROM, R. G. LANE, C. A. KELSEY, H. I. AMOLS, C. A. RICHMAN, P. A. BERARDO, J. A. HELLAND, R. S. KITTELL, M. A. PACIOTT and J. N. BRADBURY (1977) *Med. Phys.* **4**, 408-413.

J. TREMP and K. R. RAO (Zurich, Switzerland): *Survival of normal and malignant cells after irradiation with the SIN negative pion beam over the depth profile*

The proliferation and colony-forming ability of Chinese hamster fibroblast cells *in vitro* has been investigated after irradiation with a 85-MeV pion beam produced by the 590-MeV proton accelerator of the Swiss Institute for Nuclear Research (SIN). These experiments are being carried out in connection with the preclinical evaluation of pions for radiation therapy at SIN.

The method, developed by Skarsgard and Palcic (1974), is based on the embedding of

Chinese hamster cells in a cylinder-shaped tube filled with medium containing 25% gelatine. The gel/medium mixture is tissue equivalent in terms of elemental composition. The gelatine turns fluid at 37°C whereas at the radiation temperature of 17°C it remains solid. After irradiation, 2-mm-thick slices are cut off the tube at 10°C. This allows to observe the effect of pion irradiation at various points of the depth–dose profile. The slices are dissolved in 37°C medium and its cells are plated in petri-dishes whereupon their colony-forming ability can be tested.

The cells in the peak region of the pion beam have been irradiated with 200, 300, 500, 700 and 900 rad (dose rate 5 rad/min). The resulting cell-inactivation profiles show a peak which corresponds with the peak in the physical depth–dose curve and thus evidence is provided for the aptness of this method.

The cell-survival curve results from data taken from survival profiles in the peak (depth 19.0 cm) and post-peak (depth 21 cm) region. The fact that all of these survival measurements can be adequately accommodated by a single survival curve suggests that the biological effect of the beam does not change significantly throughout the peak and post-peak region.

RBE values are calculated relative to the X-ray response at surviving fractions of 0.5 and 0.1. The RBE values we found were 1.3 in both. It may be pointed out that these values are consistent with the most recent data from Skarsgard et al. at TRIUMF (1978).

It is generally acknowledged that sufficient radiobiological information on the reaction of normal and malignant tissues to negative pions be gathered before the latter is applied in medicine. In the following experiments Ehrlich ascites tumor cells were used as the biological test system and the pion source was the proton accelerator of the SIN. The dose rate was 3-5 rad/min in peak region and it was 20-30% less in plateau position. The beam contamination was about 5% muons and 10% electrons. The effect of negative pions or other radiation on the proliferative capacity of tumor cells was determined by using the method of Evans et al. (1968) which was adapted to suit the experimental conditions and the growth characteristics of the tumor culture used.

Survival curves plotted for peak and plateau pions are presented—along with those for 140 keV and 29 MeV photons and 20 MeV electrons for comparison. Under the experimental conditions used, peak pions were more effective than plateau pions by a factor of about 1.4. For 50% survival, 140-keV photons had the same effect as peak pions but the latter was more effective (factor 1.2) at 10% survival level. When 140-keV photons were taken as the standard, following are the RBE values calculated at 50% survival level: plateau pions: 0.73; peak pions: about 1.0; 29 MeV photons: 0.73; and 20 MeV electrons: 0.6. However, the highest RBE for peak pions obtained here was 1.2 at 10% survival level. The RBE values for negative pions reported so far vary between less than one for some of the test systems (Fritz-Niggli et al., 1977) and 5.4 for mouse lymphoma cell proliferation (Feola et al., 1968).

References

EVANS, T. C., R. R. HAGEMANN and D. B. LEEPER (1968) Effect of cell concentration during irradiation at low oxygen tension on survival of mouse ascites tumor cells. *Radiat. Res.* **35**, 123-131.

FEOLA, J. M., C. RICHMAN, M. R. RAJU, S. B. CURTIS and J. H. LAWRENCE (1968) Effect of negative pions on the proliferative capacity of ascites tumor cells (lymphoma) *in vivo*. *Radiat. Res.* **34**, 70-78.

FRITZ-NIGGLI, H., H. BLATTMANN, C. MICHEL, K. R. RAO and P. SCHWEIZER (1977) Preclinical experiments with the SIN negative pi-meson beam. In *Radiobiological Research and Radiotherapy*, Vol. 2, International Atomic Energy Agency, Vienna, pp. 31-47.

SKARSGARD, L. D. and B. PALCIC (1974) Pretherapeutic research programs at mesons facilities. Proc. XIIIth Int. Congr. Of Radiology. *Radiology*, **2**, 447-454.

SKARSGARD, L. D., B. PALCIC, R. M. HENKELMANN and G. K. Y. LAM (1978) RBE measurements on the π^--meson beam at TRIUMF. IAEA, Advis. Group Meeting, Dec. (1977).

The Committee on Radiation Oncology Studies plan for a program in particle therapy in the United States

J. ROBERT STEWART and W. E. POWERS

Representing CROS and its Particle Subcommittee

CROS

William E. Powers, *Chairman*
Luther W. Brady
G. Stephen Brown
Morton Kligerman
C. Ronald Koons
Seymour Levitt
Carlos Perez
Theodore Phillips
Philip Rubin
J. Robert Stewart
Herman Suit
Gordon Whitmore
H. Rodney Withers
Peter Wootton

CROS Particle Subcommittee

J. Robert Stewart, *Chairman*
Malcolm Bagshaw
Max Boone
Jack Dobson
David Hussey
Morton Kligerman
Simon Kramer
William E. Powers
Glenn Sheline
Herman Suit
Gordon Whitmore
H. Rodney Withers
Peter Wootton

Supported by the Division of Cancer Research Resources and Centers
National Cancer Institute, DHEW
Grant CA 13212

Abstract—*A review is presented of the goals of high-LET radiotherapy and the plans in the U.S.A. for determining the role of various types of particle beams in the curative treatment of patients.*

Introduction

The research plan for particle therapy in the United States was developed by the Committee on Radiation Oncology Studies (CROS) and its Particle Subcommittee, supported by a grant from the National Cancer Institute. It is based on several years of planning, and upon research results in particle therapy. The plan is published in detail elsewhere* and will be only briefly summarized here. The scientific effort has been a joint venture utilizing the talents of numerous dedicated physicians, physicists, biologists and accelerator engineers from many institutions around the world. The scientific rationale for particle therapy compared to conventional radiation has been discussed extensively at this conference and in brief consists of either or both (a) enhanced biological effect and (b) *physical* properties leading to improvement in dose distribution. Extensive physical and biological measurements clearly support this rationale and the limited clinical trials completed to date fully warrant definitive clinical study of these particle beams. The plan represents the consensus opinion of CROS and its Particle Subcommittee after extensive consultation with professionals

* *Cancer Clinical Trials:* 153–208, 1978 (Fall).

with experience and expertise in particle radiation. At intervals during the development of the plan, the United States radiotherapy community has been informed and has shown its support through presentations to and feedback from the membership of the American Society of Therapeutic Radiologists and through activities of the CROS Practice Subcommittee. The content and recommendations are based on the current status of particle technology and the scientific background now available. Contributions supporting the rationale and feasibility of the program have an international base and international cooperation in clinical evaluation is a goal to be vigorously sought in the years to come. Implicit in the plan is periodic review, updating, and resetting of priorities based on national and international progress and results.

Objectives

The goal of radiation therapy is to achieve *uncomplicated local control* of cancer in order to *cure, heal* and *palliate* the patient. Used alone or in combination with surgery or chemotherapy, radiation therapy is an effective modality in a large number of patients with cancers confined to the organ of origin or with direct or lymphatic spread to contiguous anatomical regions. Cure rates for tumors originating in many sites are high and account for a substantial fraction of living patients cured of their tumors. With improved supervoltage equipment, increase in the number of radiation therapists, and other improvements in care, there has been a substantial improvement in results over the past two decades. In addition to its role in curative therapy, radiation, because of its ability to control locally symptomatic lesions in incurable patients, has an important place in palliative therapy. Factors which frustrate our efforts to attain uncomplicated local control include the following:

- Tumors with cellular radiobiologic characteristics which make conventional radiation therapy ineffective. Examples include tumors with (a) a high fraction of radioresistant hypoxic cells, (b) great capacity for repair of sublethal damage leading to decreased effectiveness of conventional small fraction treatment, (c) rapid proliferation such that repopulation occurs between fractions, (d) cell cycle kinetics such that there are always a large number of radioresistant tumor cells present during fractionated treatment.
- Lesions located in sites involving or contiguous with critical, dose-limiting normal tissues. The site of involvement may be such that to deliver a high dose of radiation by conventional modalities, the tolerance of critical normal tissues would be exceeded, leading to unacceptable complications (radiation complications).
- Tumors which have by extensive invasion so damaged the normal structure of origin that satisfactory healing following conventional radiation is precluded, again resulting in unacceptable normal tissue sequelae (tumor complications).
- Tumors containing a large number of tumorogenic cells such that a dose of radiation resulting in high probability of killing all cells cannot be attained within the tolerance of normal tissues (tumor-radiation complications).

Concerning cure, it has been estimated in the United States that approximately 100,000 deaths occur annually due to current failure by all means of therapy to control local-regional cancer. Improved local control will also have great importance in those cancers showing promise that systemic adjuvant therapy may control distant metastases. Any new modality which will enable the radiotherapist to increase the effective dose to the tumor while limiting the effective dose to critical normal tissue will enhance control of such local lesions, and will benefit systemic therapy by complementing its effect on disseminated small metastases. Because of the unique physical and biological properties of particle radiations, it is proposed that their use will significantly enhance our ability to accomplish the goal of uncomplicated local control of cancer.

Proposal

The CROS Particle Program proposes to mount meaningful coordinated clinical studies to document the usefulness of particle radiation in the treatment of human cancers in a reasonable period of time. Compatible protocols on an international basis will greatly facilitate the evaluation. The goals differ according to particle type because of differing levels of experience and equipment technology currently available for background in planning the trials.

Because of the impressive results which have been achieved in the limited studies of fast neutron therapy there is a clear indication for a definitive evaluation of their role in radiation therapy. The physical and biological characterizations of fast neutron beams have been essentially completed; equipment design, availability and predicted reliability are good; cost is judged to be reasonable and such a study would be well supported by the medical community. Thus a major clinical investigation can be implemented which will soon provide the scientific basis for judging the clinical merit of use of high-LET radiations. This should be done concurrently with the first phase of the work with protons, negative pions and heavy ions. Our expectation is that an important clinical advantage will be shown for fast neutron therapy and, by the time results are in from the clinical trials of fast neutrons, the first phase work with the heavy ions and pions should be complete so that expanded trials could be implemented.

The equipment for the neutron study must be hospital-based, designed for clinical radiation therapy, and housed within clinical facilities. The units must be reliable, provide high dose rate, and have an isocentric gantry or equivalent for multiple angle portal set-up. The resultant neutron beam will have depth dose characteristics equal or superior to a ^{60}Co beam, i.e. an entirely reasonable beam for this definitive trial of fast neutrons. There are good prospects for collaboration among the U.S. fast neutron program and European and Japanese centers. There has already been an intercomparison of the dosimetry and radiation biology between the existing U.S. centers and the programs at Hammersmith Hospital, London, and at the National Institute of Radiological Science (NIRS), Chiba Japan. This type of collaboration will be encouraged and broadened. There are active programs in several other European hospitals. We clearly plan to keep informed of the progress at each center and whenever feasible develop collaborative programs. To the extent that this can be achieved there will be an important gain to the U.S. Particle Program.

The goal with charged particles is to assess the clinical advantage of the dose distributions which can be achieved with protons and dose distribution plus biological advantage of negative pions and heavy ion beams. The special physical characteristics of these beams: the sharp lateral cutoff, finite range and criticality of depth dependence on tissue inhomogeneities make mandatory the inclusion of computed tomography (CT) scanning in the treatment-planning process. Our experience with charged particle beams in radiation therapy is less than for fast neutrons; accordingly, the initial clinical studies will provide refined estimates of the clinical RBE(s) for human tissues, solutions to problems inherent in achieving the desired dose distribution in the patient, and assessment of the accuracy and precision with which heterogeneities of tissue density can be compensated. Once these preliminary trials are completed, prospective randomized trials at specific tumor sites can be undertaken to assess the dose distribution advantage of protons and the combined dose distribution and radiobiologic advantage of heavy ions and pions. As with the neutron program, cooperation with foreign centers involved in similar trials will be actively pursued.

We propose a Program for Particle Therapy utilizing particle generators based in hospitals and designed for clinical use to determine the role of these beams in the curative treatment of cancer patients. This recommendation is based on critical assessment of experience during the past decade in existing facilities which were designed and used mainly for basic physics research, located in areas remote from clinical facilities needed for the care of cancer patients, and housed in physics laboratories in surroundings not suitable for routine handling of large numbers of ill people. The plan proposes a transfer of technology from the laboratory to hospital-based facilities in two phases over a 10-year period, as shown in Figure 1.

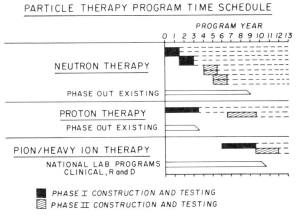

Fig. 1

Of highest priority is (a) the addition of four new hospital-based neutron generators, (b) continuation of existing programs in heavy-ion and pion therapy and their research and development programs leading toward hospital-based equipment, and (c) construction of a high-energy (approximately 250 MeV) hospital-based proton generator to make treatment of deep lesions feasible. Subsequent implementation of the planned program will require review of progress in the initial phase and status of development of components for hospital-based equipment for heavy ions and pions. Decision as to construction of either or both heavy-ion or pion facilities will be based upon feasibility, clinical expectations following initial evaluation and cost-effectiveness.

As of the time of this Third Meeting on Fundamental and Practical Aspects of Fast Neutrons and other High-LET Particles, the CROS Plan has been received and reviewed by the appropriate officials of the National Cancer Institute and by the National Cancer Advisory Board. The plan is approved in principle with implementation limited by budgetary constraints. The NCI plans initially to support a limited number of hospital-based neutron facilities and to continue support for programs in other particles. Some neutron programs not based in hospitals may be phased out. It now appears that major augmentation to a level more near the CROS Plan will require support from government or non-government sources in addition to that from the National Cancer Institute.

SESSIONS V and VI

Evaluation of present results and future of high-LET radiotherapy

Clinical experience concerning evaluation of tumour response to high-LET radiation

J. DUTREIX and M. TUBIANA

Institut Gustave-Roussy, 94800 Villejuif, France

Abstract—*The review of the presently available data from the clinical experience of neutron therapy shows a general agreement on discouraging results for* brain tumours, *and encouraging results for* tumours of head and neck *region,* salivary glands, cervical lymphnodes *and* soft-tissue sarcomas. *The number of patients and the follow-up time are limited and the conclusions can be provisional only. Clinical trials are necessary for a more precise evaluation of the therapeutic gain. Many studies have been impaired by inadequate neutron energy and the lack of availability of the beam. More numerous high-energy neutron sources specially designed and hospital based are needed for exploring the promising potentialities of the neutron beam.*

Introduction

With the exception of the pilot study carried out at Berkeley and reported in 1948 by R. S. Stone, the clinical applications of neutrons to the treatment of malignant disease are limited to the last 10 years and the follow-up of these patients is too short for a final assessment of the results. Furthermore, the number of patients for each tumour type and site is relatively too small for a statistical analysis (Tables 1 and 2).

Thus at the present time the evaluation of the tumour response can only be provisional. However, the reports presented at this meeting by the different centres provide some reliable information and cast some light on the future developments of neutron therapy.

The aim of this paper is to compare, for each tumour site, the results of the various centres in order to point out where they do and do not agree. The authors who have reported these results have kindly discussed, corrected and complemented a preliminary draft. We are much indebted to them for their help in the preparation of this review.

Brain tumours

Among the tumours with a poor response to conventional radiotherapy, *brain tumours* have raised a particular interest at several centres.

The Seattle group has reported the results on thirty-seven patients with grades III and IV *glioblastomas* (Table 3). The average survival is significantly shorter than for a group of patients previously treated with conventional photon irradiation.

Table 1. Neutron Therapy Centres.

Approximate number of patients treated until mid-1978

	Beginning	Number of patients
Hammersmith Hospital (London)	1970	800
TAMVEC (Houston)	1972	440
Seattle	1973	200
MANTA (Washington D.C.)	1973	250
NIRS (Tokyo)	1975	260
Hamburg	1976	230
Amsterdam	1976	200
Fermilab (Batavia)	1976	105
Edinburgh	1978	140
Dresden	1975	150

The results of the mixed beam appear to be the worst ones, although they are not significantly different from pure neutron irradiation. The neutron groups may be somewhat poorer than the photon group: some patients did not survive more than 2 weeks; however, the longest survival is shorter than for the photon group.

Table 2. Main localizations treated with neutrons

Approximate total number of patients treated until mid-1978

Head and neck	640
Cervical nodes	175
Brain	150
Cervix	185
Rectum	90
Soft-tissue sarcoma	160
Bone sarcoma	50
Breast	50
Oesophagus	30
Pancreas	25
Bladder	80
Lung	240
Prostate	40

At Hammersmith and Houston the treatment of *glioblastoma multiforme* has been essentially unsuccessful.

At Edinburgh the results of the first year of clinical experience show that the number of survivors is much less for neutrons than for photons.

At Manta the crude survival is not improved over photon irradiation and in some cases it would appear that neutron-irradiated patients are doing worse.

Table 3. Brain tumours

Total patient number reported: 150

Results at SEATTLE (37 patients)	
	Average survival
Neutron alone	9.4 months
Mixed beam	7.6 months
Photon	13.6 months

Average results in the literature
Average survival: 10-12 months
Survival at 6 months: 60-70%
Survival at 1 year 20-30%

With the exception of NIRS, the general conclusion is that the results with a pure neutron or mixed schedule irradiation are significantly worse than for a pure photon irradiation.

For all treatments reported above, the dose has been in the range of 5500 to 7000 equivalent rad* with a weekly dose of 900 to 1000 equivalent rad delivered with various fractionation schedules.

The autopsia performed in Seattle (on 15 out of 36 patients) show in almost all cases (14/15) that "the bulk of the tumour has been replaced by a localized mass of coagulation necrosis" without any evidence of residual malignant cells.

Diffuse gliosis and demyelination of white matter are thought to be related to the ultimate cause of death. Similar findings on the tumour and normal tissue have been reported at the Hammersmith Hospital.

The pathological findings suggest that the patients have been biologically over-dosed. This is corroborated by the observation made at MANTA that above 5500 equivalent rad there is no trend for a dose response.

It could be worthwhile using smaller doses to check whether the tumour sterilization could be achieved with less severe damage to the normal tissues; however, at Fermilab the reduction of the dose resulted in a failure to control the tumour. Another sensible investigation is to use the neutron beam for a boost on a reduced target volume. The 1976 RTOG Protocol compares photon and neutron boost of 1500 equivalent rad on a coned-down volume after 5000 rad photon irradiation of the whole brain. The present results for seventeen patients treated at Fermilab with this pattern have been reported: of nine patients with a follow-up ranging from 6 to 17 months, four are surviving. At Hamburg, seven cases have been treated with a similar technique; two survive longer than 1 year.

These last results are in the range of most of the published data for conventional irradiation which show a survival rate of 60-70% at 6 months and 20-30% at 1 year with an average survival of 10-12 months.

At MANTA the median survival on two small groups of grade IV tumours is the same (8.5 months) for boost and neutron alone. At Amsterdam the neutron boost has not achieved better results than the photons: five out of seventeen are surviving without clinical recurrence at a follow-up maximum of 10 months.

To sum up the results obtained so far in *brain tumours* are in agreement at almost all centres. They are disappointing with pure neutrons as well as mixed schedules and the preliminary results of the neutron boost fail to show any progress.

Squamous cell carcinoma of the head and neck region

The basic clinical experience is the randomized clinical trial carried out at Hammersmith Hospital. The results show a persisting control for fifty-four out of seventy-one patients treated with neutrons and twelve out of sixty-three patients treated with photons. The difference (76% to 19%) is highly significant. The

---

*We share the restrictions of many authors about the significance of the *equivalent rad*. However, we shall use it as a simplified index of the amount of irradiation. In fact the relevant figures are tumor local control rate versus complication rate; we shall quote them when available.

percentage of persisting control for neutron treatment (76%) observed in this clinical trial is approximately the same as for all head and neck cases treated from 1970 to 1978 (79%). Published data quoted by Dr Caterall show that for a similar group of tumours the expected local control by photons should be about 30%. At Houston the local control rate is the same for neutrons mixed schedules (11/19) and for photons (13/23); however, the percentage of local control without complications is higher for neutrons (11/19) than for photons (9/23).

The sets of data from these two centres show that for a same level of complications, local control is more frequently achieved with neutrons (Table 4).

Table 4. *Squamous cell carcinomas of the head and neck region*

Total number reported: 640

	Local control with no complications	Local control with 20-25% complications
Neutrons	58% mixed beam (Houston)	76% neutron alone (Ham. Hosp.)
Photons	19% (Ham. Hosp.)	61% (Houston)

Local control at other centres ranges from 40% to 84%

HAMMERSMITH. Patients treated at TDF: 110-120

	Patient number	Complete regression	Recurrence	Persisting control
Neutrons	37	30	0	30
Photons	11	7	4	3

$p: 0.02$

The comparability of the neutron and photon series in the Hammersmith Hospital study has been questioned on several occasions and even during this meeting. In particular it has been pointed out that the range of doses used in the two series is different and the average equivalent dose is larger for the neutrons than for the photons. One can question whether the photon dose is not too low; in this case the control rate at the level "no complication" would not be representative of the photons' capability. Dr Catterall has provided important additional data: when considering the subseries of patients treated at the *same dose level* (TDF 110-120) the proportion of persisting control is, as for the whole series, much higher with the neutrons than with the photons (Table 4).

Furthermore, there are some recurrences for the highest photon doses, in contrast with the absence of recurrence for the highest neutron dose. They confirm the results observed for the whole series.

At Seattle, for a mean follow-up time of 1.5 year, the local control rate is 51%. It is higher for mixed-beam irradiation—53/84 (63%), this meets the Houston data—than for neutrons only—11/40 (28%).

At MANTA the local recurrence rate is very low (16%) considering the advanced nature of the tumours.

At Amsterdam the local control amounts to 84% (36/43) with two fatal complications.

In the experience of Fermilab, the local control of *head and neck tumours is* 40% (28/70) for a follow-up maximum of 20 months.

The cases are staged "advanced cases" but the exact stage distribution is not described. Pure neutron treatment has given better results than neutron boost following photon treatment: however, the difference is not significant. It is worthwhile noting that neutron treatment has given a local control in seven out of twenty recurrences after photon treatment.

At Edinburgh a local control rate of 76% is also observed with a maximum follow-up of 1 year. This could meet the Hammersmith results. However, the same rate of local control is observed with photons and it seems to be higher than usually achieved for this type of tumour. This striking discrepancy between Edinburgh and Hammersmith raises the question of comparability between patients series which will be discussed further on.

Among the *head and neck tumours* a special mention should be made about the *salivary glands*.

Persisting control is close or even superior to 80%, at Hammersmith (29/34), Houston (8/13), Amsterdam (7/8), and Fermilab (7/11).

At Seattle, eleven high-grade tumours have been matched for comparison with tumours of similar size treated with photons:

for a diameter < 3 cm all tumours are controlled with neutrons or photons,

for a diameter 3-6 cm, 4/4 are controlled with neutrons and 2/6 with photons,

for a diameter > 6 cm all are failures, with neutrons or photons.

In general the results for *head and neck tumours* show or suggest a therapeutic gain with the neutrons. In fact various types of tumours are grouped together under the denomination *head and neck tumours* and their response to radiation might be different.

At Seattle the local control is somewhat higher for *oropharynx* (44% with neutron alone and

67% with mixed beam) than for the whole group of *head and neck tumours*.

At MANTA the specific local control is 45% for *oral cavity* and *oropharynx*, 67% for *hypopharynx* and 80% for *larynx*.

If the distribution of the *head and neck tumours* in the various centres is different, differences in response according to tumour site can explain some of the discrepancies observed between centres.

Such differences in tumour response according to site and type are also observed with photons and the therapeutic gain of the neutrons has to be evaluated in each homogeneous group. A larger number of patients is required for a significant sub-division.

Cervical lymph nodes from head and neck squamous cell carcinomas

The Seattle group has reported the result of 113 patients. Sixty patients presented with single or multiple nodes measuring from 3 to 6 cm (15 to 100 grammes) ; the percentage of patients remaining free of disease at the site of the adenopathy is:

38% for pure neutron irradiation,
60% for mixed schedules.

The average follow-up is 15 months and the maximum follow-up is approximately 4 years. The percentage can become somewhat smaller in the future; however, these results seem to be definitely better than for external photon irradiation (25% at our institution with irradiation and neck dissection).

The results of the pure neutron beam compare to the percentage of 35% obtained on the same material by Pierquin by interstitial radiotherapy with iridium implants, and the results of the mixed beam are much better.

The percentage of lymph node control for the twenty-four patients with primary tumour control rises to 82% for the pure neutron beam and 96% for mixed beam. It is higher than for external photon irradiation: 46% at the M. D. Anderson Hospital. Even for lymph nodes larger than 6 cm for which the control by external photon irradiation is very rare, the neutron beam can achieve a control for 50% of the patients.

The analysis made at MANTA on the response of cervical nodes is in agreement with Seattle: for 50% of the patients (30/62) a full response has been achieved without recurrence; out of twenty-eight patients with fixed nodes eleven (39%) were controlled.

Breast

Advanced *breast adenocarcinomas* (T4) have given different results at the Hammersmith Hospital and at Houston.

At Hammersmith all (11/11) the tumours have been controlled without any long-term complication.

At Houston the conclusion is that *advanced breast carcinomas* can be locally controlled but at the expense of a high complication rate particularly breast fibrosis: the local control (for 26 patients) amounts to 69% with 35% complications; this does not seem to be better than with photon irradiation for which 60% local control is achieved with 7% complications.

The contradiction between the conclusions of the two centres is particularly striking. It is not explained by the difference between the two beams which are used nor by the extent of the treated volume. It calls for a more precise specification of the complications, and a more detailed comparison of the time distribution of the dose.

Sarcomas

There is a general agreement on the high rate of local control of *soft-tissue sarcomas* (Table 5).

At Hammersmith Hospital a complete regression is observed in 24/28 patients and in twenty-one cases (75%) a persisting control with a complication rate of 32%.

Table 5. Soft-tissue sarcoma

	Patient number	Local control	Complications
Hammersmith	28	75%	32%
Houston	23	70%	13%
NIRS	12	58%	33%
Hamburg	51	88%	not analysed
MANTA	20	71%	not analysed

The same figures are obtained at Tokyo for a group of twelve patients.

At Amsterdam, seven out of sixteen tumours remained uncontrolled but four of them died within 6 months and a full regression of soft-tissue sarcomas may take longer.

At Houston the local control is 16/23 (70%) with 13% complications.

The experience at MANTA on twenty patients has shown "extremely favourable results" in spite of the complications related to the extended size of the region which had to be treated.

At Hamburg on forty-six soft-tissue sarcomas a complete regression has been observed with very few recurrences but follow-up and complications are not analysed. The series of *osteosarcomas* are much less numerous. At Tokyo local control has been obtained in twelve out of thirteen cases, markedly improved over photon treatment. At MANTA a persisting control has been achieved in seven out of nine patients with grades I or II *chondrosarcoma*. The chondrosarcoma is a slow-growing tumour: at our Centre the average survival is 6 years (to be compared to 16 months for osteogenic sarcoma). Its response to neutrons could corroborate Breur's observation of the relationship between neutron efficiency and doubling time. However, one should also point out that a very long follow-up is necessary for the final assessment of the results.

Deep-seated tumours

Deep located tumours of the *pelvis, abdomen* and *thorax* require a convenient penetration of the radiation beam. For this reason their treatment has been mainly considered at the centres where very high-energy neutrons are available.

Ninety-six *gynaecological cancers* have been analysed at Houston among a total number of 108 patients treated with neutrons (Table 6).

The mixed beam (59 patients) has given the best results: 61% local control with 7% complications, which is significantly better than for a photon control group (48% local control with the same level of complications). A neutron boost gave a local control rate equal to that achieved with the mixed beam but the complication rate was higher (20%). Experience with the pure neutron beam is limited to a twice-weekly fractionation schedule.

Table 6. Cervix carcinoma

HOUSTON. Ninety-six patients analysed

	Local control	Complications
Mixed beam	61%	7%
Boost	60%	20%
(Photons)	48%	7%

NIRS. Sixty-nine patients
Results for 20 grade III-IVa patients

	Local control	Complications
Neutrons	80%	25%
(Photons)	66%	25%

The results obtained at Tokyo are in agreement with these data. A mixed-beam irradiation followed by an endocavitary application has provided a higher rate of local control than a photon beam: out of twenty cases of stages III and IVa squamous cell carcinoma, 80% are alive without evidence of disease (instead of 66% for the photon group) with an incidence of 25% rectal complications in both groups. In contrast to squamous cell carcinoma, the adenocarcinoma of the cervix (in fact a small series of nine patients) did not respond to neutron therapy.

The response of *bladder carcinoma* seems encouraging in the series of twenty patients treated at Amsterdam. The local cure rate reaches 80% versus 33% with ^{60}Co γ-rays without severe skin and intestine complications when a six-field technique is used.

At the same centre the local control of *rectal cancer* (24 patients) appears much better than with X-ray therapy at the expense of a high rate of intestinal complications.

At Houston the treatment of a series of sixteen recurrent or inoperable *colo-rectal cancers* "has met with moderate success", in that seven of these very unfavourable tumours were locally controlled. At Edinburgh a complete tumour regression has been observed only in one out of fifteen cases.

A high proportion of local control has been observed at Houston without complication on a series of eighteen *prostatic cancers*, most of them (16) treated with mixed schedules.

On account of the poor response to conventional radiation *pancreatic carcinoma* is candidate for high-LET radiation. Seventeen cases have been treated at MANTA. The response has been very poor for neutrons alone; a better response has been achieved in combination with chemotherapy. However, there is no evidence of any significant benefit with the neutrons.

Sixteen patients have been treated at Berkeley with helium ions; five patients survive without evidence of disease with an average survival of 17 months. This result can be considered as a modest success related to favourable dose distribution achieved with the helium ions.

The largest series of *lung cancer* has been treated at Dresden with neutrons of 6.2 MeV mean energy combined to ^{60}Co γ-rays. The tumour dose is approximately 5500 equivalent rads in 7 weeks with a contribution of 20-40% by neutrons, given either before, after or before and after the ^{60}Co irradiation.

A non-randomised comparative trial has shown that with neutrons the proportion of "totally destroyed tumour" in 116 dissections is significantly higher than for 149 cases treated with ^{60}Co only. This pathological finding was confirmed in a randomised trial.

However, in contrast to this the survival rate is higher for pure γ-treatment than when neutrons are added: at 1 year 13/18 versus 3/29. The autopsy could not provide any interpretation for the discrepancy between survival rate and local control rate.

At most centres the bad prognosis of the *lung cancer* is not considered improved with the neutron therapy. However, a neutron boost has been found of some help at Houston and NIRS for the pain relief in Pancoast tumours.

At Fermilab the combination of neutrons with *chemotherapy* has given an interesting result: two out of three patients remain without overt disease at 8 months.

For carcinoma of the *oesophagus* treated at NIRS by radiation therapy alone with either mixed beam or fast neutron boost, six out of ten patients were locally controlled, with a follow-up longer than 6 months. This result may be better than with the X-ray irradiation: at our centre, the survival rate at 6 months is 60%. At Houston, nine out of eighteen patients remained without local disease, but none was surviving at 18 months (the average survival at our centre is 10% at 18 months.)

It is difficult to reach definitive conclusions about tumour responses from this available clinical data. The total number of patients treated with neutrons is less than 3000 in the world and the follow-up is short for many of them.

The age of neutron therapy is similar to that of roentgentherapy at the beginning of the century. Neutron therapy has been in use for a shorter time than was electron therapy at the time of the conference on high-energy electrons held at Montreux in 1964: at that time, the clinical experience of electron therapy was 20 years old and large series of patients had been treated during the preceding decade.

In his presidential address A. Zuppinger defined the aims of that meeting as follows: "we would like to know for which tumours the chances of success would be definitely improved by using the electron therapy, in which cases it is unnecessary, and whether there are circumstances in which its use is to be discouraged. . . . We would like to hear about all cases which have suffered complications . . ." These aims were not yet fulfilled during the Montreux Conference and we cannot yet expect to reach similar aims for high-LET radiation at the present time.

The conclusions which can be drawn from the present clinical experience with neutron therapy are tentative only. Let us summarize them.

The results concerning *brain tumours* are very discouraging. The therapeutic benefit with respect to conventional low-LET radiation is doubtful for adenocarcinoma of the *breast* and for carcinoma of the *gastro-intestinal tract*. However, the lack of success and the detrimental effects may not be due to the biological properties of high-LET radiation but related to dosimetry, namely dose distribution and energy absorption in the tissues at stake. Therefore, the conclusions should not be extended to all high-LET radiation, in particular heavy ions and pi-mesons.

Encouraging results appear to be available for carcinomas of the *upper respiratory and digestive tract*, in both primary tumours and cervical adenopathy, for advanced carcinomas of the *cervix* and *soft-tissue sarcoma*.

An important point emphasized at Seattle and Houston is the superiority of the mixed beam with respect to pure neutron treatment. This may be due partly to an improvement of the dose distribution when high-energy photons are added to neutrons. However, some radiobiological experiments (Raju and Jett, 1974; Railton et al., 1974; Nelson et al., 1975) suggest that this advantage might possibly be related to a radiobiological mechanism. However, for the clinical material which is considered here one must point out that the fractionation is five sessions per week for the mixed beam while it is frequently only three or two sessions per week for the pure neutron treatment.

The optimum time dose distribution for neutron treatment remains at present open for discussion. However, for all series when different *fractionation* schedules can be compared, reduced fractionation always appears to be as detrimental for neutron as for conventional low-LET radiation. This assertion has raised criticism from some authors and should not be considered valid for all kinds of tumours.

As far as the *overall time* is concerned, the difference between the Hammersmith schedules and the schedules of the American centres has been considered as a possible cause of some discrepancies in these results. However, if the role of the overall time of irradiation is related to the kinetics of the tissues at stake, the optimum overall time should not be different for high- and low-LET radiation. The *total dose* is not yet optimized. In the clinical results which have been presented the doses range approximately from 5000 to 7000 equivalent rad. When comparing the effects of different doses, as it was possible in particular for *head and neck tumours*, the conclusion is trivial: higher doses in this range increase the percentage of control at the expense of more complications. What is considered "complications" is in great part an individual judgement and this in itself explains some discrepancies in the therapeutic ratios observed at different centres. However, when photon and neutron results are compared at the same centre this bias should not occur.

An obvious conclusion of this data is that neutrons cannot achieve miracles. They may improve the results for some tumour sites but the progress evaluation requires careful trials.

Unfortunately, up to now most neutron therapeutic studies have been handicapped by many factors and this is why, in spite of considerable efforts, we are facing so many question marks. Let us consider some of these handicaps.

1. Suboptimal dose distribution due to the too low energy of neutrons and/or too large penumbra.
2. Suboptimal fractionation resulting from lack of cyclotron availability.
3. Suboptimal setting up of the patients caused by fixed horizontal beams and too low dose rates. This is particularly hazardous during treatment of the pelvic tumours.
4. Uncertainties about RBE values which are demonstrated by unsatisfactory matchings of biological neutron and photon doses, in particular for late radiation effects.

The RBE concept is unambiguous in experimental radiobiology but in clinical radiotherapy the choice of the relevant RBE is difficult. In clinical work RBE is influenced by physical and biological parameters, such as dose distribution and fractionation, it may vary with the critical tissue and the biological effect taken into consideration. One constant value of the RBE cannot be sufficient to match the doses delivered with the two types of beams. Furthermore, some of the studies which we are reviewing started at a time when there was still little clinical experience. Moreover, difficulties involved in intercomparison between clinical centres hampered exchange of useful information.

We must be thankful for the pioneers who, in spite of these difficulties, have provided many interesting data, but it is important to stress that, due to these handicaps, the possible improvement achieved by neutron therapy is probably underestimated for many tumour sites; unfortunately we cannot assess the extent to which it is underestimated.

Many of these handicaps could be now eliminated through the use of specially designed and hospital-based medical cyclotrons. Radiotherapists should no longer accept these handicaps and they must not compromise.

In summary, the present clinical experience suggests that neutron therapy can give either better or worse results than photon therapy, depending on tumour types and sites.

This has been observed for most new modalities which have been introduced in radiotherapy. Each of these new procedures has met with partial success and has added something to therapeutic potentialities.

References

RAJU, M. R. and J. H. JETT (1974) RBE and OER variations of mixtures of plutonium alpha particles and X-rays for damage to human kidney cells (T-1). *Radiat Res.* **60**, 473-481.

RAILTON, R., D. PORTER, R. C. IRWSON and W. J. HANNAN (1974) The oxygen enhancement ratio and relative biological effectiveness for combined irradiations of Chinese hamster cells by neutrons and gamma-rays. *Int. J. Radiat. Biol.* **25**, 121-127.

NELSON, J. S. R., R. E. CARPENTER, and R. G. PARKER (1975) Response of mouse skin and the C3HBA mammary carcinoma of the C3H mouse to X-rays and cyclotron neutrons: Effects of mixed neutron–photon fractionation schemes. *Europ. J. Cancer* **11**, 891-901.

Evaluation of normal tissue responses to high-LET radiations

KEITH E. HALNAN*

Glasgow Institute of Radiotherapeutics and Oncology, Glasgow, Scotland

Abstract—*Clinical results presented have been analysed to evaluate normal tissue responses to high-LET radiations. Damage to brain, spinal cord, gut, skin, connective tissue and bone has occurred. A high RBE is probable for brain and possible for spinal cord and gut but other reasons for damage are also discussed. A net gain seems likely. Random controlled trials are advocated.*

Introduction

The severity of normal tissue response to high-LET therapy is of crucial importance, just as much today as it was in 1940. It scarcely needs emphasis that if fast neutrons and other high-LET radiations are to be of significant value, better than low-LET radiation, there must be an improved therapeutic ratio or "gain factor"; improved tumour sensitivity, whether or not because of lower oxygen enhancement ratio, must not be counterbalanced by equally increased severe early or late normal tissue reactions.

The results presented at this meeting have, therefore, been analysed for evidence on this point and great thanks are due to all the authors concerned for all the information given so well and so promptly. For convenience work reported will be referred to by geographical site—Amsterdam (J. J. Battermann and K. Breur), Berkeley (J. R. Castro, C. A. Tobias, J. M. Quivey, G. T. Y. Chen, J. T. Lyman, T. L. Phillips and E. L. Alpen), Edinburgh (W. Duncan and S. J. Arnott), Fermilab (G. A. Lawrence), Hamburg (H. Franke), Hammersmith (M. Catterall), Houston (L. Peters and D. H. Hussey), Japan (H. Tsunemoto), Los Alamos (M. Kligerman), MANTA (R. D. Ornitz and C. C. Rogers) and Seattle (T. W. Griffin, J. C. Blasko and G. E. Laramore).

Doses quoted are in units used at the centre concerned, such as neutron rads, neutron plus gamma rads, sometimes with a scaling factor, or cobalt or photon equivalent doses, and where possible are given with fractionation and overall time—e.g. 1600 rad/20f/4w means 1600 rads (as reported) given in 20 fractions over 4 weeks.

Normal tissue effects

The major normal tissue effects reported are severe late effects and this analysis will mainly summarise these. Most of them are in very small numbers and of low statistical significance—it seems foolish to attempt reliable calculation of RBE therefore from these data—the main exception being the important results for gut and for skin from Amsterdam.

Brain

Brain damage is analysed first since it seems so far that this tissue is especially at risk. Evidence on this comes from several centres treating cerebral tumours (like most work in this field the first evidence on this coming from Hammersmith but not yet published in detail). In Edinburgh there are only three out of ten neutron-treated patients surviving, compared with eleven of thirteen photon treated, though it is suggested that this may be explained by imbalance in randomization; however, the acute early dementia reported from Hammersmith has *not* been seen and this is possibly related to the Edinburgh dosage which was only 1300 rad/20f/4 weeks. At MANTA three cases of brain necrosis are very clearly reported, arising respectively after 1784 rad to the frontal lobe, 1930 rad to the mid-brain and 1675 rad to the whole brain with a smaller boost of 480 rad. In

* Now at Hammersmith Hospital, Ducane Road, London W12 0HS, England.

251

Seattle also neutron-treated patients fared worse than controls, after doses of 1550-1850 rad and all the patients have died. However, at Amsterdam, at the Fermilab and perhaps at Houston results from neutron therapy are no worse than those from low-LET therapy and at Hamburg and in Japan they might even be *better*—the significant point seems to be that in all these centres either mixed treatment schedules have been given, or the control results have already been poor, or shrinking fields have been used with the whole brain not given full dosage. Histological post-mortem findings from Hammersmith, from Edinburgh, from Seattle and from MANTA, all show extensive gliosis and demyelination.

These bad results are all highly disturbing and suggest either a very high RBE or overdosage. A high RBE is obviously possible, but alternatively dosage calculations perhaps should take more account of the higher neutron absorption in hydrogen-rich brain tissue or specifically in fatty myelin. Bewley's original calculations (Bewley, 1963) were very important, detailed calculations should be made again for brain tissue. I would like to suggest, however, that simple over-dosage may also be a major reason, particularly when the *whole* brain is treated. The old standard textbooks such as Paterson (1963) state "for brain tissue a wise limit for irradiation of large volumes is of the order of 4000 rads (in 3 weeks)" which is equivalent to about 4400 rad/25 fractions/5 weeks or only perhaps 1400 neutron rad in, say, 20 fractions/4 weeks, or even less. Some evidence can be found to support this. Todd (1963) analysed Manchester patients some years ago and found that irradiation of gliomas with a dose of <4500 rad/3 weeks to the whole brain gave better results than higher dosage to smaller volume. Urtasun et al. (1976) recently showed benefit from Metronidazole, treating gliomas to a dose of 3000 rad/9f/3 weeks and he has indeed been criticized for giving too low a dose! Attention is directed to a comprehensive recent review (Jones, 1978).

The conclusion we are bound to reach, however, is certainly that as given so far neutron dosage to the brain has often been too high, the reason for the damage is much less certain but the suggestion is that there may be special sensitivity to high-LET radiation and that there may be physical overdosage.

After such dismal results it was very helpful to be reminded by Phillips that radiotherapy *is* of value for gliomas, with added value from BCNU. If we continue to cause so much brain damage with neutrons the whole neutron programme may suffer. I suggest that neutron schedules should be very carefully reviewed and that it might be better to continue for the present with "neutron boost" lower dosage schedules.

Spinal cord

It is obviously interesting therefore to analyse next the treatment of the spinal cord. Several cases of damage have now been reported, from nearly all centres treating "head and neck" cases, all seem to be therefore in the cervical cord. Clear and important descriptions came first from Hammersmith with five patients developing cord myelopathy at doses of more than 1000 rad and 800 rads now being wisely taken as the upper limit for cord tolerance. Four cases are reported from MANTA, with doses of 1543 rad/28f/7 weeks causing damage at 4 and 8 months, 1462 rad/30f/7 weeks at 12 months after L'Hermitte's syndrome at 5 months, and 1600 rad/20f/5 weeks at 23 months. Fermilab keep cord dosage below 1250 rad and have seen transient damage after 700 neutron rad plus 2000 photon rad but see no cord damage 1 year after 1300 and 1500 rad to two patients. Seattle has observed two cases of cervical cord damage. No cases have been seen at Houston.

These seem to fit in reasonably well with known tolerance to low-LET radiation if one takes myelopathy risks as serious above 3300 rad/5 weeks to lengths of cord of over 20 cm and above 5000 rads/5 weeks to shorter lengths, but it would be very valuable to know the length of cord irradiated in all neutron cases reported. Cord RBE is probably higher than skin RBE but much better data are needed before this supposition could be given any numerical confidence. Animal data with fractionated treatment continue to be important for this as for other normal tissues.

Gut

Gut is another very relevant normal tissue, of increasing importance as we begin to use higher penetration high-LET radiations and to treat tumours in the abdomen and pelvis. Much valuable data has been reported at this meeting and well deserves analysis.

At Amsterdam conventional bladder treatment is well tolerated, under 1% of patients have severe gut damage after tumour dosage of 5000 rad ^{60}Co/4 weeks; and yet nine of fifty patients treated by neutrons for pelvic tumours—bladder, rectum and uterus—have died of gut damage leading to peritonitis. These patients have been treated by parallel opposed fields with central minimum dosage of 1720 rad/20f/4 weeks. This has meant that even higher doses of 1900-2200 neutron rad will have been given to some pelvic organs, plus an additional 200-300 gamma rad.

It was particularly helpful to hear from Amsterdam that pelvic gut complications have been reduced by two factors—lowering the central dose and using a six-field plan instead of a parallel pair treatment, it was also valuable to hear the suggestion that the RBE for gut (3.0) may be higher than for skin (2.4) and that there seems to be a steep dose–response curve for both tissues.

Next, it was particularly interesting to have data from Houston on advanced cervical cancer and on prostatic tumours; the incidence of complications was 10 of 96 cervical cancers and 2 of 18 prostatic cancers, compared with 3 of 40 and none of 7 control patients respectively. The very informative actuarial survival curves comparing mixed-beam treatment and controls seem to show similar complication rates with improved local control for cervical cancer and this would be very important if continued for larger numbers in a random control trial. Japanese results for stages III and IV cervical cancer seem to show slightly better results with more complications for neutron treatment but numbers are at present low for statistical significance. Good data came from MANTA again. The first interesting point is the demonstration for pancreatic cancer that chemotherapy with fluorouracil can combine with neutron therapy to cause bleeding from the stomach, which did not occur with the same radiation dose without chemotherapy. There were 3 very severe gut complications when there was surgery before or after whole pelvis irradiation of 1550 rad/20f/4 weeks or higher doses. Hamburg experience also suggested that early gut reactions could be very severe after 800-1200 rad and mixed-beam treatments are therefore being given.

In general it seems extremely clear that whole pelvis or large-volume irradiation at the kind of dose levels outlined above can be hazardous. Assessment of these findings, however, is difficult when the neutron treatment may be mixed with X-rays or with intracavitary caesium treatment and random control trials seem desirable. The incidence of gut complications from photon irradiation can be high also, as reported, for example, from Houston (Strockbane, et al., 1970).

The larynx and pharynx

"Head and neck" tumours have been among the most important groups treated, because of accessibility and suitability for lower-energy neutrons. The extremely important pioneer random-control trial from Hammersmith does not need further detailed discussion and analysis here, other than to repeat that there are both better results and more complications on the neutron side—complications probably both from poor penumbra and difficulties in adequate dose distribution, and also from high biological dosage—the really important point, however, seems to be that there is a net gain justifying, I suggest, further work with higher energy neutrons with a flexible beam from an isocentric head, and with treatment of earlier stage tumours. Reports from other centres do not conflict with this. Edinburgh's well-designed random trials have not yet gone on long enough but should be highly productive. In Amsterdam there have been three deaths from complications out of fifty-one patients, with one laryngeal necrosis in nine patients with laryngeal cancer—from doses of 1890 rad/23f/5 weeks.

At Houston, interestingly, there are more complications in the control series—24% (12/49)—than from neutrons—14% (16/116); this very good analysis shows a net gain from a "less aggressive" (i.e. lower dosage) mixed beam. At Seattle there was a very high 73% complication rate for surgery after neutron radiation but again better results are coming from newer mixed schedules.

The general conclusion is that a case is beginning to be made for better results from high-LET radiation but that random control trials should continue with dosage needing adjustment so that complications are similar in the two arms, the larynx and pharynx both clearly being vulnerable organs and relatively slight reduction of dosage by perhaps 5% being advisable in some centres.

Skin, connective tissue and bone

Choice of dosage and of RBE is still considerably based on skin reactions. The classical early work reported from Hammersmith (e.g. Field, 1969) reassured us about long-term damage relative to acute reactions, and yet we now begin to hear suggestions yet again that high-LET radiation has a peculiar propensity towards insidious long-term normal tissue damage. Several new descriptions have been given of damage to skin, to connective tissue and to bone. Three skin necroses have been caused in Amsterdam by doses of more than 2000 rad. Three cases come from MANTA of soft tissue and bone necrosis after even higher doses of as much as 2240 rad with four cases of "unacceptable fibrosis"—three from 2240 rad to the neck and one from 1720 rad to the right acetabulum. In Seattle there are three bone necroses and three in cartilage, all but one from pure neutron treatment, with the higher neutron dosage. The significant report from Houston is of high dose effects in breast cancer. There great efforts have been made to give acceptable high dose X-ray therapy to the breast, using twice-daily fractionation (Montague, 1968). "Neutron only" dosage has been as high as 1920 to 2560 rad and apparently similar biological dosage has been attempted with neutron boost or mixed-beam schedules. These have caused high dose effects in 50% (8/16) of neutron-only treatments and 35% (9/26) overall. In contrast, the breast treatments given at Hammersmith, with relatively long-term observation now on several patients, have not led to permanent damage after lower dosage of 1560 rad.

The suggestion of special long-term damage to these tissues from neutrons does not seem to be firmly supported, from clinical evidence so far reported, if dosage used is appropriate and adjustment is made for large volume treatments.

Other tissues

There are many other normal tissues on which we have good low-LET data, including kidney, lung, liver and gonads, but with little or no high-LET data in man yet available. As an example to follow there has been a very fine paper from Hammersmith of effects on the eye (Roth *et al.*, 1976) which merits detailed study. It is suggested that doses over 1510 rad cause "ocular destruction" and that 1405 rad should be the "tolerance dose". A similar megavoltage X-ray dose is surely about 5500-6000 rad and the RBE seems nearer 4 than 3.

Finally, the elegant early studies at Berkeley and at Los Alamos deserve reference. So far one cannot find any significant general differences in normal tissue effects from heavy ions or pions. One very interesting finding with helium ions was that there was upper gastro-intestinal bleeding in two of sixteen patients given 5000 rad cobalt equivalent but if the area irradiated was reduced for the last 1000 rad no bleeding occurred—volume effects again. The Los Alamos pion work has also begun well and sensibly with dosage gradually being increased and no serious complications therefore yet reported. The suggested relatively high tolerance of rectum and other gut to pions is interesting and deserves further study and confirmation, perhaps at Villigen, from where we hope to have further reports soon.

It has been a great pleasure firstly to read and then to hear so much good work done so well on so many patients and presented at this meeting.

Discussion, Conclusions, Recommendations

The first important question is whether high-LET radiation has any special or peculiar harmful effect on normal tissue. Certainly harm has been done to brain tissue and several cases of damage or necrosis have been seen in spinal cord, gut, bone, cartilage, connective tissue and skin. I suggest the verdict is the Scottish "not proven" except for brain. However, there are several very important factors that deserve more attention.

The first is dose distribution and volume irradiated. The importance of volume is well known in normal skin reactions to superficial and orthovoltage radiation. Typical historical views come from Paterson (1963) and von Essen (1963), and similarly we know very well the importance of length of spinal cord irradiated. However, very few of the reports on neutron therapy detail the volume treated, and the dose given often seems unaltered whatever size fields are used. And yet the facilities available with most neutron therapy, such as poor penumbra and fixed beams, often lead to large-volume irradiation from parallel opposed field plans. Similarly, the large advanced tumours treated require correspondingly large fields and in addition normal tissues are invaded and damaged by tumour—one cannot give the same dose to a large T3 or T4, N3, laryngeal tumour requiring 12-cm-long fields as to a T1 N0 tumour requiring only 5 × 5-cm fields and expect the same normal tissue effects. Account must be taken of volume and of dose distribution. Several authors recommend mixed or "neutron boost" schedules. Surely if these do give better results this may be due to the better dose distribution obtained from the photon part of the treatment, rather than from any intrinsic radiobiological advantage.

The next factor is time, fractionation especially. RBEs of 3 or 3.1 have been used by many workers irrespective of whether treatment is given in 12, 20, 30 or even 50 fractions over 4, 6 or 8 weeks. X-ray fractionation may be very considerably extended, even using twice-daily treatment, or overall time increased to 8 weeks with considerable benefit to normal tissue reactions. With high-LET radiation, however, surely we cannot similarly increase dosage with impunity. There is very much less sublethal recovery, and if 1500 neutron rad/12f/4 weeks is taken as a standard dose suitable for small volume treatment it should only be increased very slightly if we double the number of fractions and overall time. Further discussion on the time factors is needed, we do not yet know whether fractionation improves the therapeutic ratio for neutrons, especially for late effects.

Dosage itself needs more consideration. The higher-energy neutrons now coming into use obviously have more skin-sparing effect and, as with megavoltage X-rays, skin reactions will not be comparable to those in unseen deeper tissues. It is known already that the hydrogen-rich fatty tissues such as myelin may absorb up to 20% more dose from elastic interactions than skin, this is obviously one possible factor in brain damage, and perhaps in fatty subcutaneous tissue also; new calculations are desirable. When comparing and assessing results it is very helpful if full details are given of the composite rad dose (centigray if one prefers) in both neutrons or other high-LET radiation and gamma or X-rays. It is understandably convenient to translate this into a photon, megavoltage X-ray or cobalt-60 equivalent dose but this necessitates use of an RBE or scaling factor, which may or may not be correct, and may well be different from what is used in another centre. Constant dose checks *in vivo*—as in the oesophagus shown by Catterall—are very important.

Random controlled trials seem undoubtedly to be needed, and to be quite ethical, now, aiming at a similar high dose normal tissue effect, or complications, in the control and high-LET arms. The complications must be assessed in the same kind of way, with a numerical index like that of Karnofsky, and with an attempt at assessing net gain. Fowler's suggestion of two randomly allocated dose levels of, say, 2.5% above and below the dose one thinks is most correct has much to commend it—this is very difficult for us clinicians to accept but we ought to try.

Conclusions

1. Several centres report undesirably high numbers of high dose normal tissue damage, brain being the clearest but not the only example. A high RBE is likely for brain tissue, and possible for spinal cord and gut. This is unlikely to be the sole cause of damage and further reasons include:
   (a) Difficulties in planning and dose distribution.
   (b) Large volumes treated.
   (c) Physical errors in dose calculation.
   (d) Incorrect adjustment for fractionation and time.
   (e) Additive effects from chemotherapy and from surgery.

2. A net gain is likely, in some sites at least.

3. Random controlled trials are essential, with very careful assessment of normal tissue as well as of tumour response. Adequate evaluation of high-LET radiotherapy requires better treatment units, with isocentric beams, good penetration energies, and as good dose distributions as from megavoltage X-rays.

References

BEWLEY, D. K. (1963) Physical aspects of the fast neutron beam. *Brit. J. Radiol.* **36**, 81-85.

FIELD, S. B. (1969) Early and late reactions in skin of rats following irradiation with X-rays and fast neutrons. *Radiology* **92**, 381-384.

JONES, A. (1978) Cerebral astrocytoma—trends in radiotherapy and chemotherapy: a review. *J. Roy. Soc. Med.* **71**, 669-674.

MONTAGUE, E. (1968) Experience with altered fractionation in radiation therapy of breast cancer. *Radiology* **90**, 962-966.

ROTH, J., J. BROWN, M. CATTERALL and A. BEAL (1976) Effects of fast neutrons on the eye. *Brit. J. Ophthal.* **60**, 236-244.

STROCKBANE, M. F., J. E. HANCOCK and G. H. FLETCHER (1970) Complications in 831 patients with squamous carcinoma of the intact uterine cervix treated with 3000 rads or more whole pelvis irradiation. *Amer. J. Roentgenol.* **108**, 293-304.

TODD, I. D. H. (1963) Choice of volume in the X-ray treatment of supratentorial gliomas. *Brit. J. Radiol.* **36**, 645-649.

URTASUN, R. C., P. BAND, J. D. CHAPMAN, M. L. FELDSTEIN, B. MIEKLE and C. FYER (1976) *New Engl. J. Med.* **294**, 1364.

VON ESSEN, C. F. (1963) A spatial model of time–dose–area relationship in radiation therapy. *Radiology* **81**, 881-885.

The application of RBE values to clinical trials of high-LET radiations*

H. R. WITHERS and L. J. PETERS

*Section of Experimental Radiotherapy,
The University of Texas System Cancer Center,
M. D. Anderson Hospital and Tumor Institute,
6723 Bertner Drive, Houston, Texas 77030, U.S.A.*

Abstract—*RBE $_{n/x}$ values for tumor control and normal tissue injury vary (Barendsen, these proceedings; Field and Hornsey, these proceedings). This variation is determined in part by the inherent radiosensitivity of the component cells, as well as by their proliferative status and/or oxygenation. A therapeutic gain would be expected from treating tumors with neutrons or other high-LET beams if tumor cells were characterized by an X-ray survival curve with an initial slope which was shallower and/or a shoulder which was broader than those characteristic of the cells of dose-limiting normal tissues. The presence of hypoxic tumor cells increases the RBE for tumor control, especially if the X-ray treatment is given as a few large doses. However, reoxygenation must be relatively inefficient if there is to be a significant therapeutic advantage over conventional treatment which uses a large number of small dose fractions. The survival of a cell as a function of its position within the division cycle is less variable with high-LET irradiation than it is with X-irradiation: hence, changes in the age distribution of critical tumor and normal tissue cells may lead to large variations in RBE, especially when small dose fractions are used. If surviving tumor cells do not progress through the division cycle during a course of fractionated radiotherapy, their resulting parasynchronization into resistant phases would lead to a therapeutic gain from the use of high-LET radiations. Also, since potentially lethal injury is repaired in slowly proliferating cells after X-, but not after neutron irradiation, the RBE $_{n/x}$ may be further increased for an inadequately redistributing population of tumor clonogenic cells. Conversely, if tumor cells redistribute efficiently, or if G_1, S or G_2 delay induces parasynchrony in relatively sensitive phases of the division cycle, there could be a therapeutic disadvantage from using high-LET radiations.*

At present, there are no human data permitting the correlation of a poor response to conventional fractionated X-ray therapy with intrinsic radioresistance, inefficient reoxygenation or poor cell cycle redistribution. Such data are needed for proper selection of the subpopulations of patients most likely to benefit from high-LET radiotherapy.

Introduction

A therapeutic gain from substituting high-LET beams for conventional X- or γ-rays will only be obtained if the RBE for tumor control is greater than the RBE for the dose limiting normal tissue, i.e. if the therapeutic gain factor (TGF) is > 1. Clinical experience indicates that the relevant normal tissue is usually (though not exclusively) the fibrovascular stromal tissue whose injury results in late effects such as contraction, fibrosis, or necrosis. In the instance of the $E_d = 50$ MeV/Be neutrons generated at TAMVEC, the best estimate of the RBE for these late effects in humans is 3.1 when fractional doses equivalent to 200 rad of X-rays are used.

Thus, the RBE for tumor control by that beam must be greater than 3.1 for a therapeutic gain to be realized in the MDAH–TAMVEC trial. The RBE for a number of different assays of experimental animal tumor responses varies considerably (Barendsen, these proceedings; Howlett, et al., 1975; Mason and Withers, 1977). It has not been possible to make reliable correlations between this variation and the biology of the tumors.

The clinical data presented at this symposium and previously (Van Peperzeel et al., 1974) suggest that the RBE for control of human tumors also varies and that some method is needed for selecting patients in whom a

*This investigation was supported in part by grant numbers CA-12542 and CA-11138, awarded by the National Cancer Institute, DHEW.

therapeutic gain might be expected from using high-LET radiotherapy.

Three factors could cause substantial variations in the RBE for tumor control by a course of fractionated radiotherapy.

1. Inadequate reoxygenation.
2. Inadequate redistribution of surviving tumor cells throughout the division cycle between dose fractions.
3. Inherent variations in the radiation sensitivity of tumor cells to the relatively low fractional doses used in clinical radiotherapy.

Inadequate reoxygenation

Figure 1 shows theoretical isoeffect curves (Withers and Peters, in press) calculated on the assumption that 1% of the tumor clonogens were hypoxic at the beginning of treatment and that the efficiency of reoxygenation may vary from tumor to tumor. If reoxygenation is sufficient to maintain the proportion of hypoxic cells at 1%, the response to any scheme other than a few large fractions is essentially that of a well-oxygenated tumor (dashed line). This reflects the fact that the 99% well-oxygenated tumor cells dominate the response to low fractional doses. If the tumor clonogens surviving the early treatments are less efficiently reoxygenated and their proportion increases to 10%, or even 20%, and remains at those levels, there is still only a modest effect on the total dose required for a

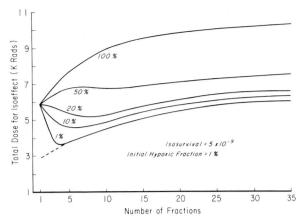

Fig. 1. Theoretical isoeffect curves (Withers and Peters, in press) showing the influence of different efficiencies of reoxygenation on the dose required to achieve a constant probability of control of a tumor which was composed initially of 1% hypoxic cells, using single or fractionated dose radiotherapy. (The dashed line is the isoeffect curve for a completely well-oxygenated tumor.) Note that with large numbers of small dose fractions, the influence of hypoxia is less marked than with a few large doses and the efficiency of reoxygenation is less critical.

given probability of tumor control by a conventional 30 to 35 fraction course of X-rays using fractional doses of about 200 rad. When larger proportions of surviving tumor clonogens remain hypoxic through inefficient reoxygenation, there is an appreciable increase in the total dose required for tumor control by multifraction X-ray therapy.

Replacing X-rays with neutrons would reduce the separation of the isoeffect curves (Fig. 1) by

Table 1. Hypoxic gain factor at 200 rad X-rays

X-rays	50 MeV$_{d \rightarrow Be}$ neutrons
OER = 2.3[a]	OER = 1.65[b]
Effective D_o for 200 rad fractions = 320 rad (oxic)[c]	Effective D_o(oxic) = 103 rad

$RBE_{n/x}$ = 3.1
Effective OER (mixed oxic-hypoxic)

% Cells hypoxic	X-rays	Neutrons	Gain factor
100	2.3	1.65	1.39
80	1.92	1.48	1.30
60	1.60	1.32	1.21
40	1.35	1.20	1.13
20	1.15	1.09	1.05
10	1.07	1.04	1.03
5	1.04	1.02	1.02
0	1.00	1.00	1.00

[a] N. J. McNally, 1975.
[b] E. J. Hall, 1977.
[c] Effective D_o is used to describe the slope of a multifraction dose survival curve which is a simple exponential, on the assumption of equal survival decrements from each dose fraction.

an amount determined by the hypoxic gain factor*: that is, the separation of the curves would be reduced but not eliminated. The HGF value is affected not only by the respective OERs, but also by the proportion of surviving tumor clonogens which are hypoxic and the size of the fractional doses to which the tumor is exposed. Table 1 shows calculated HGFs, at a dose of 200 rad of X-rays, for various mixtures of oxic and hypoxic cells. The OERs at doses equivalent to 200 rad of X-rays were assumed to be 2.3 (McNally, 1975) and 1.65 (Hall, 1977) for X-rays and neutrons, respectively. If 100% of the cells were hypoxic and remained so, the HGF would be $2.3/1.65 = 1.4$, but this value decreases rapidly as the proportion of hypoxic cells decreases and if 20% or fewer were hypoxic, the HGF would be negligible. Thus, it is only in tumors which reoxygenate poorly that a significant gain would accrue from the use of neutrons or other high-LET particles or, for that matter, hypoxic cell sensitizers.

Inadequate redistribution

The therapeutic gain which may result from the reduced variation in division cycle related fluctuations in radiosensitivity when high-LET particles are substituted for X-rays could be greater than that obtained through the reduced oxygen effect (Withers and Peters, in press).

From data of Gragg et al. (1978) we have estimated the $RBE_{n/\gamma}$ at 200 rad of γ-rays to be 5.1 for S-phase CHO cells and 1.75 for cells at the G_1S boundary. This large cell-age-dependent variation in RBE reflects primarily (but not exclusively) the large difference between the initial slope of the X-ray and neutron survival curves for S-phase cells. The significance of these data is that if a tumor were composed largely of cells with a γ-ray response similar to that of S-phase cells, the RBE for tumor control would be large, whereas it would be low if most tumor cells demonstrated a response similar to that of G_1S cells.

RBE values for a series of theoretical tumors composed of two compartments corresponding in their radiation survival characteristics to G_1S or S-phase CHO cells may be calculated. As may be seen from column 3 of Table 2, net RBE values would vary between 5.1 for a tumor composed of 100% resistant-phase cells and 1.75 for a tumor composed entirely of sensitive-phase cells. For example, if a tumor were composed of 40% resistant and 60% sensitive phase cells, the net RBE would be $(0.4 \times 5.1) + (0.6 \times 1.75) = 3.1$.

The ratio of the RBE for these theoretical tumors to the RBE for the dose-limiting normal tissues gives us an estimate of what may be called the kinetics gain factor (KGF). Using the RBE value of 3.1 found to be relevant for late contraction in patients treated with $E_d = 50$ MeV/Be neutrons, KGF values were calculated for hypothetical tumors containing various mixtures of resistant and sensitive phase cells (Table 2, column 4). For example, if 100% of the tumor cells were in a resistant phase of the cycle, the RBE for tumor control would be 5.1 and the KGF would be $5.1/3.1 = 1.6$. This means that for the same late injury to normal tissues from neutrons and γ-rays, the biologically effective tumor dose with neutrons would be, in this extreme condition, 1.6 times greater than that from γ-rays.

Table 2. *Influence of age distribution on RBE at 200 rad γ-rays*[a]

% Resistant	% Sensitive	Net RBE	KGF[b]
0	100	1.75	.57
10	90	2.09	.68
20	80	2.42	.78
30	70	2.76	.89
33	67	2.85	.92
40	60	3.09	1.00
50	50	3.43	1.11
60	40	3.76	1.22
70	30	4.1	1.33
80	20	4.43	1.43
90	10	4.77	1.54
100	0	5.1	1.65

[a]$RBE_{n/\gamma}$ at 200 rad—resistant phase 5.1,
—sensitive phase 1.75,
—late injury 3.1.
[b]KGF = Kinetics gain factor. (= Net RBE/3.1).

The KGF values shown in Table 2 vary by a factor of ~ 3. This contrasts with a 1.4-fold variation in HGF (Table 1). However, it is important to note that all of this variation does not imply a therapeutic gain from high-LET radiotherapy: all the theoretical tumors with a KGF of 1.0 would be equally well treated by γ-rays or high-LET particles, while those with a KGF value less than 1.0 would be better treated with γ-rays. Such a therapeutic disadvantage to neutron therapy could result from efficient redistribution of surviving cells to a predominance of sensitive phases or could follow parasynchrony in sensitive phases induced, for example, by prolonged G_1S or G_2 delay.

*Hypoxic Gain Factor, HGF $= \dfrac{\text{oxygen enhancement ratio (OER) X-rays}}{\text{oxygen enhancement ratio neutrons}}$ (Alper, 1963).

The importance of a reduced cycle-related variation in cell killing with high-LET beams may not be evident in data derived from experiments using large single doses or even a small number of large dose fractions. The KGF values in Table 2 were calculated for dose fractions equal to 200 rad (γ-rays): at higher doses the relative influence of cycle-related variations in RBE is less. Therefore, RBE values commonly derived for the various responses of animal tumors in experiments using a small number of large dose fractions may not be relevant to the clinical situation.

When tumors are treated with a large number of small fractions the effective KGF may vary greatly depending upon the efficiency with which surviving tumor cells redistribute through the cell cycle between doses. In a slowly redistributing tumor, surviving tumor cells could become increasingly synchronized in resistant phases of the division cycle with a resultant increase in the RBE and, hence, KGF. This could be one explanation for the high RBE values reported for slowly growing pulmonary metastases in man (Van Peperzeel et al. 1974). Such a phenomenon may also affect the non- or slowly proliferating normal tissues, but the KGF should nevertheless increase because the RBE value for the normal tissue would already reflect any influence of parasynchrony of surviving cells.

A second phenomenon could increase the RBE for control of poorly redistributing tumors. If repair of potentially lethal damage (PLD) is greater in slowly dividing than in rapidly dividing cells (Hahn and Little, 1972) and is less after neutron irradiation than after exposure to X-rays (Urano et al. 1976), then the RBE for control of a tumor will be greater if its clonogenic cells divide infrequently than if they cycle rapidly.

Variation in intrinsic radiosensitivity at low doses of X-rays

The responses of different normal tissues to low doses of fractionated X-irradiation probably vary considerably and it is reasonable to assume that the same is true for tumors. If the initial slope of the survival curve for tumor cells is steep, or the shoulder narrow, the *effective* survival curve for multifraction irradiation will be steeper than if the survival curve had a shallow initial slope and/or a broad shoulder. It could be anticipated that the RBE for leukemia cells would be lower than that for neoplastic epithelial cells. At present, however, there are insufficient data on the low dose responses of various epithelial tumors to permit accurate prediction of the types most likely to demonstrate a high RBE, although some trends are beginning to emerge from clinical trials.

Tumor characteristics which may aid selection of patients for high-LET radiotherapy

There seems little doubt that a proportion of tumors will be better treated by high-LET beams than by X-rays and, conversely, that some will be better treated by X-rays. It is also predictable that a proportion of tumors will lie between these two poles and will be treated equally well by either high- or low-LET beams. The practicality of high-LET radiotherapy depends upon the relative incidence of these three classes of tumor response. It seems, from preliminary clinical evidence, that the proportion of tumors better treated by high-LET beams will not be large enough to warrant the widespread deployment of cyclotrons or heavy particle accelerators. Rather, the optimal usage of high-LET machinery may be achieved by establishing a relatively small number of facilities jointly supported and managed by groups of radiotherapy centers to which those patients most likely to benefit from such therapy would be referred.

What characteristics of tumors would help us to select those which will be better treated by high-LET beams

Selection of patients for high-LET radiotherapy may be better if it were possible to:
1. Identify tumors which reoxygenate poorly.
2. Identify tumors which redistribute poorly.
3. Identify tumors whose cells are relatively resistant to low doses of X-rays, i.e. those tumors whose cells are characterized by an X-ray survival curve with a shallow initial slope and/or a broad shoulder.

There are no established methods for achieving these goals. The third goal will most likely become apparent only from careful comparison of the clinical results of neutron and X-ray therapy for a wide variety of tumors. The first goal may be achieved by using electrodes to make serial measurements of average tumor pO_2 during the course of fractionated X-ray therapy and correlating these findings with the outcome of treatment. The second goal seems even more unattainable, but the correlation of a pre-radiation proliferation profile of the tumor with the result of X-ray therapy may be an initial approach. Both these latter suggestions pose substantial conceptual, technical, and logistic problems, but are proposed because they have not been tested and other approaches are even less appealing.

Conclusions

1. Tumors, as well as normal tissues, vary in their relative responses to neutrons and X-rays. Animal tumors may not be very appropriate models for the human situation because of their characteristically rapid growth (Barendsen, these proceedings).

2. The proliferation pattern of the clonogenic cells in a tumor may be a more important determinant of RBE than generally supposed and may be more important than hypoxia in determining a therapeutic gain from the use of high-LET radiotherapy. Slowly-proliferating tumors may be those which are best treated by high-LET radiotherapy because of inadequate redistribution of surviving tumor cells through the division cycle between doses resulting in their parasynchrony in phases of the cycle relatively more resistant to X-rays than to neutrons.

3. There is a need for predicting poor responses to X- or γ-rays and thereby to become more proficient in selecting patients for whom high-LET radiotherapy may offer an advantage.

References

ALPER, T. (1963) Comparison between the oxygen enhancement ratios for neutrons and x-rays, as observed with *E. coli B. Brit. J. Radiol.* 36:97-101.

BARENDSEN, G. W. Review of RBE data of fast neutrons for tumour responses. These proceedings, pp. 175-179.

GRAGG, R. L., R. M. HUMPHREY, H. D. THAMES and R. E. MEYN (1978) The response of Chinese hamster ovary cells to fast neutron radiotherapy beams. III. Variations in RBE with position in the cell cycle. *Radiat. Res.* **76**, 283–291.

HAHN, G. M. and J. B. LITTLE (1972) Plateau-phase cultures of mammalian cells: An *in vitro* model for human cancer. *Curr. Top. Radiat. Res. Q.* **8**, 39-83.

HALL, E. J. (1977) Radiobiological intercomparisons *in vivo* and *in vitro*. *Int. J. Radiat. Oncol. Biol. Phys.* **3**, 195-201.

HOWLETT, J. F., R. H. THOMLINSON and T. ALPER (1975) A marked dependence of the comparative effectiveness of neutrons on tumor line, and its implications for clinical trials. *Brit. J. Radiol.* **48**, 40-47.

MASON, K. A. and H. R. WITHERS (1977) RBE of neutrons generated by 50 MeV deuterons on beryllium for control of artificial pulmonary metastases of a mouse fibrosarcoma. *Brit. J. Radiol.* **50**, 652-657.

MCNALLY, N. J. (1975) The effect of repeated small doses of radiation on recovery from sublethal damage by Chinese hamster cells irradiated in oxic and hypoxic conditions in the plateau phase of growth. In T. ALPER (Ed.), *Cell Survival After Low Doses of Radiation*, John Wiley & Sons, Inc., pp. 119-125.

PETERS, L. J., D. H. HUSSEY, G. H. FLETCHER and J. T. WHARTON. Second preliminary report of the M. D. Anderson study of neutron therapy for locally advanced gynecological tumors. These proceedings, pp. 3–10.

URANO, M., N. NESUMI, K. ANDO, S. KOIKE and N. OHNUMA (1976) Repair of potentially lethal radiation damage in acute and chronically hypoxic tumor cells *in vivo*. *Radiology* **118**, 447-451.

VAN PEPERZEEL, H. A., K. BRUER, J. J. BROERSE and G. W. BARENDSEN (1974) RBE values of 15 MeV neutrons for responses of pulmonary metastases in patients. *Europ. J. Cancer* **10**, 349-355.

WITHERS, H. R. and L. J. PETERS. Biological basis of radiotherapy. In G. H. FLETCHER, *Textbook of Radiotherapy*, 3rd edition, Lea and Febiger, Philadelphia, PA (in press).

WITHERS, H. R. and L. J. PETERS. Radiobiology of high LET irradiation (neutrons). *International Meeting for Radio-Oncology*, Baden/Vienna, Austria, May 17-21, 1978 (in press).

Doses and fractionation schemes to be employed in clinical trials of high-LET radiations

J. F. FOWLER

Gray Laboratory of the Cancer Research Campaign, Mount Vernon Hospital, Northwood, Middlesex HA6 2RN, England

Abstract—*An improvement in local tumour control of 20% would require 2 × 130 cases in order that nine trials out of ten should show the difference at a significant level. If the improvement is 30% only 2 × 55 patients are required. However, the detection of increased normal tissue complications requires larger numbers of patients than this, e.g. 2 × 150 if the increase is from 5% to 15%.*

The use of two dose levels on the neutron side in a clinical trial would enable definitive data to be obtained in, say, 6 years with 300 patients, instead of 10 years with 400 patients if only one dose level were used.

Introduction

If Dr. Catterall's results are confirmed in other centres, a bigger improvement in local control will be observed for neutrons than for hyperbaric oxygen. HBO yielded an increase in local control of about 20% in cancer of the head and neck and uterine cervix. This difference requires between 200 and 300 cases altogether (2 × 130) in order that 9 trials out of 10 should show the difference at a significant level, i.e. $P < 0.05$ (Boag et al., 1971). Table 1 shows that for the larger improvement of 30% in local control, less than half these numbers of patients are needed. Therefore fast neutron trials should indicate relatively soon whether such large improvements in local control will occur in other trials.

Normal tissue complications

However, this is only half of the problem. If the normal tissue complications rise to an unacceptable proportion, the new method will not be judged useful. The numbers of normal tissue reactions are, in an ethical trial, naturally fewer than the number of good tumour responses. Therefore we may need more patients to demonstrate whether normal tissue reactions have really been increased, with reliable statistical accuracy, than are needed for proof of improved tumour response. Table 1 illustrates this point.

An increase in normal tissue complications from 5% to 25% is certainly too high. An increase from 5% to 15%, whilst possibly acceptable in the early stages, provided that the new method gives very definitely better tumour response, is still high enough to require further developments in the method to bring it down.

The results of the first clinical trial of advanced head and neck cancer (Catterall et al. 1975, 1977) show an improvement in local control from 19% to 76% together with an increase in serious complications from 4% to 17%. Ideally, the trial should be repeated with everything the same except total neutron dose. This should be reduced with the object of bringing normal tissue complications down to about 4% again. It is difficult to estimate how much to reduce the dose to achieve this. Perhaps 5%? As much as 10%? It is a genuine question for a clinical trial and both reductions—perhaps others—would be ethical.

Future clinical trial

It is easier, but still inaccurate, to estimate how much loss of tumour control such dose reductions would bring. Even assuming that tumour response versus dose curves are steeper for neutrons than for X-rays, a 5% reduction is unlikely to bring the 76% local control all the way down to 19%. A reduction of 10% is also rather unlikely to do so, but it is not beyond the bounds of possibility. These considerations lead to the suggestion of a modest decrease in neutron dose only, for the repeated clinical trial. If only half the improvement in local control is then found that is reported for the first trial, the numbers of patients need not exceed 2 × 100.

More patients will, however, be needed to ensure that 4% instead of 17% complications are caused; perhaps about 320 patients in all, 2 × 160, simply to exclude a level as high as 15% (Table 1).

But it does seem that such a repeated trial is sufficiently likely to show a statistically clear result, to be worth doing. We shall then be able to answer the proper question: "How much increase in local control is obtained with neutrons, when no signficant increase in normal tissue complications is caused?"

Two dose levels

If *two* dose levels could have been chosen for such a trial at the outset, this question could have been answered without repeating the trial. The repeated control group would obviously have been unnecessary. One and a half times the number of patients (and $1\frac{1}{2}$ times the intake period) would have been needed; but this is much less than *twice* the number of patients and intake time, plus the follow-up time of the first trial, that we are now faced with.

being treated with neutrons and 100 as controls, requires, say, 2 years' intake and 3 years of follow up, as minimum values. In 5 years half the

Table 2.

| ONE DOSE LEVEL IN NEW TREATMENT |

100 patients in each arm = 200
100 patients per year   = 2 years
Plus 3-year follow up = 5 years

Result: Better local control but more complications
Conclusion: Promising, but does not prove that new method is different.
Repeat using lower dose level:
   200 more patients, and 5 years

| TOTAL: 400 patients, 10 years |

Table 3.

| TWO DOSE LEVELS IN NEW TREATMENT |

Levels chosen during the discussions to choose the best level for any trial.
±2 to 5%, within real uncertainty.
   100 patients in each arm = 300
   100 patients per year   = 3 years
   Plus 3-year follow up = 6 years

Result: By interpolation, the local control and survival can be assessed FOR THE SAME COMPLICATION LEVEL as in the 100 control patients.
Conclusion: New treatment is or is not better, by x%, with prob. P.

| TOTAL: 300 patients, 6 years |

Table 1. Numbers of patients required in clinical trials

Prospective chances of demonstrating an increase at $P = 0.05$	Local control of tumour		Increase of normal tissue complications		
	40-60%	40-70%	5-25%	5-15%	5-10%
1 chance in 2	2 × 50	2 × 20	2 × 25	2 × 60	2 × 200
3 chances in 4	2 × 90	2 × 35	2 × 40	2 × 105	2 × 360
9 chances in 10	2 × 130	2 × 55	2 × 65	2 × 160	2 × 550

From Boag *et al.* (1971).

Tables 2, 3 and 4 illustrate the economy of using two dose levels (with the new modality only). The doses should be spaced apart by only a small amount—5 to 10%—so that both doses are within the ethically acceptable range of genuine uncertainty that is discussed whenever dose levels are chosen for a new modality.

From Tables 2 to 4 the choice seems clear. The conventional two-arm method, 100 patients answer is obtained, exactly as now with the Hammersmith neutron trial. Another 5 years, and another 200 patients, are needed to obtain the answer required.

Total: 400 patients and 10 years (Table 2).

If, however, a three-arm trial is planned, with two levels of neutron dose from the start, only 300 patients and 6 years are needed to obtain the same answer (Table 3).

The choice is therefore between a full answer in 6 years or half an answer in 5 years, with 5 more years essential (Table 4). It is only if there is overwhelming virtue in getting the half-answer first, as an aid in planning the second half, that the more wasteful and slower method would be contemplated.

Table 4. Summary.

For 100 patients in each arm.

One dose level in new treatment:
400 patients, 10 years

Two dose levels in new treatment:
300 patients, 6 years

Note: Same principle holds for any other number of patients required:
1 level: 4N + 2FU periods
2 levels: 3N + 1FU period

There must, therefore, be some indications, preferably from phase 1 trials (pilot studies to estimate toxicity), to guide the choice of the two levels of dose. Clinical impressions during the early stages of a trial resulting in adjustments to the dose levels would obviously not be preferred, but these could certainly be used to prevent enormously high levels of complication from occurring.

Choice of dose level

Table 5 gives suggestions for a possible clinical trial at Hammersmith, if such a repeat with the same type of advanced head and neck cancer cases can be done. The total dose should be reduced by at least 5%. The overall time should preferably not be lengthened; that will only save the acute mucosal reaction. It will not reduce the incidence of late complications because these depend on very slowly turning-over tissues.

It may, of course, be necessary to reduce the acute mucosal reaction level and then an increase in overall time might be added. But this should be kept to a minimum, so as not to allow the tumours time to proliferate more during treatment. Such an increase in overall time should on no account be employed as a substitute for the reduction in total dose. There might indeed be some clinical temptation to do this, because the late reactions will not be observable for a long time, to enable the prediction of no reduction in them to be confirmed.

It is important not to let the overall time rise to as long as the 7 weeks employed in the U.S.A. neutron therapy trials. The tumours have too much time to repopulate during the treatment.

Table 5.

First trial of advanced head and neck cancer (Catterall et al 1975, 1977)	Proposed future trial 16 MeV d on Be cyclotron neutrons	Proposed daily schedule Two dose levels
1560 rad	1476 rad[a]	19 d and 21 d rad[c]
12 × 130 rad	12 × 123 rad	20 × 0.95 d 20 × 1.05 d
4 weeks	4 weeks[b]	4 weeks[b]

[a]The total dose should be reduced by at least 5%.
[b]The overall time should not be lengthened.
[c]The mean dose per fraction d has to be chosen using the experience of the clinicians who have treated patients daily.

Shorter overall times

Indeed, it is one of the potential advantages of methods of coping with hypoxic cells that shorter overall times (than the conventional 6 weeks or so) should then be reliable. This was demonstrated by our earlier mouse tumour results. Mammary tumours in C3H mice were subjected to ten different X-ray fractionation schedules, some of which were repeated with X-rays plus misonidazole and some with cyclotron neutrons (Fowler *et al.* 1976). The poor (short) régimes were improved up to the level of the best results by either misonidazole or neutrons.

The reason why such long overall times as 6 or more weeks are popular is probably the relief from acute symptoms. The reason why good tumour results have been obtained is, in my view, that reoxygenation in the tumours has had time to occur. Very prolonged schedules make no kinetic sense for tumour cell killing. If the hypoxic cells are dealt with by neutrons (or misonidazole or hyperbaric oxygen) there is then no advantage *for tumour eradication* in prolonging the treatment for longer than is absolutely necessary to avoid severe acute reactions. In principle, it is obvious from the simplest kinetic considerations that the sooner a lethal dose can be accumulated in the tumour the better the chance of cure. Therefore the 4-week overall time used by Dr. Catterall at Hammersmith should preferably not be lengthened (Table 5), unless acute reactions are too severe. Even then the overall time should not be lengthened too much, because it saves only the acute reaction, not the late reaction, and is therefore misleading.

Size of dose per fraction

Concerning size of dose per fraction, it is not as important as whether the tumours reoxygenate or not, provided that no individual dose fraction is too large. This critical size is not well known, but may be about the equivalent of 400 to 500 rad of mega-voltage X-rays. Although larger doses per fraction may show a bigger gain factor in the killing of hypoxic cells, the starting-point might be worse than with multiple small fractions. Many small doses may show a smaller gain in hypoxic cell killing, but their starting-point for X-rays-only may be better than for a few large fractions and the final result is still better. Therefore I have included in Table 5 a plan for daily fractionation using two dose levels, and an overall time that is not too extended.

References

BOAG, J. W., J. L. HAYBITTLE, J. F. FOWLER and E. W. EMERY (1971) The number of patients required in a clinical trial. *Brit. J. Radiol.* **44**, 122-125.

CATTERALL, M., I. SUTHERLAND and D. K. BEWLEY (1975) First results of a randomised clinical trial of fast neutrons compared with X-or gamma rays. In *Treatment of Advanced Tumours of the Head and Neck. Brit. Med. J.* **2**, 653–656.

CATTERALL, M., D. K. BEWLEY and I. SUTHERLAND (1977) Second report of a randomised clinical trial of fast neutrons compared with X or gamma rays. In *Treatment of Advanced Tumours of the Head and Neck. Brit. Med. J.* **1**, 1642-1643.

FOWLER, J. F., P. W. SHELDON, J. DENEKAMP, and S. B. FIELD (1976) Optimum fractionation of the C3H mouse mammary carcinoma using X-rays, the hypoxic-cell radiosensitizer Ro-07-0582, or fast neutrons. *Int. J. Radiat. Oncol. Biol. Phys.* **1**, 579-592.

Review of protocols for high-LET radiotherapy in the United States*

SIMON KRAMER

*Thomas Jefferson University Hospital,
Philadelphia, Pennsylvania, U.S.A.*

Abstract—*The need for a relatively large number of patients to establish the value of particle therapy is emphasized. Therefore, a cooperative clinical trials program has been initiated in the United States. A system of regional networks at each of the particle facilities has been, or is being, established. Current protocols for neutron beam therapy and proposed protocols for pi-meson and heavy-ion therapy are described. A registry of all patients treated by high-LET radiation has been initiated so that the incidence and extent of late and normal tissue morbidity can be determined.*

Problems in clinical treatment research

The question of which and how many patients need to be entered into clinical treatment research studies of particle therapy remains of the utmost importance. Consider the path of clinical treatment research. The initial investigation (Fig. 1) will lead to one of three conclusions. Rarely the results are so convincing that the method can be accepted as standard practice. More often initial results are promising but not conclusive and this is where a great deal more work has to be undertaken. Occasionally the results are no better or even worse than current standard practice so that further efforts are abandoned. Of course, these conclusions may be fallacious in which case quite often such methods may surface again at some future date. If promising results are obtained we now enter the second phase of treatment research. More data are needed. These may show (Fig. 2) that indeed the method works well for everything, or almost everything, and again now becomes standard practice. More likely, a method will be found to be working well for a number of types and stages of disease. Now it must be defined which those are, i.e. more patient data are needed. Next there is the possibility that it works well for only a very few patients; then the cost effectiveness must be established and appropriate systems developed. For instance, if only two or three relatively rare disease states are favorably affected, it might be more cost effective to bring all such patients to one or two national centers for therapy, rather than establish multiple treatment centers. Thirdly, it may not be cost effective and will have to be abandoned. Finally, worst of all, the initial promising results are not confirmed by further investigation and again this line of research is abandoned.

In neutron beam therapy we appear to have reached the second block of Fig. 1 (Catterall, 1974; Eichhorn *et al.* 1977; Hussey *et al.* 1974; Parker *et al.* 1977) and are entering the second phase of treatment research and indeed, Dr. Catterall's work has already confirmed that neutron therapy is excellent for some disease states (Catterall *et al.*, 1977; Catterall, 1977). But, clearly there is need for much more data at a variety of sites and ultimately all possible sites may well have to be investigated before the full value of neutron therapy is defined, and questions such as neutron beam alone versus a mixed beam of neutron and photons are settled. The total number of patients needed is likely to be very high. The Committee for Radiation Oncology studies has estimated that 7600 patients will have to be investigated, one-half in randomized, one-half in nonrandomized prospective clinical trials, fully to establish the value of neutron beam therapy in cancer treatment. All patients should be involved in registries as suggested by Duncan (1974) so that long-term normal tissue morbidity may be assessed. Our past experience with clinical trials

---

*This investigation was supported by Grant Number CA23113, awarded by the National Cancer Institute, DHEW.

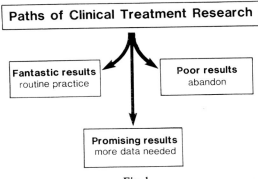

Fig. 1.

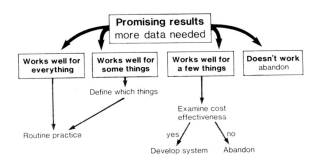

Fig. 2.

in the United States leads us to believe that we must have cooperative clinical trials and the establishment of regional networks at most, if not all, particle facilities in the United States if that number of patients is to be accessioned into studies in a reasonable space of time.

The problems of co-operative clinical trials are well understood. The establishment of networks adds new problems. Not only must the principal investigators at the neutron facilities agree on their program, but now each principal investigator must develop such agreement within his network. Most of the control patients in randomized trials will be treated at the home facility of the network members and thus quality control of the management of the control patient becomes a major task as does the mechanism of involving the network participants in the management of the neutron-beam patient. The need for networks in other countries is perhaps less compelling, but the proliferation of facilities in the United States and the consequent dissipation of patient material makes it essential for us.

To put matters in perspective, it might be well to consider some figures. Table 1 gives the initial patient number estimates for the first five cooperative neutron trials in the United States. Improvement in loco-regional control of 20% with a probability of detection of 85% or greater and a p value 0.05 will require 1135 patients.

Table 1. *Initial estimate of patient numbers*

240 Ca cervix
280 H & N (radiation + surgery)
240 H & N (radiation alone)
150 Gliomas
225 Prostate

1135 Total

Improvement in loco-regional control = 20%.
Probability of detection $\geq 85\%$ ($p = 0.05$).

Let us now look at the availability of appropriate patients in the United States. I have previously made some estimates on patients with head and neck cancer for clinical trials (Kramer, 1977). This is summarized in Table 2. It will be seen that of 41,400 patients only 1%, or about 500, can probably be entered into all trials addressing locally advanced or regional disease in this area. Similar estimates have been made for patients with locally advanced regional carcinoma of the cervix uteri (Table 3). Here I have assumed that 20% of potentially available patients are entered into trials, perhaps an optimistic estimate. Again we come to a total of 500 patients. These figures may well be underestimates by a factor of 2 or even 3 but considering that there is great competition for these patients within the research-groups in the United States and that the total number of neutron beam facilities will obviously be quite small, the number of patients to enter neutron studies will clearly be modest. If clinical trials are to be completed in a reasonable space of time, say 2 to 3 years of accession and another 2 to 5 years of follow-up, the patient material must be utilized optimally and consideration should be given to such areas as paired rather than randomized controls and to prognostic predictive factors allowing for better stratification.

Table 2. *Estimate of patient numbers for clinical trials.*

Head and neck cancer, U.S.A., 1977

Total H & N cancer (6% of 690,000)			41,400
Localized	36%		
Regional	44%	=	18,216
Disseminated	20%		
Uncommon or different biology			6000
To be considered			12,216
In non-participating institutions	60%		7326
Potentially available			4800
Entered into trials	10%		500
Neutron trials			?

Table 3. Estimate of patient numbers for clinical trials[a]

Cancer of the cervix, U.S.A., 1978

Total number		20,000
Localized	65%	
Regional (IIIA, IIIB, IVA)	30%	6000
Disseminated	5%	
In non-participating institutions	60%	3600
Potentially available		2400
Entered into trials	20%	480
Neutron trials		?

[a] 1977 *Cancer Facts and Figures*, American Cancer Society.

The U.S.A. experience

The clinical neutron trial experience in the United States falls into three phases. The first two were initiated on the modified physics machines; the third awaits the installation of hospital optimized machines. The first phase is the initial work and can be described as human biology; this has been described in detail before. The second phase consisted of the development of controlled clinical trials. These were developed by the participants through extensive discussions. Drs. Hussey and Peters from M. D. Anderson, Houston, Texas, Drs. Parker and Griffith from the University of Washington in Seattle, Drs. Rogers and Ornitz from the Mid-Atlantic Neutron Treatment Association in Washington D.C., and Dr. Lionel Cohen of the Fermi Laboratory in Chicago were the principal contributors, and more recently, Dr. Rodriguez-Antunez of the Great Lakes Neutron Treatment Association, Cleveland, Ohio. In 1976 the Radiation Therapy Oncology Group undertook the management of these trials. And in 1977 the first of these protocols began to accession patients. At present there are nine protocols with a total accession of patients of 235 (Table 4). Schemata of these protocols are given in detail in the protocol abstracts and will be described here only briefly. Because of the different constraints at the participating institutions a number of options are available for most of the protocols. The choice of such options must be made by each participating institution before any patient is entered and adhered to throughout that trial.

Table 4. *RTOG case accession in neutron protocols 15 August, 1978*

52 Cervix	42 Malignant gliomas
114 Head and Neck	9 Prostate
12 RT and surgery	5 Bladder
99 RT alone	13 Esophagus
3 Neutron boost	0 Lung
	235 TOTAL

Table 5.
Cervix

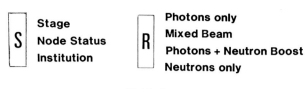

Table 6.
Head & Neck – RT & Surgery

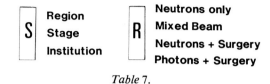

Table 7.
Head & Neck – RT only

S	Region Stage Institution	R	Photons only Mixed Beam Neutrons only

Table 8.
Head & Neck – Neutron Boost

S	Region Stage Institution	R	Photons only Photons Neutron Boost

In carcinoma of the cervix (Table 5) patients with stages IIIA, IIIB and IVA are eligible. They will be stratified by stage, nodal status and institution and randomized to two or more treatment arms. In the head and neck three protocols are available. The first (Table 6) is for potentially operable patients with T2, T3 and T4 primary tumors and nodes of any N staging except IIIA. They will be stratified by region, stage and institution and randomized into two or more of the four treatment arms. Patients with inoperable lesions of the head and neck (Table 7) will be stratified by region, stage and institution and entered into two or more of three treatment arms. A third protocol for head and neck patients has recently been introduced to determine the value of the neutron boost versus a photon boost, after initial primary management by photons (Table 8). Patients will be stratified by region, stage and institution, and randomized to photons plus photon boost, or photons plus neutron boost. The study for patients with malignant gliomas (Table 9) will include grade III and grade IV biopsied lesions and radiologically diagnosed deep-seated unbiopsied tumors. They will be stratified by grade, whether or not they are biopsied and by institution and randomized to photons versus photons with neutron boost.

In carcinoma of the prostate (Table 10), stage C, patients will be stratified according to well differentiated or poorly differentiated histology, whether or not previous orchiectomy was performed and pelvic node status and then randomized to two or more of the treatment arms. In carcinoma of the bladder (Table 11) a nonrandomized study has been agreed upon in patients with stage BI, grades III or IV and stages B2, C and DI any grade, with a choice of three treatment arms. Similarly a nonrandomized study has been undertaken in patients with squamous cell carcinoma of the esophagus (Table 12) of either the upper, middle or lower third with lesions of less than 15 cm in length with a choice of three treatment options. Finally, there is a study for patients with nonresectable, non oat-cell cancer of the lung, without evident distant metastases (Table 13), they will be stratified by size of primary tumor (greater or less than 6 cm), histology (adeno-carcinoma versus squamous or undifferentiated) and whether or not post-irradiation chemotherapy is planned, and then randomized to one of two treatment arms.

Table 9.
Malignant Gliomas

S: Grade, Not Biopsied, Institution
R: Photons only, Photons + Neutron Boost

Table 10.
Prostate

S: Histology, Orchiectomy, Node Status
R: Photons only, Mixed Beam, Neutrons only

Table 11.
Bladder

C: Mixed Beam & Surgery, Mixed Beam only, Neutrons only

Table 12.
Esophagus

C: Neutrons - 2 fractions per week, Neutrons - 4 fractions per week, Mixed Beam

Table 13.
Lung

S: Primary size, Histology, Chemotherapy
R: Photons only, Neutrons only

As regards the clinical trials with pi-meson therapy the human biology studies are well advanced and thirteen controlled clinical trials are proposed as shown in Table 14.

Table 14.
Proposed pi meson protocols

Brain	Stomach
Oral cavity	Pancreas
Oropharynx	Cervix
Hypopharynx	Prostate
Larynx	Bladder
Thoracic	Rectum
Superior sulcus	

Heavy ion therapy has not yet reached the stage of controlled clinical trials. Protocols are proposed for sites as shown in Table 15.

Table 15. Proposed stripped nuclei protocols

Pancreas
Esophagus
Cervix
Choroid of eye
All sites
    Locally advanced or recurrent

As mentioned earlier, information on normal tissue effects and particularly late normal tissue morbidity resulting from particle therapy is urgently needed. To this end, a particle registry has been established by the Radiation Therapy Oncology Group which, hopefully, will include all patients treated. At present, 912 patients have been registered (Table 16), data are now available for 147 patients that have survived 6 months, 79 that have survived 9 months and 49 patients that have survived 15 or more months (Table 17). As more potentially curable patients are treated by particle radiation this data base will, of course, be expanded so that appropriate information can be obtained.

Table 16. RTOG Particle Registry
Cases registered by particle type

30	Pion
831	Neutron
51	Heavy ion
912	Total

Table 17. RTOG Particle Registry

Month	No. cases
3rd	280
6th	147
9th	79
12th	49
15th	49

Conclusion

There is a need for a large number of patients to be entered into clinical trials in order fully to establish the value of fast neutron and other high-LET particle therapy in clinical radiation therapy. To this end a national clinical cooperative trials program has been established in the United States and regional networks of collaborating radiation oncologists and their facilities have been or are being established at each of the high-LET facilities. The management of the clinical trials program, as well as a high-LET registry established to determine late normal tissue morbidity, has been undertaken by the Radiation Therapy Oncology Group.

References

CATTERALL, M. (1974) A report on three years' fast neutron therapy from the Medical Research Council's cyclotron at Hammersmith Hospital, London. *Cancer* **34**, 31-25.

CATTERALL, M., D. BEWLEY, I. SUTHERLAND (1977) Second report on results of a randomized clinical trial of fast neutrons compared with X- or gamma-rays in treatment of advanced tumors of the head and neck. *Brit. Med. J.* **1**, 162.

CATTERALL, M. (1977) The results of randomized and other clinical trials of fast neutrons from the Medical Research Council, Cyclotron, London. *Int. J. Radiat. Oncol. Biol. Phys.* **3**, 247-253.

DUNCAN, W. (1974) Registration of clinical data and criteria for selection of patients for fast neutron trials. *Europ. J. Cancer* **10**, 381-383.

EICHHORN, H. J. and A. LESSEL (1977) Four years' experience with combined neutron-telecobalt therapy. Investigations on tumor reaction of lung cancer. *Int. J. Radiat. Oncol. Biol. Phys.* **3**, 277-280.

HUSSEY, D. G. FLETCHER, J. CADERAO (1974) Experience with fast neutron therapy using the Texas A & M variable energy cyclotron. *Cancer* **34**, 65-77.

KRAMER, S. (1977) Cancer of the head and neck: A challenge and a dilemma. *Sem. in Oncology* **4**, 353-355.

PARKER, R. H. BERRY, J. CADERAO, J. GERDES, D. HUSSEY, R. ORNITZ and C. ROGERS (1977) Preliminary clinical results from the United States fast neutron teletherapy studies. *Int. J. Radiat. Oncol. Biol. Phys.* **3**, 261.

Clinical trials with fast neutrons in Europe

K. BREUR and J. J. BATTERMANN

Department of Radiotherapy, Antoni van Leeuwenhoek Hospital, Amsterdam, The Netherlands

Abstract—*There is an increasing number of European centres with facilities for fast neutron treatment. To prepare protocols for multi-centre trials and to make treatment results intercomparable, many obstacles have to be overcome. A provision for two dose levels in the neutron treatment arm of a trial could enable the estimation of the local cure rate for an acceptable late complication rate. The considerations on which future trials are based as well as the main characteristics of trials in progress in Europe are reported.*

Introduction

After pilot studies during a number of years, the first randomized trial of fast neutrons was started in London in 1972, using the 7-MeV neutron beam from the MRC cyclotron in Hammersmith Hospital. The results of this trial have been reported on several occasions by Dr. Catterall *et al.* (1975, 1977). They indicated that fast neutrons can yield a substantial increase in local cure rate for advanced tumours in the head and neck region.

Among therapists some reluctance remained to accept these results as a final proof for the therapeutic advantage of fast neutrons over high-energy photons. The criticism raised repeatedly was based on the reported high complication rate in the neutron series. Another disturbing factor was considered to be the variation in X-ray doses applied to the patients in the control arm of the trial. While the total neutron dose was kept constant for all patients at 1560 rad in twelve sessions over 4 weeks, the total dosages for the X-ray treatments, which were partly carried out in other participating hospitals in London, varied a great deal. It was questioned therefore whether at least part of the treatments of the control group had not been sub-optimal or whether, by accepting a similar high complication rate, a fixed excessively high total dose of megavoltage X-rays given to all patients in the control arm could not have achieved a local cure rate similar to that obtained with fast neutrons. Dr. Catterall and Dr. Bewley could refute some of these criticisms. They remained convinced of the therapeutic advantages of fast neutrons, even in spite of the relatively inferior dose distribution.

From 1972 on H. J. Eichhorn and co-workers were able to use 6.2-MeV neutrons from a Soviet cyclotron in Dresden. They found at autopsies a higher percentage of bronchial carcinomas to be histologically "tumour negative" in the treated areas for the group treated partially with neutrons compared to a group treated solely by telecobalt radiation. This was found at first in a pilot series and later on it was confirmed in a randomized trial (Eichhorn *et al.* 1979).

It was generally felt that in coming years still more data and, if possible more convincing data, should be gathered. During the last 5 years the number of centres in Europe which obtained facilities for the clinical application of fast neutrons has increased to ten and will increase still more in the near future.

About 5 years ago an EORTC Study and Project Group on Fast Particle Therapy was formed, in which representatives of all European centres actively engaged in this field participated. During yearly meetings the developments and progress in the various centres and relevant physical and biological results were discussed. The ultimate aim of the group was to analyse whether pooling of all efforts by means of multi-centre trials could be achieved in order to accumulate, within a relatively short period, data that could demonstrate unequivocally the real value of fast neutrons in the treatment of cancer. For drafting protocols for controlled clinical trials in which many centres can participate, many difficulties of all kinds have to be overcome to arrive at a common agreement, even more so if centres in different countries are concerned. Protocols for neutron trials offer still more problems because a number of specific factors have to be taken into account.

Factors influencing neutron-treatment results

Treatment results will be influenced by the degree of adequacy of the *dose distribution* that can be obtained with the available beam. Physical parameters such as the relative depth dose, the penumbra and the dose build-up in superficial layers are determining factors. These are dependent on several characteristics, e.g. the energy spectrum of the neutron beam and the collimator design. The relative depth dose is also dependent on the source–skin distance. This distance is in turn influenced by the neutron yield per minute, because with a low output one is forced in most cases to treat with an SSD of less than 1 metre. The penumbra is also dependent of the dimensions of the neutron source, which especially in DT generators can be fairly large. The EORTC Task Group decided that for each patient the actual achieved dose distribution should be well documented.

To equalize daily and total dose levels for the various types of beams, *biological factors* have to be taken into account. The main one, of course, is the RBE which is correlated with the mean energy of the neutron beam and in addition is dependent on the percentage gamma-ray contribution. The RBE is also influenced by the size of the dose given per fraction. The OER and the skin-sparing effect probably will not differ significantly for the mean energies mostly used in Europe at the moment, i.e. about 7 MeV for cyclotron neutrons and about 14 MeV for DT neutrons (see Table 1).

Of the utmost importance for the design of the trials is of course the decision to be taken by the Cooperative Group on the fractionation scheme and the overall time of treatment.

The number of fractions that can be given per week is restricted in some centres by the availability of the machine. Up till now the Hammersmith centre could only use the beam 3 days a week. The same situation exists in Louvain. All other European centres have or will have hospital-based machines, enabling the therapist to give 5 daily fractions per week.

As most available data from Hammersmith and the centres in the United States came from treatments with 2 or 3 fractions per week, it seemed opportune to investigate the effects of treatments with 5 daily fractions per week, the more so, because it appears that the results improve if more than 2 fractions per week are applied. Although the preliminary results in the U.S.A. could suggest an advantage in the use of a mixed neutron–X-ray schedules (Peters and Hussey, 1979), it still has to be sorted out whether this could be the effect of a 5-days-a-week schedule, in combination with a better dose distribution, rather than due to a biological phenomenon.

The overall treatment time to be chosen is also of importance. If the aim of the trials is to compare the effects of neutrons with those of high-energy photon radiation, it would seem logical to keep the overall time in both arms of the trial approximately similar. This would eliminate the differences with regard to the repopulation rate of the tumours and the

Table 1. Centres in Europe with facilities for fast neutron radiotherapy.

	Type	Mean neutron energy (MeV)
London	Cyclotron	7.5
(Manchester)[a]	DT-generator	14.7
(Glasgow)[a]	DT-generator	14.7
Berlin–Dresden	Cyclotron	6.2
Amsterdam	DT-generator	14.1
Hamburg	DT-generator	14.1
Edinburgh	Cyclotron	6.7
Essen	Cyclotron	6.3
Heidelberg	DT-generator	14.1
Louvain	Cyclotron	20
(Zürich)[b]	DT-generator	14.1
(Orleans)[c]	Cyclotron	
(Nice)[c]	Cyclotron	

[a] Until recently insufficient yield for clinical use.
[b] Installation of DT apparatus in near future.
[c] Plans for clinical application of cyclotron neutrons.

regeneration rate of the normal tissues. Theoretically, however, the reduced shoulder in the survival curve and the reduced effect of reoxygenation could permit shorter overall treatment times with fast neutrons.

Unfortunately the practice of radiotherapy with X-rays or gamma-rays with regard to overall time varies a great deal in Europe. In Great Britain many centres use overall times of 3 or 4 weeks for tumours in the head and neck region, while in the other countries overall times of 5 to 7 weeks are more commonly used. As a compromise it was decided in the Cooperative Group to aim at an overall time of 5 weeks for the neutron treatments and to allow for a slightly longer overall time for the photon treatments, depending on the practice in the particular centre. Edinburgh, however, having a great deal of experience with the 4-weeks' overall treatment time for the head and neck region, will use this period for both arms of the trial.

Although the dose distribution that can be obtained with fast neutron beams is always inferior to that obtainable with megavoltage X-rays, the use of multiple fields and wedge filters can reduce this factor to some extent. It is clear that neutron facilities providing for an iso-centric beam positioning have an advantage above those with only a fixed horizontal or vertical beam.

If collaborating centres could manage to overcome all the difficulties mentioned and could come to an overall agreement, one is still faced with the problem of estimating the optimal total dose, in fact being the maximum tolerance dose for each region. In view of the many differences in relevant factors, such an estimation should first of all be based on pilot studies over a few years in each centre.

For the *intercomparison* of treatment results and eventually the pooling of data, a careful intercomparison of dose measurements in the various centres is necessary. Though in principle this would appear a task that could be executed rather easily by our physicists, it has proved to be an area full of pitfalls, mainly by the lack of uniformity in methods of measurement and the instruments used. For example: it had been established that skin and mucosal reactions were practically the same in Edinburgh and Amsterdam for the dosages used in these centres. RBE comparison by Dr. E. Hall (1979) indicated that the RBE was almost the same for both beams, while the mean energy differed considerably (Table 1). These findings stimulated a re-examination of the dose measurements. It now seems likely that the Amsterdam neutron dose in fact is relatively higher by about 8% and the gamma contribution lower, making the total effective dose (TED = $rad_{neutron\ dose}$ + 1/3 $rad_{gamma\ dose}$) about 7% higher than the values reported earlier for the Amsterdam beam. This correction would give a better agreement with the difference in RBE of about 10% that one would expect between the two beams of different energies on the basis of experimental studies. This example shows that accurate dose measurements have to be performed for intercomparison between the participating centres and emphasizes the need for an internationally accepted neutron dosimetry protocol.

Two neutron dose levels

If in the final results of future trials the neutron group should again demonstrate a relatively higher complication rate than the X-ray group, the criticism could be raised again that an apparently higher local cure rate obtained with neutrons might be the result of a relatively higher effective dose given in the neutron series. To provide for an answer to such criticisms the EORTC group discussed at length the following proposal, advanced, among others, by J. Fowler (1979). Starting from the dose level which, from the experience obtained in pilot studies, appears to be a reasonable tolerance dose, one group of patients would receive a total dose 3–5% higher and the other group a dose 3–5% lower than the estimated optimal dose. Since it can be expected that the dose–effect curve for the complication rate and possibly also the curve for local cure rate will be rather steep in the region between those two levels, it is not unlikely that a distinct difference in complication rate could be established. An interpolation between the two dose levels could indicate the local cure rate obtainable with a dose that will give a complication rate similar to the one in the X-ray treatment arm based on extensive experience.

There are already clear indications that with increasing dose a steep rise in complication rate might be expected (Battermann and Breur, 1979). If during the early phase of such a trial with two doses the highest dose level should prove to give too high a complication rate, this dose would have to be lowered. If on the other hand at final evaluation there would be shown to be no significant difference between the two dose levels with regard to the complication rate, the results of both groups could be pooled.

Though the proposal for the two levels in neutrons arm was accepted by the majority of the members of the group, the method has hardly been put to practice up till now. In Amsterdam two dose levels are applied for the neutron radiation in the pelvic area with 1680 rad and 1800 rad TED, respectively.

Proposals for trials in Europe

After a summing up of all the obstacles mentioned, it is less surprising that there are still no well-organized multi-centre trials in progress in Europe at this moment. Drafts for protocols of such trials have been accepted in principle for the squamous cell carcinomas in the head and neck region, the bladder carcinomas stages III and IV, inoperable rectal carcinomas, inoperable salivary gland tumours and carcinomas of the prostate. Furthermore, studies are being undertaken on malignant gliomas, "Pancoast" tumours and soft tissue sarcomas.

The general objective of future trials by the EORTC group is to compare the effectiveness of:
(a) fast neutron radiation (using agreed fractionation schedules and
(b) megavoltage X-rays and electrons (current techniques).

For the time being the group refrains from studies on mixed neutron + X-ray treatments and combination of neutron treatment with chemotherapy.

Head and Neck tumours

Since the beginning of this year randomized clinical trials were started in Edinburgh and Amsterdam on squamous cell carcinomas in the head and neck region. Both centres use the same set of criteria. A few other centres also proceed along the same lines (Duncan and Arnott, 1979). The same is planned for randomized studies on inoperable T_3 and T_4 (M_0) bladder cancer and on inoperable rectal cancer.

Results

For the few trials already in progress, it is much too early for a first evaluation. The preliminary experience in the various regions has already been reported in papers presented earlier during this meeting. Several years of joint efforts are still required to reach meaningful conclusions.

References

BATTERMANN, J. J. and K. BREUR (1979) Results of fast neutron radiotherapy in Amsterdam. In G. W. BARENDSEN, K. BREUR and J. J. BROERSE (Eds.), *Proc. of 3rd Meeting on Fundamental and Practical Aspects of Fast Neutrons and other High LET Particles in Clinical Radiotherapy.* Pergamon Press Ltd., Oxford, pp. 17-22.

CATTERALL, M., I. SUTHERLAND and D. K. BEWLEY (1975) First results of a randomised clinical trial of fast neutrons compared with X- or gamma-rays. In treatment of advanced tumours of the head and neck. *Brit. Med. J.* 2, 653-670.

CATTERALL, M., D. K. BEWLEY and I. SUTHERLAND (1977) Second report of a randomised clinical trial of fast neutrons compared with X- or gamma-rays. In treatment of advanced tumours of the head and neck. *Brit. Med. J.* 2, 1642-1649.

DUNCAN, W. and S. J. ARNOTT (1979) Results of clinical applications with fast neutrons in Edinburgh. In G. W. BARENDSEN, K. BREUR and J. J. BROERSE (Eds.) *Proc. of 3rd Meeting on Fundamental and Practical Aspects of Fast Neutrons and other High LET Particles in Clinical Radiotherapy.* Pergamon Press Ltd., Oxford. pp. 31-35.

EICHHORN, H. J., A. LESSEL and K. DALLÜGE (1979) Five years of clinical experience with a combination of neutrons and photons. In G. W. BARENDSEN, K. BREUR and J. J. BROERSE (Eds.), *Proc. of 3rd Meeting on Fundamental and Practical Aspects of Fast Neutrons and other High LET Particles in Clinical Radiotherapy.* Pergamon Press Ltd., Oxford, pp. 79-81.

FOWLER, J. F. (1979) Doses and fractionation schemes to be employed in clinical trials of High LET radiations. In G. W. BARENDSEN, K. BREUR and J. J. BROERSE (Eds.), *Proc. of 3rd Meeting on Fundamental and Practical Aspects of Fast Neutrons and other High LET Particles in Clinical Radiotherapy.* Pergamon Press Ltd., Oxford, pp. 263-266.

HALL, E. J. (1979) Review of RBE data for cells in culture. In G. W. BARENDSEN, K. BREUR and J. J. BROERSE (Eds.), *Proc. of 3rd Meeting on Fundamental and Practical Aspects of Fast Neutrons and other High LET Particles in Clinical Radiotherapy.* Pergamon Press Ltd., Oxford, pp. 171-174.

PETERS, L. and D. H. HUSSEY (1979) Current status of the MDAH–TAMVEC fast neutron therapy study at Houston. In G. W. BARENDSEN, K. BREUR and J. J. BROERSE (Eds.), *Proc. of 3rd Meeting on Fundamental and Practical Aspects of Fast Neutrons and other High LET Particles in Clinical Radiotherapy.* Pergamon Press Ltd., Oxford, pp. 3-10.

Summary of discussion on multi-centre collaboration in clinical trials with high-LET radiations

G. W. BARENDSEN

Radiobiological Institute of the Organization for Health Research TNO, Lange Kleiweg 151, Rijswijk, The Netherlands

Introduction

The general discussion on problems in multi-centre clinical trials was chaired by Dr. Duncan and various topics were introduced by members of a panel.

In a short introduction the chairman pointed out that two main topics should be distinguished:

(a) General aspects of clinical co-operation related to principles and practice of clinical trials, e.g. language used to describe reactions of tumours and normal tissues, quality control, and

(b) Specific aspects of co-operative trials with various high-LET radiations.

General aspects of clinical co-operation

As a first question Dr. Duncan asked whether the panel thought that it would be worthwhile to establish a small working party consisting of representatives of the American and European groups to outline in a report the principles of clinical trials and rules for the application of fast neutrons in randomized as well as non-randomized studies. Dr. Kramer supported this idea suggesting that, in particular, recommendations should be made about assessment of equivalence of the standard practices with high-energy X-rays in various countries. In the U.S.A., for instance, frequently longer overall times are employed as compared to some of the European countries. Therefore the problem of identification of "equivalent dose" must be discussed. Dr. Withers expressed some doubts about whether equivalence of treatments could be attained in different centres in different countries, because this would require agreement about a very large number of details that could only be attained by frequent interchange of participants in the trials. Dr. Kramer suggested that this detailed agreement would not be required to attain a common strategy. Dr. Catterall suggested that in addition to general recommendations a world-wide registration of all neutron-treated patients might be desirable for tumours at different sites. Dr. Tubiana stated that a registration of results of treatments would require a common language. In the review that Dr. Dutreix and he had carried out of clinical data presented at this symposium, they had encountered important differences, e.g. with the meaning of "complication". Dr. Halnan pointed out that what is meant, for example, by "unacceptable fibrosis" is not the same in all centres. Dr. Phillips supported the suggestion that a uniform system of scoring responses of normal tissues and tumours would be required in order to prepare a registration system. He suggested that a world-wide scoring of late normal tissue reactions might be based on a five level scoring system proposed by Dr. Cohen in Chicago. Dr. Kligerman suggested that this system should be tried out for a period of perhaps half a year in order to evaluate whether it is a suitable system for a variety of centres. Dr. Wiernik expressed some less optimistic views of the possible value of a registration system, if it involved many questions to be answered. Dr. Halnan responded by suggesting that with a suitable computerized system even answers to part of the questions could be evaluated.

Dr. Breur suggested that if registration of all patients treated with neutrons would be too large a task, it could be restricted to those patients that did not fit in established trials. Dr. von Essen stated that in spite of differences in the types of radiation used, e.g. fast neutrons, pions and heavy ions, it would be useful to have at least some co-operation on criteria of response and criteria of damage, to be related to dose distribution and total dose.

Dr. Dische commented on the basis of experience with trials with hyperbaric oxygen, that it is extremely important to assess the comparability of results by using a common language. In some centres a higher rate of complications is considered acceptable. In addition, differences in overall metastases rates are observed and these might make assessment of late reactions impossible. Dr. Duncan concluded this part of the discussion, stating that a careful analysis of the problems was still required.

Specific aspects of co-operative trials with high-LET radiations

(a) Dose specifications

Dr Barendsen started this discussion off by pointing out problems involved in reporting of the doses given, the neutron energy spectrum of the beam, the dose per fraction applied and the overall time. Considerable differences exist between the centres in Europe and the U.S.A. These differences pertain to proposed randomized trials for cancer of the head and neck, of the bladder, the rectum, the uterine cervix as well as the brain. In particular the use of the concept of photon-rad equivalent, which uses implicitly RBE values which are not mentioned, provides problems. It might be better to state the neutron dose and gamma dose separately and if necessary for purposes of presentation and discussions to combine these to a "tumour treatment effective dose" $= D_n + \frac{1}{3}D_\gamma$.

Dr. Catterall stated that to obtain comparability among centres all neutron generators should provide beams suitable for good clinical treatments. As long as this is not achieved it might not be too great a disadvantage that the doses and schedules used in different centres are not all equal. Dr. Phillips pointed out that, although the expression of doses of neutrons in photon equivalent rad, i.e. $D_n \times \text{RBE} + D_\gamma$, as given in the U.S.A., is not very satisfactory, in Europe the problem is not solved by quoting D_n and D_γ separately, because the RBE depends on the neutron energy spectrum. Thus it is necessary to prescribe doses of neutrons for each neutron beam used with corrections for the doses possibly to be derived from such data as were presented by Dr. Hall and Dr. Field during this meeting. Dr. Kramer added that not only is it necessary to consider the neutron energy spectrum but also to take into account the normal tissue that is dose limiting in the treatment planned. Dr. Burgers made the statement that in addition to the influence of the neutron energy spectrum, the type of photon radiation that is taken as a standard radiation should be considered. It is certainly important to state whether the RBE quoted for a given neutron beam and a given effect in normal tissue is derived relative to 250-kV X-rays or to 4-MeV X-rays. Furthermore, it is not unlikely that volume effects have to be taken into account because of the poor dose distribution for fast neutrons.

(b) Fractionation schemes and mixed schedules

Dr. Barendsen pointed out that in past trials of fast neutrons two fractions per week had been used in some centres, three fractions per week in others and four or five fractions per week in another few centres. A discussion appears to be required about whether all future protocols should require at least three or at least four fractions per week. The latest U.S.A. protocols specify mostly four or five fractions per week. If availability of the machine limits the number, mixed neutron-photon schedules might be considered. Dr. Kramer commented that although four fractions per week might be preferable, fewer fractions must at present be used if machine availability is limited.

Dr. Field asked whether mixed neutron–photon schedules were considered to be of advantage because of physical characteristics, i.e. to make up for the poor depth dose and penumbra of neutrons by adding better collimated high-energy X-rays, or whether a biological advantage might be present. Dr. Peters answered that in different centres the rationale for using mixed schedules might be different. At M. D. Anderson Hospital the use of mixed schedules is considered purely as a technical expedience to improve dose distributions. Dr. Field asked what neutron energy would be required to provide adequate depth–dose characteristics. Dr. Peters answered that neutrons in his view would never be equivalent to a high-energy X-ray beam in that respect. He added that in cases of head and neck cancer the mixed schedule allowed him to compensate for the poor penumbra characteristics of neutrons and they had not had spinal-cord problems. Dr. Lawrence indicated that he used mixed schedules based on a similar rationale. Dr. Withers, however, suggested that the absence of repair of potentially lethal damage after neutron irradiation might render normal tissues more responsive than tumours and this might limit the tolerance dose. Thus addition of photons as part of the treatment might help to eliminate this problem of late reactions of slow turnover tissues. Dr. Raju added that his experiments with cultured cells demonstrated that repair of potentially lethal damage in G_1 cells and in non-cycling cells is more effective after photons than after high-LET radiation. Dr. Barendsen commented that the significance of this repair of potentially lethal damage in normal tissues required further studies.

Dr. Joslin asked Dr. Breur whether the increasing RBE with decreasing dose would be important in the penumbra region. The dose in the penumbra region although lower than in the centre of the beam, might be more effective than judged from the physical dose distribution because of the larger RBE at smaller doses. This might account for part of the complications observed in the treatment of head and neck cancer and of tumours in the pelvic area. Dr. Breur answered that with multiple fields the complications encountered earlier with two fields in the pelvic area were no longer observed. Dr. Duncan added that in Edinburgh the reactions of normal tissue were seen exactly where they were expected on the basis of dose distribution and that one cannot see a difference whether neutrons or photons are used. Dr. Barendsen later suggested that, although the RBE in the penumbra might be larger due to the smaller dose, this does not imply a greater effectiveness for damage to normal tissue than estimated from the physical dose distribution. The effectiveness per unit dose is for high-LET radiation more constant as a function of the dose than for X-rays. This is compatible with an approximately exponential dose–survival curve and the absence of accumulation of damage for fast neutrons. However, in the case of low-LET radiation the effectiveness per unit dose decreases with decreasing dose and the "biologically effective" beam profile for X-rays is slightly narrower than would be expected on the basis of the physical dose distribution of photons. Dr. Fowler added that the therapist should certainly take account of the fact that at the edge of the beam the dose does not decrease as rapidly as with the high-energy X-ray beams. For X-rays the biologically effective beam might be slightly smaller than measured with dosimeters.

Dr. Catterall illustrated, in answer to questions, the method of compensating bolus for air gaps which appear as a result of regression of tumours in the head and neck region treated with neutrons.

Dr. van Peperzeel brought up the problems associated with control groups of patients in randomized trials, if these are to be treated in another hospital than the centre where neutron treatments are given. She felt that this procedure might cause problems in the interpretation of results. This might especially apply if, in addition to radiotherapy, chemotherapy is added, which may also differ among institutes. Dr. Kligerman commented that at Los Alamos, where pion treatments are given, the control patients are staged and treatment planning is done at Los Alamos, even though the actual treatment is carried out elsewhere. In addition, the radiotherapists who refer patients are asked to come to Los Alamos and co-operate even though for a brief time. Dr. Phillips added that in northern California a regional co-operative cancer treatment group runs the trials in all areas. The protocols include chemotherapy. He felt that it is possible to exercise good quality control in this organization by strictly following protocols.

Concluding remarks

Dr. Duncun closed this discussion session with the remark that it is evidently impossible as yet to identify and evaluate the great number of problems involved in co-operative clinical trials with high-LET radiations. He urged that as a first arrangement a common and well-defined language should be agreed upon among all centres that want to co-operate. This can probably best be attained in small groups which should draft discussion documents for various types of treatments and sites. It is evident that this symposium should get a follow up soon, since this is in the interest of all concerned.

Concluding remarks on the future of high-LET radiotherapy

G. W. BARENDSEN, K. BREUR and J. J. BROERSE

Organizers of the meeting

THE presentation of papers on clinical experience during the first day of the meeting showed that a large amount of data has been accumulated. It was very stimulating to hear the excellent reviews of these data by Dr. Dutreix and Dr. Tubiana, who evaluated tumour responses, and by Dr. Halnan, who evaluated normal tissue reactions observed in all centres. These evaluations showed clearly that for some types of tumours encouraging results have been obtained, but that for other tumours the results were disappointing.

Throughout the meeting many participants stated the requirement for isocentric beams with good penetration, small penumbra and high dose rate, comparable to megavoltage X-ray beams. Furthermore, international agreement on dosimetry methods is required.

For the future collaboration in clinical trials with high-LET radiations between centres in Europe and the U.S.A. coherent protocols will have to be developed especially with respect to the assessment of normal tissue damage and analysis of tumour responses. It is likely that the advantage of high-LET radiotherapy can only be established definitely by randomized trials in which many centres participate.

As a result of this meeting multi-disciplinary contacts have been intensified and the contributions of all participants were highly appreciated. Continuation of the discussion, probably in part in small working parties, will be necessary to establish the merits of the high-LET radiation modalities for the cure of different types of tumours in various sites.

List of participants

of the 3rd meeting on fundamental and practical aspects of the application of fast neutrons and other high-LET particles in clinical radiotherapy.

Sponsored by the Radiobiological Institute of the Organization for Health Research TNO, 151 Lange Kleiweg, P.O. Box 5815, 2280 HV Rijswijk, The Netherlands, from September 13-15, 1978

G. J. M. J. VAN DEN AARDWEG, Laboratory for Celbiology and Histology, Nic. Beetsstraat 22, Utrecht, The Netherlands.

M. W. AARNOUDSE, Laboratory for Radiopathology, Bloemsingel 1, Groningen, The Netherlands.

J. AINSWORTH, Lawrence Berkeley Laboratory, Berkeley, Cal. 94720, U.S.A.

E. L. ALPEN, Donner Laboratory, University of California, Berkeley, Cal. 94270, U.S.A.

L. ARMSTRONG, Atomic Energy of Canada Ltd., P.O. Box 6300, Postal Station "J", Ottawa, Ontario, Canada.

S. J. ARNOTT, Department of Radiation Oncology, Western General Hospital, Edinburgh EH4 2XU, Scotland.

J. A. ATEN, Laboratory for Radiobiology, c/o Antoni van Leeuwenhoek Hospital, Plesmanlaan 121, Amsterdam, The Netherlands.

J. BAARLI, CERN, Geneva, Switzerland.

F. M. BACON, Tube Development Division, Sandia Laboratories, Albuquerque, N.M, 87131, U.S.A.

G. W. BARENDSEN, Radiobiological Institute TNO, P.O. Box 5815, Rijswijk, The Netherlands.

D. H. W. BARNES, MRC Radiobiology Unit, Harwell, Didcot, Oxon OX11 0RD, England.

H. H. BARSCHALL, Engineering Research Building, University of Wisconsin, Madison, Wisconsin 53706, U.S.A.

H. BARTELINK, Antoni van Leeuwenhoek Hospital, Plesmanlaan 121, Amsterdam, The Netherlands.

J. P. BATAINI, Curie Foundation, 26 Rue d'Ulm, 75231 Paris, France.

J. J. BATTERMAN, Antoni van Leeuwenhoek Hospital, Plesmanlaan 121, Amsterdam, The Netherlands.

A. D. R. BEAL, Department of Medical Physics, King Faisal Specialists Hospital and Research Center, P.O. Box 3354, Riyadh, Saudi Arabia.

D. W. VAN BEKKUM, Radiobiological Institute TNO, P.O. Box 5815, Rijswijk, The Netherlands.

M. BELLI, Instituto Superiore di Sanità, Radiation Laboratory, Viale Regina Elena 299, 00161 Roma, Italy.

A. P. VAN DER BERG, Rotterdam Radiotherapeutical Institute, P.O. Box 5201, Rotterdam, The Netherlands.

J. A. VAN BEST, Radiobiological Institute TNO, P.O. Box 5815, Rijswijk, The Netherlands.

D. K. BEWLEY, MRC Cyclotron Unit, Hammersmith Hospital, Ducane Road, London W12 0HS, England.

W. BINDER, Allgemeines Krankenhaus der Stadt Wien, Universitätsklinik f. Strahlentherapie und Strahlenbiologie, Alser Strasse 4, A-1090 Vienna, Austria.

H. BLATTMANN, Strahlenbiologisches Institut der Universität Zürich, August Forelstrasse 7, CH-8008 Zürich, Switzerland.

E. A. BLEHER, Klinik f. Strahlentherapie, Inselspital Bern, Brückfeldstrasse 13, CH-3012 Bern, Switzerland.

P. BLOCH, Hospital of the University of Pennsylvania, Department of Radiation Therapy, 3400 Spruce Street, Philadelphia, PA 19104, U.S.A.

L. BONKE-SPIT, Rotterdam Radiotherapeutical Institute, P.O. Box 5201, Rotterdam, The Netherlands.

D. BONNETT, MRC Cyclotron Unit, Western General Hospital, Crewe RD, Edinburgh EH4 2XU, Scotland.

J. BOOZ, Institut f. Medizin, der Kernforschungsanlage Jülich GmbH, Postfach 1913, 517 Jülich 1, Federal Republic of Germany.

A. BRAHME, Department of Radiation Physics, Karolinska Institute, S-10410 Stockholm, Sweden.

J. T. BRENNAN, Department of Radiation Therapy, Hospital University of Pennsylvania, 3400 Spruce Street, Philadelphia, PA 19104, U.S.A.

K. BREUR, Antoni van Leeuwenhoek Hospital, Plesmanlaan 121, Amsterdam, The Netherlands.

J. J. BROERSE, Radiobiological Institute TNO, P.O. Box 5815, Rijswijk, The Netherlands.

G. S. BROWN, 1211 W. La Palm, Suite 100, Anaheim, CA. 92801, U.S.A.

C. A. BULL, King Faisal Specialist Hospital and Research Center, P.O. Box 3354, Riyadh, Saudi Arabia.

G. BURGER, Institut f. Strahlenschutz, Ingolstädter Landstrasse 1, D-8042 Neuherberg, Federal Republic of Germany.

J. M. V. BURGERS, Antoni van Leeuwenhoek Hospital, Plesmanlaan 121, Amsterdam, The Netherlands.

R. CALLE, Fondation Curie, Institut du Radium, 26 Rue d'Ulm, 75231 Paris, France.

J. R. CASTRO, University of California, Lawrence Berkeley Laboratory, Building 55, Room 106, Berkeley, CA 94720, U.S.A.

M. CATTERALL, MRC Cyclotron Unit, Hammersmith Hospital, Ducane Road, London W12 0HS, England.

M. CHEMTOB, SPSN—C.E.A.—DPr/SPS, Centre d'Etudes Nucleaires de Fontenay aux Roses, B.P. no. 6, 99260 Fontenay-aux-Roses, France.

J. K. E. COLDITZ, N.V. Philips Gloeilampenfabrieken, Neutron Generation Development Department, Building EAV, Eindhoven, The Netherlands.

M. COPPOLA, Divisione Radioprotezione CNEN, CSN Casaccia, S.P. Anguillarese, C.P. 2400, 00100 Roma, Italy.

P. L. COVA, Clinica S. Ambrogio, 16 Via Faravelli, 20149 Milano, Italy.

J. J. CURRY, American College of Radiology, 925

Chestnut Street, 7th floor, Philadelphia, PA 19107, U.S.A.

S. B. CURTIS, Lawrence Berkeley Laboratory, Building 74, Room 15913, Berkeley, CA 94720, U.S.A.

J. VAN DAM, Centrum voor Gezwelziekten, Academisch Ziekenhuis St. Rafaël, 3000 Leuven, Belgium.

L. DAVIS, American College of Radiology, 925 Chestnut Street, 7th floor, Philadelphia, PA 19107, U.S.A.

J. DENEKAMP, Gray Laboratory, Mount Vernon Hospital, Northwood, Middlesex HA6 2RN, England.

B. F. DEIJS, Laboratory for Radiobiology, c/o Antoni van Leeuwenhoek Hospital, Plesmanlaan 121, Amsterdam, The Netherlands.

J. F. DICELLO, Los Alamos Scientific Laboratory, MP-3 MS 844, P.O. Box 1663, Los Alamos, N.M. 87545, U.S.A.

G. DIETZE, Physik-Technische Bundesanstalt, 6.5, Bundesallee 100, 33 Braunschweig, Federal Republic of Germany.

S. DISCHE, Marie Curie Research Wing for Oncology, Regional Radiotherapy Department, Mount Vernon Hospital, Northwood, Middlesex HA6 2RN, England.

B. DIXON, Radiobiology Department, Regional Radiotherapy Centre, Cookridge Hospital, Leeds LS16 6QB, England.

W. DUNCAN, Department of Radiation Oncology, Western General Hospital, Edinburgh EH4 2XU, Scotland.

J. DUTREIX, Institut Gustave-Roussy, 16 bis Avenue Paul-Vaillant Couturier, 94800 Villejuif, France.

M. M. ELKIND, Division of Biological and Medical Research, Argonne National Laboratory, Argonne, Illinois 60439, U.S.A.

C. F. VON ESSEN, SIN, 5234 Villigen, Switzerland.

M. FAES, Studiecentrum voor Kernenergie, S.C.K./C.E.N., B-2400 Mol, Belgium.

A. FELIUS, Rotterdam Radiotherapeutical Institute, P.O. Box 5201, Rotterdam, The Netherlands.

P. FESSENDEN, Department of Radiology, Stanford University, School of Medicine, Stanford, CA 94305, U.S.A.

S. B. FIELD, MRC Cyclotron Unit, Hammersmith Hospital, Ducane Road, London W12 0HS, England.

J. R. FIKE, c/o Radiobiological Institute TNO, P.O. Box 5815, Rijswijk, The Netherlands.

J. F. FOWLER, Gray Laboratory, Mount Vernon Hospital, Northwood, Middlesex HA6 2RN, England.

H. D. FRANKE, Radiotherapy Department, University Hospital Hamburg-Eppendorf, Martinistrasse 52, 2000 Hamburg 20, Federal Republic of Germany.

U. FRANKE, Radiotherapy Department, University Hospital Hamburg-Eppendorf, Martinistrasse 52, 2000 Hamburg 20, Federal Republic of Germany.

E. FREIBERG, HAEFELI, Basel, Switzerland.

J. GAHBAUER, The Cleveland Clinic, 9500 E Euclid Avenue, Cleveland, Ohio, U.S.A.

J. L. GARSOU, Radiotherapie/Controle Physique des Radiations, Université de Liège, Hospital de Baviere, B-4010 Liège, Belgium.

M. GIL-GAYARRE, Catedra de Radiologia y Medicina Fisica, Hospital Clinico de San Carlos, Ciudad Universitaria, Madrid 3, Spain.

A. E. GILEADI, KMS Fusion Inc., 3941 Research Park Drive, Ann Arbor, Michigan 48106, U.S.A.

L. S. GOLDSTEIN, University of California, Division of Radiation Oncology, San Francisco, CA 94143, U.S.A.

H. J. GOMBERG, KMS Fusion Inc., 3941 Research Park Drive, Ann Arbor, Michigan 48106, U.S.A.

D. T. GOODHEAD, MRC Radiobiology Unit, Harwell, Didcot, Oxon OX11 0RD, England.

R. L. GOODMAN, Hospital of the University of Pennsylvania, Department of Radiation Therapy, 3400 Spruce Street, Philadelphia, PA 19104, U.S.A.

D. GREENE, Physics Department, Christie Hospital, Withington, Manchester M20 9BX, England.

T. W. GRIFFIN, Division of Radiation Oncology, University of Washington Hospital, 1959 NE Pacific Street, Seattle, Washington 98105, U.S.A.

K. GÜNTHER, Zentralinstitut f. Molekularbiologie der Akademie der Wissenschaften der DDR, Lindenbergerweg 70, 1115 Berlin, D.D.R.

M. GUICHARD, Institut Gustave-Roussy, 16 bis Avenue Paul-Vaillant Couturier, 94800 Villejuif, France.

E. J. HALL, Department of Radiology, Radiological Research Laboratory, 630 West 168th Street, New York, NY 10032, U.S.A.

K. E. HALNAN, Institute of Radiotherapeutics and Oncology, Western Infirmary, Glasgow G11 6NT, Scotland.

A. T. HAN, Division of Biological and Medical Research, Argonne National Laboratory, Argonne, Illinois 60439, U.S.A.

H. HARINGA, Antoni van Leeuwenhoek Hospital, Plesmanlaan 121, Amsterdam, The Netherlands.

G. HARTMANN, DKFZ, Abt. Strahlenschutz und Dosimetrie, Im Neuenheimer Feld 280, 6900 Heidelberg, Federal Republic of Germany.

S. HELLMAN, Joint Center for Radiation Therapy, Department of Radiation Therapy, Harvard Medical School, 50 Binney Street, Boston, MA 02115, U.S.A.

J. H. HENDRY, Paterson Laboratories, Christie Hospital, Withington, Manchester M20 9BX, England.

J. D. HEPBURN, Accelerator Physics Branch, Atomic Energy of Canada Ltd., Physics Division, Chalk River Nuclear Laboratories, Chalk River, Ontario, Canada.

A. F. HERMENS, Radiobiological Institute TNO, P.O. Box 5815, Rijswijk, The Netherlands.

A. HESS, Radiobiologische Universitätsklinik, Abt. Strahlentherapie, Martinistrasse 52, D-2000 Hamburg 20, Federal Republic of Germany.

B. HOGEWEG, Radiobiological Institute TNO, P.O. Box 5815, Rijswijk, The Netherlands.

J. W. HOPEWELL, Research Institute, Churchill Hospital, Oxford, England.

S. HORNSEY, MRC Cyclotron Unit, Hammersmith Hospital, Ducane Road, London W12 0HS, England.

J. HORTON, The Cleveland Clinical Foundation, Division of Radiology, 9500 Euclid Avenue, Cleveland, Ohio 44106, U.S.A.

K. HÖVER, DKFZ, Abt. Strahlenschutz und Dosimetrie, Im Neuenheimer Feld 280, 6900 Heidelberg, Federal Republic of Germany.

R. HÜNIG, Department of Radiotherapy, University Hospital, CH-4031 Basel, Switzerland.

G. ILIAKIS, Institut f. Biologie, Abt. f Biophysikalische Strahlenforschung, Paul-Ehrlichstrasse 20, D-6000 Frankfurt/Main, Federal Republic of Germany.

W. J. L. JACK, Department of Radiation Oncology, Western General Hospital, Edinburgh EH4 2XU, Scotland.

R. JAHR, Physikal. Technische Bundesanstalt, Bundesallee 100, 33 Braunschweig, Federal Republic of Germany.

R. E. JONES, Marconi Avionics Ltd., Neutron Division, Elstree Way, Borehamwood, Hertfordshire, England.

J. DE JONG, Katholieke Universiteit, Geert Groote Plein, Nijmegen, The Netherlands.

C. A. F. JOSLIN, Leeds University Department of Radiotherapy, Regional Radiotherapy Centre, Cookridge Hospital, Leeds LS16 6QB, England.

H. JUNG, Institut f. Biophysik und Strahlenbiol. der

Universität Hamburg, Martinistrasse 52, 2 Hamburg 20, Federal Republic of Germany.
H. B. KAL, Radiobiological Institute TNO, P.O. Box 5815, Rijswijk, The Netherlands.
A. H. KEIJSER, Radiotherapeutisch Instituut Limburg, H. Dunantstraat 5, Heerlen, The Netherlands.
J. B. A. KIPP, Laboratory for Radiobiology, c/o Antoni van Leeuwenhoek Hospital, Plesmanlaan 121, Amsterdam, The Netherlands.
M. M. KLIGERMAN, Cancer Research and Treatment Center, 900 Camino de Salud, N.E., Albuquerque, N.M. 87131, U.S.A.
E. A. KNAPP, Accelerator Technology Division, M.S. 811, University of California, Los Alamos Scientific Laboratory, Los Alamos, N.M. 87545, U.S.A.
A. J. VAN DER KOGEL, Radiobiological Institute TNO, P.O. Box 5815, Rijswijk, The Netherlands.
H. D. KOGELNIK, Universitätsklinik f. Strahlentherapie und Strahlenbiologie, Alserstrasse 4, A-1090 Vienna, Austria.
G. KOK, Westeinde Hospital, P.O. Box 432, The Hague, The Netherlands.
S. KRAMER, Thomas Jefferson University Hospital, Department of Radiation Therapy and Nuclear Medicine, 10th and Walnut Streets, Philadelphia, PA 19107, U.S.A.
H. KRAUS, Radiologische Universitätsklinik, Abt. Strahlentherapie, Martinistrasse 52, 2000 Hamburg, Federal Republic of Germany.
R. LAVINE, Varian S.A., Zug, Switzerland.
G. LAWRENCE, Cancer Treatment Facility, Fermi Laboratory, P.O. Box 500, Batavia, Illinois 60510, U.S.A.
M. O. LEACH, Department of Physics, University of Birmingham, P.O. Box 363, Birmingham B15 2TT, England.
V. E. LEWIS, Division of Radiation Science and Acoustics, National Physical Laboratory, Teddington, Middlesex, England.
M. LINDGREN, Radiofysiska Centrallaboratoriet, Lasarettet, S22185 Lund, Sweden.
C. E. LÜCKE-HUHLE, Kernforschungszentrum Karlsruhe GmbH, Institut f. Genetik and f. Toxikologie von Spaltstoffen, Postfach 3640, D-7500 Karlsruhe 1, Federal Republic of Germany.
G. LUTHARDT, NUKEM GmbH, Postfach 110080, 645 Hanau 1, Federal Republic of Germany.
E. MAIER, Institut f. Medizinische Strahlenphysik und Strahlenbiologie, Universitätsklinikum Essen, Hufelandstrasse 55, D-4300 Essen 1, Federal Republic of Germany.
E. P. MALAISE, Institut Gustave-Roussy, 16 bis Avenue Paul-Vaillant Couturier, 94800 Villejuif, France.
D. W. A. MCCREADIE, Scottish Home and Health Department, St. Andrews House, Edinburgh, Scotland.
J. C. MCDONALD, Sloan-Kettering Institute for Cancer Research, 425 East 68th Street, New York, NY 10021, U.S.A.
N. J. MCNALLY, Gray Laboratory, Mount Vernon Hospital, Northwood, Middlesex HA6 2RN, England.
J. MEDER, Instytut Onkologii, Wawelska 15, 02-034 Warszawa, Poland.
H. G. MENZEL, Institut f. Biophysik, Universität des Saarlandes, D-6650 Homburg/Saar, Federal Republic of Germany.
J. P. MEULDERS, Laboratoire du Cyclotron, Chemin du Cyclotron 2, B-1348 Louvain-la-Neuve, Belgium.
M. MOHIUDDIN, Thomas Jefferson University Hospital, Department of Radiation Therapy, 11th & Walnut Streets, Philadelphia, PA 19103, U.S.A.

D. MUSORROFITI, Via Carlo Botta 59, 10081 Castellamonte (Torino), Italy.
B. J. MIJNHEER, Antoni van Leeuwenhoek Hospital, Plesmanlaan 121, Amsterdam, The Netherlands.
A. H. W. NIAS, Richard Dimbleby Department of Cancer Research, St. Thomas Hospital Medical School, London SE1, England.
U. B. NORDBERG, Radiofysiska Centrallaboratoriet, Lasarettet, S-22185 Lund, Sweden.
B. NORDELL, Department of Radiation Physics, Karolinska Institute, Fack, S-10401 Stockholm, Sweden.
M. OCTAVE-PRIGNOT, Unité de Radiothérapie et de Neutronthérapie Exp., Centre des Tumeurs, Avenue Hippocrate 54, 1200 Brussels, Belgium.
R. M. OLIVER, Department of Health and Social Security, Hannibal House, Elephant and Castle, 10 John Adam Street, London WC2N 6HN, England.
M. OLSEN, Department of Radiotherapy, University of Texas, Medical Branch, Galveston, Texas 77550, U.S.A.
R. D. ORNITZ, The George Washington University, Medical Center, Division of Radiation Oncology and Biophysics, 901 Twenty-third Street N.W., Washington, D.C. 20037, U.S.A.
J. OVADIA, Michael Reese Hospital and Medical Center, Department of Medical Physics, 20th Street and Ellis Avenue, Chicago, Ill. 60616, U.S.A.
C. J. PARNELL, MRC Cyclotron Unit, Hammersmith Hospital, Ducane Road, London W12, England.
H. A. VAN PEPERZEEL, Department of Radiotherapy, Academisch Ziekenhuis, Catharijnesingel 101, Utrecht, The Netherlands.
J. PETERS, High Voltage Engineering Europe B.V., Amsterdamseweg 61, Amersfoort, The Netherlands.
L. J. PETERS, Section of Experimental Radiotherapy, The University of Texas System, Cancer Center, M. D. Anderson Hospital, 6723 Bertner Avenue, Houston, Texas 77030, U.S.A.
R. PETIT, UCL, Clinique Universitaires St. Luc, B-1200 Brussels, Belgium.
G. PFISTER, Institut f. Kernenergetik und Energiesysteme der Universität Stuttgart, Pfaffenwaldring 31, D-7000 Stuttgart 80, Federal Republic of Germany.
T. L. PHILLIPS, M330, Section of Radiation Oncology, University of California, San Francisco, CA 94143, U.S.A.
B. PIERQUIN, Centre Hospitalo, Universitaire Henri Mondor, Service de Carcinologie, Radiotherapie, 51, Avenue de Maréchal de Lattre de Tassigny, 94000 Créteil, France.
J. A. VAN DER POEL, Rotterdam Radiotherapeutical Institute, P.O Box 5201, Rotterdam, The Netherlands.
R. C. S POINTON, Christie Hospital and Holt Radium Institute, Withington, Manchester M20 9BX, England.
P. C. VAN DER POL, Department of Radiotherapy, Leyenburg Hospital, Leyweg 275, The Hague, The Netherlands.
W. PORSCHEN, KFA, Institut f. Medizin, Postfach 1913, 517 Jülich, Federal Republic of Germany.
W. E. POWERS, Cancer Research Center, P.O. Box 1268, Columbia, Missouri 65201, U.S.A.
L. M. VAN PUTTEN, Radiobiological Institute TNO, P.O. Box 5815, Rijswijk, The Netherlands.
M. R. RAJU, H-10 Group, Los Alamos Scientific Laboratory, Los Alamos, N.M. 87545, U.S.A.
K. R. RAO, Universität Zürich, Zürich, Switzerland.
J. S. RASEY, Division of Radiation Oncology, University of Washington, Seattle, Washington 98195, U.S.A.
J. RASSOW, Institut f. Medizinische Strahlenphysik und

Strahlenbiologie, Universitätsklinikum Essen, Hufelandstrasse 55, D-4300 Essen 1, Federal Republic of Germany.

G. REINHOLD, HAEFELI, Baden, Switzerland.

H. S. REINHOLD, Radiobiological Institute TNO, P.O. Box 5815, Rijswijk, The Netherlands.

F. RICHARD, UCL Clinique Universitaires, St. Luc, B-1200 Brussels, Belgium.

M. P. RICHTER, Department of Radiation Therapy, Hospital of the University of Pennsylvania, 3400 Spruce Street, Philadelphia, Pennsylvania 19104, U.S.A.

D. J. RUBIN, National Institute of Health, Building 31, Room 3A49, Bethesda, Maryland 20014, U.S.A.

P. E. RÜEGSEGGER, Institut f. Biomed. Technik, Universität Zürich, Moussonstrasse 18, CH-8044 Zürich, Switzerland.

A. L. DE RUITER-BOOTSMA, Department of Histology and Cell Biology, Nic. Beetsstraat 22, Utrecht, The Netherlands.

G. SCARPA, CNEN–CSN, Casaccia, P.O. Box 2400, 00100 Roma, Italy.

E. SCHERER, Strahlenklinik, Radiologisches Zentrum, Universitätsklinikum Essen, Hufelandstrasse 55, 43 Essen 1, Federal Republic of Germany.

R. SCHITTENHELM, Siemens A.G., Medical Group, Henkestrasse 127, D-8520 Erlangen, Federal Republic of Germany.

K. J. SCHMIDT, Kernforschungszentrum Karlsruhe GmbH, Institut f. Genetik und f. Toxikologie von Spaltstoffen, Postfach 3640, D-7500 Karlsruhe 1, Federal Republic of Germany.

G. SCHMITT, Strahlenklinik, Radiologisches Zentrum, Klinikum Essen, Hufelandstrasse 55, 43 Essen 1, Federal Republic of Germany.

E. VAN DER SCHUEREN, Wilhelmina Gasthuis, Department of Radiotherapy, 1e Helmersstraat 104, Amsterdam, The Netherlands.

H. SCHUHMACHER, Institut f. Biophysik der Universität des Saarlandes, D-665, Homburg/Saar, Federal Republic of Germany,

J. G. SCHWADE, Landow Building, Room C808-CTEB, National Institute of Health, 7910 Woodmont Avenue, Bethesda, Maryland 20014, U.S.A.

H. G. SEYDEL, Thomas Jefferson University Hospital, Department of Radiation Therapy, 11th & Walnut Streets, Philadelphia, PA 19103, U.S.A.

J. B. SMATHERS, Bioengineering Program, Texas A&M University, College Station, Texas 77843, U.S.A.

F. SMIT, Department of Radiotherapy, University Hospital, Bloemsingel 59, Groningen, The Netherlands.

A. R. SMITH, Cancer Research and Treatment Center, The University of New Mexico, 900 Camino de Salud Ne, Albuquerque, N.M. 87131, U.S.A.

J. SPIRA, Boston University School of Medicine, Department of Radiological Physics, 80 East Concord Street, Boston, MA 02118, U.S.A.

W. STAR, Department of Clinical Physics and Bio-Engineering, West of Scotland Health Boards, 11 West Graham Street, Glasgow, Scotland.

R. STEWART, 50 North Medical Drive, Salt Lake City, Utah 84132, U.S.A.

R. STRANG, Department of Clinical Physics and Bio-Engineering, West of Scotland Health Boards, 11 West Graham Street, Glasgow, Scotland.

Ch. STREFFER, Institut f. Med. Strahlenphysik und Strahlenbiologie, Universitätsklinikum Essen, Hufelandstrasse 55, 4300 Essen 1, Federal Republic of Germany.

N. SUNTHARALINGAM, Department of Radiation Therapy, Thomas Jefferson University Hospital, 11th & Walnut Streets, Philadelphia, PA 19107, U.S.A.

A. SUTTLE, The University of Texas, Medical Branch, Graduate School of Biomedical Sciences, Galveston, Texas 77550, U.S.A.

M. L. SUTTON, Christie Hospital and Holt Radium Institute, Wilmslow Road, Withington, Manchester M20 9BX, England.

C. R. THOMAS, Marconi Avionics Ltd., Neutron Division, Elstree Way, Borehamwood, Hertfordshire, England.

J. L. TOM, The Cyclotron Corporation, 950 Gilman Street, Berkeley, CA 94611, U.S.A.

D. T. TRAN, CGR—MeV, B.P. no. 34, Route de Guyancourt, 78530 BUC, France.

E. L. TRAVIS, Gray Laboratory, Mount Vernon Hospital, Northwood, Middlesex, England.

A. TREURNIET-DONKER, Rotterdam Radiotherapeutical Institute, P.O. Box 5201, Rotterdam, The Netherlands.

K. R. TROTT, Strahlenbiologisches Institut, Bavariaring 19, München 2, Federal Republic of Germany.

H. TSUNEMOTO, National Institute of Radiological Science, 9-1, 4-Chome, Anagawa, Chiba-Shi, Japan.

M. TUBIANA, Institute Gustave-Roussy, 16 bis, Avenue Paul-Vaillant Couturier, 94800 Villejuif, France.

T. T. TURKAT, 215 Est Janss Road, Thousand Oaks, California 91360, U.S.A.

A. T. M. VAUGHAN, Department of Physics, University of Birmingham, P.O. Box 363, Birmingham B15 2TT, England.

P. C. VERAGUTH, Inselspital Bern, Abt. Strahlentherapie, Brückfeldstrasse 13, Ch-3012 Bern, Switzerland.

J. VERMEIJ, Department of Radiotherapy, University of Groningen, Bloemsingel 1, Groningen, The Netherlands.

P. A. VISSER, Antoni van Leeuwenhoek Hospital, Plesmanlaan 121, Amsterdam, The Netherlands.

D. D. VONBERG, MRC Cyclotron Unit, Hammersmith Hospital, Ducane Road, London W12 0HS, England.

A. WAMBERSIE, Unité de Radiothérapie et de Nutronthérapie Experimentale, Centre des Tumeurs, UCL 54.69, Avenue Hippocrate 54, 1200 Brussels, Belgium.

B. VAN DER WERF-MESSING, Rotterdam Radiotherapeutical Institute, P.O. Box 5201, Rotterdam, The Netherlands.

C. F. WESTERMANN, Department Clinical Physics, Westeinde Hospital, P.O. Box 432, The Hague, The Netherlands.

TJ. WIEBERDINK, Antoni van Leeuwenhoek Hospital, Plesmanlaan 121, Amsterdam, The Netherlands.

G. WIERNIK, Research Institute, Churchill Hospital, Headington, Oxford, England.

J. R. WILLIAMS, Department of Medical Physics and Medical Engineering, Western General Hospital, Crewe Road, Edinburgh, Scotland.

M. WINANT, UCL Clinique Universitaire, St. Luc, B-1200 Brussels, Belgium.

H. R. WITHERS, Section of Experimental Radiotherapy, M. D. Anderson Hospital, Houston, Texas 77030, U.S.A.

G. WOLBER, Deutsches Krebsforschungszentrum, Institut f. Nuklearmedizin, Im Neuenheimer Feld 280, 6900 Heidelberg, Federal Republic of Germany.

P. WOOTTON, NN 158, University Hospital RC-08, University of Washington, Seattle, Washington 98195, U.S.A.

TH. WIJSHIJER, Brown Boveri A.G., Abt. EKR, Ch-5401 Baden, Switzerland.

H. YOSEF, Department Radiobiology and Oncology, Belvidere Hospital, Glasgow, Scotland.

J. ZOETELIEF, Radiobiological Institute TNO, P.O. Box 5815, Rijswijk, The Netherlands.

A. ZUPPINGER, Abt. Radiologie, Universität Bern, Alpenstrasse 17, Ch-3006 Bern, Switzerland.

F. ZYWIETZ, Institut f. Biophysik und Strahlenbiologie, Martinistrasse 52, D-2 Hamburg 20, Federal Republic of Germany.